A FIELD GUIDE
TO COASTAL FISHES

FROM MAINE TO TEXAS

TO CHRIS —
HOPE YOU
CATCH THE
"BIG" ONE!.

Val Kells
6/18/11

A Field Guide to Coastal Fishes

FROM MAINE TO TEXAS

Val Kells

and Kent Carpenter

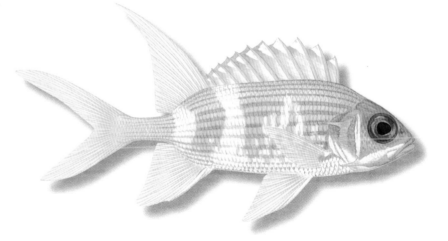

The Johns Hopkins University Press
Baltimore

The Johns Hopkins University Press
2715 North Charles Street
Baltimore, Maryland 21218-4363
www.press.jhu.edu

Library of Congress Control Number: 2010928198

ISBN-13: 978-0-8018-9838-9 (pbk. : alk. paper)
ISBN-10: 0-8018-9838-2 (pbk. : alk. paper)

A catalog record for this book is available from the British Library.

Special discounts are available for bulk purchases of this book.
For more information, please contact Special Sales at 410-516-6936
or specialsales@press.jhu.edu.

The Johns Hopkins University Press uses environmentally friendly book
materials, including recycled text paper that is composed of at least
30 percent post-consumer waste, whenever possible.
All of our book papers are acid-free, and our jackets and covers
are printed on paper with recycled content.

Contents

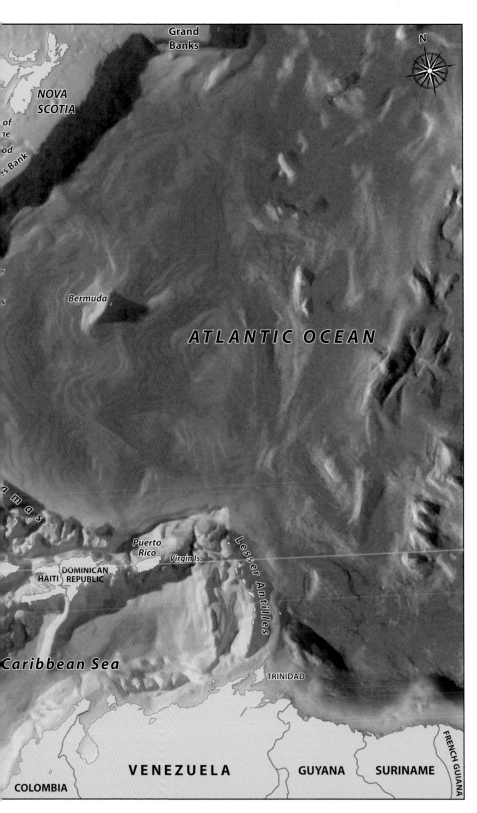

Grand
Banks

NOVA
SCOTIA

of
e
od
s Bank

7

s

Bermuda

ATLANTIC OCEAN

m a s

Puerto
Rico

Virgin Is.

HAITI

DOMINICAN
REPUBLIC

Lesser Antilles

Caribbean Sea

TRINIDAD

VENEZUELA

GUYANA

SURINAME

FRENCH GUIANA

COLOMBIA

N

Acknowledgments

We faced great challenges in bringing many complex elements together to produce this book. Although it was three and a half years in full-time production, this book is actually the result of many more years of dedicated work. Thousands of hours were spent traveling, researching, collecting, photographing, drawing, painting, designing, and writing to bring this book to fruition. Although we, as authors, can claim credit for its content, this book could not have been possible without the assistance of many other people.

We wish to acknowledge the following persons for their generous contributions of information, reference photographs, research, assistance, and knowledge—and much more. Without their help, this book would not be complete.

Kathleen Cole, University of Hawaii at Manoa; Bruce Collette, National Marine Fisheries Service; Matthew Craig, University of Puerto Rico; James Dooley, Aldelphi University; William Eschmeyer, California Academy of Sciences; Mary Fabrizio, Virginia Institute of Marine Science; Ron Fricke, Staatliches Museum für Naturkunde Stuttgart; John Graves, Virginia Institute of Marine Science; Mark Hall and Rebecca Bray, Biomes Marine Biology Center; Aimee Halvorson, Virginia Institute of Marine Science; Emma Hickerson, National Oceanographic and Atmospheric Administration; Jeffrey M. Leis, Australian Museum; Phil Lobel, Boston University; Keiichi Matsuura, National Science Museum, Tokyo; Thomas Monroe, National Marine Fisheries Service; Edward Murdy, National Science Foundation; John Musick, Virginia Institute of Marine Science; Joseph Nelson, University of Alberta; John Randall, Bishop Museum; David Smith, Smithsonian Institution; James Tyler, Smithsonian Institution; James Van Tassell, Hofstra University; Benjamin Victor, Ocean Science Foundation; Douglas Weaver, Texas A&M University, Corpus Christi; Jeffrey Williams, Smithsonian Institution.

We extend our gratitude to David Dooley, Wayne Starnes, Arnold Postello of the South Carolina Aquarium, and especially Louis Johnson, for providing photo reference. Special thanks to Leigh Middleditch Jr. and Robert Yates III for copyright advice and direction. To Mary Murray for sharing invaluable design knowledge. To Patterson Cunningham for fine-tuning the manuscript. To Georgia Middleditch for help editing and fact-checking. To Rosemary and Charles Robinson for translating foreign text. To Jeff Haas for reliable computer support. And to Captain Norman Miller for his help seining, collecting, photographing, and fishing around and off Ocracoke Island.

We thank everyone who gave their support, encouragement, and friendship throughout the long production of this book. Many people touched this project in ways small to very large. They include not only: Bob Baydush and Susan Smith; Lane Becken; Wendy Browning-Lynch, Randy, and Will Lynch; Don and Dorothy Chan; Joasha and John Dundas; Jessie Grove; Jess and Kathleen Haden; Carol Hurt; Andy Kells; Katherine, Kate, and Elizabeth McDonald; Sue McKenna and Lew Flanders; Betty, Casey, Jo, and Leigh Middleditch; Elizabeth Neff; Hunter Palmer; Jean and David Patteson; Louise Satterfield; and Heidi Sonen. Cheers!

Acknowledgments

The initial concept for this book was formulated approximately fifteen years ago. When all the pieces were finally and properly in place, the project was presented to the Johns Hopkins University Press. A partnership and common goal were quickly established, and this book soon took shape. Thus, we owe an enormous amount of gratitude to the directors, editors, and staff of the Press. Collectively, they had the foresight to see the benefit this book would bring our readers.

In particular, we thank our editor, Vincent J. Burke, Ph.D, and our copyeditor, Debby Bors. Debby spent many hours carefully refining the manuscript. Her keen eyes and hard work added a final, priceless polish. Vince had the faith, vision, trust, and patience to champion our project. With unwavering support, he helped to nurture it from concept to reality. Thank you, Vince—this book is a dream come true.

Finally, we would like to express our deepest thanks to our families for their patience, encouragement and understanding. They have supported us throughout our careers, in ways large and small, to meet our many professional goals. This book took an enormous amount of time that might otherwise have been spent with them.

Thank you, Andrew, Dave, and Drew. Thank you, Cecilia, Don, and Nicole. Now that it is complete, we can all go fishing . . . and diving.

Preface

About This Book

We developed this book in response to a need for a comprehensive, current, and compact field guide to fishes of the Atlantic and Gulf coasts. Our hope is that scientists, educators, students, naturalists, fishermen, divers, aquarists, and fish-watchers will find this work useful and informative.

Area and Species Covered

The species included are brackish and marine water fishes that are encountered from the northern coast of Maine to the southern border of Texas. This area generally extends from the intertidal zone to depths of about 660 feet (200 meters). We provide identification and natural history information for most fishes we know to have stable populations within this range of coverage. This includes fishes that spend all or part of their adult lives in marine waters and several non-native species.

We also describe species that are predominantly freshwater inhabitants but are also found in low-salinity waters. For comprehensiveness, some deep-water species and their families are included. Rare species and those generally occurring below 660 feet have been mostly omitted. Other species were excluded for lack of information such as specimen photographs or video clips for live color reference.

Many of the fishes that occur from Maine to Texas also occur at other, often distant, locations. Strays and waifs may be found in areas outside of the species' typical range. Wherever possible, those locations are noted in the text. Most of the fishes found in the southern waters of the United States also occur in the Caribbean and adjacent waters; thus the range map includes this broader area.

Names and Sequence of Species

The Latin and common names of the families in this book follow those presented in *Fishes of the World*, fourth edition, by Joseph S. Nelson. The common names of the individual species are taken from *Common and Scientific Names of Fishes from the United States, Canada, and Mexico*, sixth edition, published by the American Fisheries Society (AFS), Special Publication 29. With a few exceptions, Latin names, authority, and date follow Eschmeyer's most recent online *Catalog of Fishes*. The orders and families of fishes are organized in phylogenetic sequence following Nelson's fourth edition of *Fishes of the World*. The sequence of the species within each family follows the alphabetical order of the Latin genus and species names, rather than the alphabetical order of the common names. Seahorses are the only exception due to layout difficulties. For example, Gobies are listed in the following sequence:

Yellowline goby, *Elacatinus horsti*
Tiger goby, *Elacatinus macrodon*
Neon goby, *Elacatinus oceanops*

We have capitalized all single-word common names and the first word of multiple-word common names, as presented in *FAO Species Identification Guide for Fishery Purposes; The Living Marine Resources of the Western Central Atlantic*, published by the Food and Agriculture Organization (FAO) of the United Nations, with the support of the American Society of Ichthyologists and Herpetologists. Although we elected to use the accepted AFS common names for the individual species, other commonly used local names or those accepted by the FAO are also mentioned where possible. Whenever we encountered errors or conflicting information, we made appropriate corrections and inclusions based on the most recently published documentation.

Organization and Presentation

We have arranged this book into three primary sections: Introduction, Families, and Species. These sections are supported by supplemental materials, which include a glossary of terms, a list of additional resources, and an index. A full bibliography and descriptions of similar species and species not illustrated in this book can be found online at www.press.jhu.edu.

The Introduction provides an overview of the evolution, diversity, and features of fishes. It also includes information that will help the user in his or her identification of fishes. Each family section describes, in concise terms, each of the 76 families of fishes that are found along the coast from Maine to Texas.

The species section is the largest in this book and includes descriptions of 937 individual species based on the most recently published scientific information available. An additional 72 species are described in the online Appendix. A condensed summary of range and habitat is provided for each species. To save room, names of the states are abbreviated. North, South, East, and West are also abbreviated and read, respectively: N, S, E, W. For example, south Florida reads: S FL. The biologic description provides a brief summary of the species' behavior, diet, and/or ecology. The depths provided are approximate maximum recorded depths. The lengths given for each species are the approximate maximum recorded adult total length.

Each account is accompanied by a large, full-color illustration of the species. Each fish is shown in living color, as it would appear in hand, or at the surface in clear water. While no two fish of the same species are exactly alike, the illustration intends to closely represent the species as one might observe it. Great care was taken to accurately portray the correct placement and proportion of anatomical features. All illustrations show the adult fish unless otherwise noted. The illustrations are presented in a 'size relative' fashion, meaning that those in the same genus and on the same page are shown at a size relative to the longest fish in the genus. Each illustration provides identifying anatomical and color features. All are shown facing left or from above with all fins displayed. The only exception is the presentation of right-eyed flatfishes, which are shown facing right.

Diversity and Classification

Fish are the most diverse group of vertebrates on the planet. There are currently over 29,000 known living species of fishes and many thousands of others that have become extinct in the 500 million years since the early fish ancestors first swam in the seas. Over 1,000 marine fish species live along the Atlantic and Gulf coasts of the United States, inhabiting bays, inlets, estuaries, seagrass beds, coral reefs, and rocky, sandy, and mangrove-lined shores. These species range in size from the Whale shark, which may grow to 50 feet, to the Island goby, which as an adult grows to less than 1 inch in length. Atlantic and Gulf coast fishes also include species that represent the most primitive of ichthyofauna—like the Atlantic hagfish—to the most highly evolved forms, such as the Balloonfish.

The scientific classification of fishes is, and will likely always be, an ongoing process subject to debate and change as new information unfolds. However, many scientists divide fishes into five recognized classes: Myxini, the Hagfishes; Petromyzontida, the Lampreys; Chondrichthyes, the Cartilaginous fishes; Actinopterygii, the Ray-finned fishes; and Sarcopterygii, the Lobe-finned fishes. Of these five classes only the Lobe-finned fishes do not occur along the Eastern United States.
The table below shows how three representative fishes are classified in the three classes that occur along the Eastern United States:

	Hagfish	Whale shark	Island goby
Kingdom:	Animalia	Animalia	Animalia
Phylum:	Chordata	Chordata	Chordata
Class:	Myxini	Chondrichthyes	Actinopterygii
Order:	Myxiniformes	Orectolobiformes	Perciformes
Family:	Myxinidae	Rhincodontidae	Gobiidae
Genus:	*Myxine*	*Rhincodon*	*Lythrypnus*
Species:	*glutinosa*	*typus*	*nesiodes*

The "jawless" fishes—Hagfishes and Lampreys—are truly primitive. They lack true jaws, do not have paired fins, and exhibit a simple, cartilaginous skeletal structure. They have a single nostril located on the top of the head. Their form of locomotion is simple and eel-like. The simplicity of the jaws in this class of fishes limits them to rasping prey. However, this characteristic has not prohibited them from succeeding. Through scavenging and parasitizing, jawless fishes have survived, evolved, and prospered for hundreds of millions of years.

The Cartilaginous fishes of the Class Chondrichthyes—the familiar sharks, skates, and rays as well as the less familiar chimaeras—are more structurally advanced than jawless fishes, even while lacking true bones. They have true jaws and their nostrils are located on both sides of the head, generally under the snout. The skull and jaws are constructed of large, single units, rather than multiple pieces as seen in the Ray-finned fishes. Cartilaginous fishes lack a swim bladder and most rely on large, oily livers for buoyancy. Unlike jawless fishes, they possess paired fins: the pectoral and pelvic fins.

Another differentiating feature is that all cartilaginous species practice internal fertilization and produce either egg cases or live young.

The largest and most diverse class of fishes is, by far, the familiar Ray-finned fishes of the Class Actinopterygii. This group possesses a bony, rather than cartilaginous, skeleton. Like the Cartilaginous fishes, bony fishes also possess true jaws. However, their jaws are composed of many small bones rather than large, cartilaginous units. The skull is also a complex structure of small bones. The wide array of jaw and tooth types in this class has spawned a large variety of feeding systems, including biting, crushing, filter-feeding, sucking, picking, and scraping. Nostrils are found on the upper part of both sides of the head. Bony fishes usually have swim bladders, many of which are complex in structure. They also possess paired fins, but in some the pectoral and pelvic fins may be absent. Their methods of reproduction are wide and varied.

Adaptations to Life in Water

Living completely submerged in water presents a host of challenges. Fishes need to regulate the amount of salt and water in their bodies, extract dissolved oxygen from the water, and possess senses adapted to aquatic life. Beyond the basics of sight and smell, most fishes have a lateral line, a sensory organ that is highly developed to detect the minutest motions in the water. Fishes also have developed many ways to communicate with each other under water. Some grind their teeth, while others grunt by manipulating the air bladder. Some fishes have light-producing organs which they may use to locate each other in the darkness.

Water can be over 900 times more dense than air. Many fishes cruise through this dense solution by undulating the body, caudal peduncle, and tail to create forward thrust. Others move by flapping, fanning, or sculling the fins. Eels move much like snakes, winding their way through the water and over the bottom. Skates undulate their pectoral-fin lobes, whereas rays flap them. Many fishes have developed ingenious forms of locomotion. Searobins 'crawl' across the bottom by 'walking' with their free pectoral-fin rays. Remoras, while able to swim freely, have an adapted dorsal fin that forms a suction disk. This disk allows the Remora to 'hitch a ride' on its host. Some fishes hardly swim at all. The Sargassumfish spends most of its life clinging to and crawling through mats of floating *Sargassum* seaweed.

Identifying Fishes

All fishes change shape and most change color as they develop. Juveniles can be drastically different from their adult counterparts. Adults of the same sex and species often have subtle differences. Numerous fishes are sexually dimorphic, meaning males and females are different in color and form. Many fishes change color and pattern depending on the time of day, time of migration, diet, depth, mood, or breeding phase. Some change color when they are hunting; others change color to appear intimidating. In addition, almost every fish changes color when it is caught, in distress, or after it has died.

Introduction

Depending on the subject at hand, identification can be an easy or daunting task. Some species of fishes are so unique that they do not resemble any other and are thus easy to identify. Others are so similar in appearance that only subtle nuances distinguish one from another. Even though the variety and changes in the color of fishes is great, observed colors and patterns are the most common tools used for identification.

Below are several examples of commonly observed color patterns in fishes.

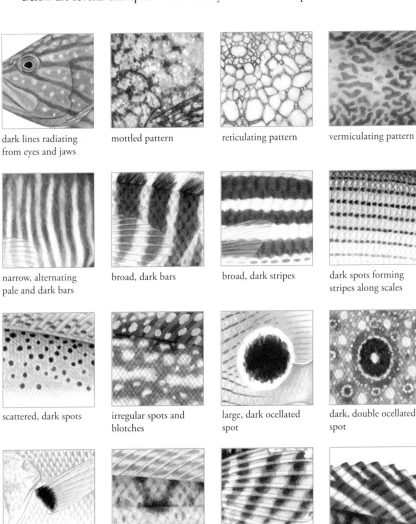

dark lines radiating from eyes and jaws

mottled pattern

reticulating pattern

vermiculating pattern

narrow, alternating pale and dark bars

broad, dark bars

broad, dark stripes

dark spots forming stripes along scales

scattered, dark spots

irregular spots and blotches

large, dark ocellated spot

dark, double ocellated spot

dark blotch on pectoral-fin base

dark, V-shaped saddle

banded pectoral fin

banded dorsal fin

When a fish is not identifiable by either color or pattern, the fish's anatomy can help to secure an identification. Shape, size, and placement of anatomical features vary from one species to the next and thus distinguish one from another. The illustrations on this page show the primary external features of several cartilaginous and bony fishes that are commonly used as tools for identification.

Cartilaginous Fishes

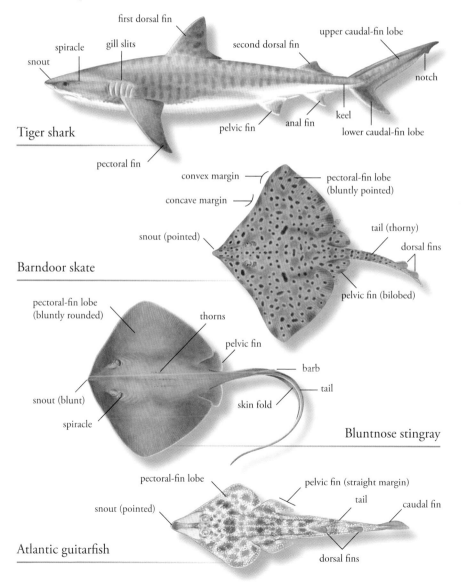

17

Bony Fishes

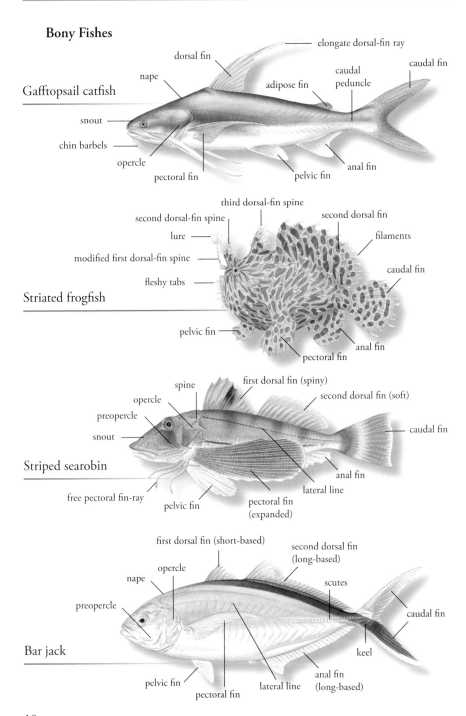

Gafftopsail catfish

- elongate dorsal-fin ray
- dorsal fin
- nape
- caudal peduncle
- caudal fin
- adipose fin
- snout
- chin barbels
- opercle
- pectoral fin
- pelvic fin
- anal fin

Striated frogfish

- third dorsal-fin spine
- second dorsal-fin spine
- lure
- modified first dorsal-fin spine
- fleshy tabs
- second dorsal fin
- filaments
- caudal fin
- pelvic fin
- anal fin
- pectoral fin

Striped searobin

- spine
- first dorsal fin (spiny)
- opercle
- preopercle
- snout
- second dorsal fin (soft)
- caudal fin
- anal fin
- lateral line
- free pectoral fin-ray
- pelvic fin
- pectoral fin (expanded)

Bar jack

- first dorsal fin (short-based)
- second dorsal fin (long-based)
- nape
- opercle
- scutes
- preopercle
- caudal fin
- keel
- pelvic fin
- pectoral fin
- lateral line
- anal fin (long-based)

Lengths and proportions will also help in differentiating one fish from another. Total lengths are used in this book. To determine total length, measure from the tip of the snout or the tip of the lower jaw to the tip of the caudal fin. If the caudal fin is forked, press the upper and lower tips toward each other. The lengths provided in the accounts give the reader a general idea of how big a species may become. The illustration below shows positions of specific identifying features and indicates lengths and depth.

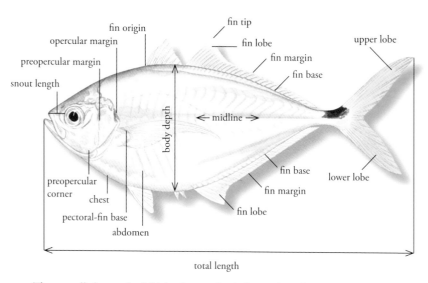

total length

The overall shape of a fish's body can also help in identification. Shapes of the fishes presented in this book are in profile, as a fish would appear from the side. Proportions of a fish's depth relative to its length are important identifiers. A fish is said to be 'deep' when the measurement of depth is great relative to length, such as the Deep-bodied boarfish. A fish is said to be elongate when its depth is small relative to length, such as the Lancetfish. Some examples of profile are shown below.

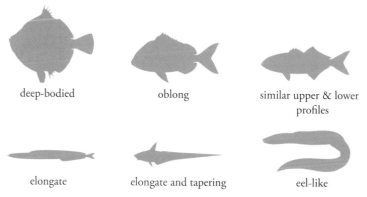

deep-bodied

oblong

similar upper & lower profiles

elongate

elongate and tapering

eel-like

The cross-sectional shape of a fish is the shape of the body as it appears head on. A fish that is flattened from side to side is said to be laterally compressed. A fish that is flattened from top to bottom is described as flattened or depressed. Below are some simplified cross-sectional views.

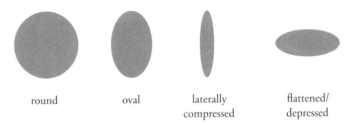

round   oval   laterally   flattened/
                           compressed   depressed

Sometimes it is necessary to go further and count spines, rays, or scales to determine the identity of a fish. When counting spines and rays, the norm is to count anteriorly (front) to posteriorly (toward tail). Even if a spine is very small or directed forward, it is still counted. Additionally, if the last ray on the dorsal or anal fin is split to a unified base, it is still counted as a single ray.

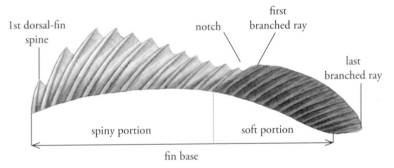

1st dorsal-fin spine   notch   first branched ray   last branched ray

spiny portion   soft portion

fin base

There are times when lateral-line scales need to be counted. These are counted from the first pored, lateral-line scale behind the opercle to the last pored scale that corresponds to the crease in the caudal peduncle when the caudal fin is moved from side to side. At other times, scales above the lateral line are counted. These are counted from the highest arch of the lateral line diagonally backward to the base of the dorsal fin.

The habitat, geographic area, and depth range may also help in identification. For example, the Striped blenny and the Stretchjaw blenny have very similar appearances and features. However, their ranges barely overlap. When using depth as an identifier, it should be noted that many recorded depths are those taken by trawl or line. These records reflect the deepest point of the trawl or line and not necessarily the deepest level at which a fish may swim.

Conservation

Many people have devoted countless hours to the conservation of our precious marine environments. It cannot be overstated how important it is to reduce, and possibly reverse, the harm due to pollution, destruction of habitat, and overfishing. If each person takes one small step toward conservation and preservation, the overall affect could be tremendous.

Many fishes included in this book are currently Threatened or Endangered. The IUCN (International Union for Conservation of Nature) Red List of Threatened Species contains an evaluation of the extinction risk of global plants and animals. The information plays a significant role in guiding conservation activities and serves to monitor changes in the conservation status of species. Of the many fish species that are monitored, sharks are one of the most vulnerable. They have been heavily exploited and are largely misunderstood and misrepresented. Sharks grow slowly, have a long gestational period, and do not produce a large quantity of young. Therefore, it is very difficult for them to recover from the depletion they have suffered.

On the positive side, there are several examples of population recovery due to conservation. One is the Goliath grouper. This mammoth fish was severely overfished in the 1980's. It also suffered loss of spawning habitat. Its decline was compounded by its slow growth and late maturity. Laws were enacted to help its recovery, and in 1990, it became illegal to take or possess a Goliath grouper in US waters. Today, there are promising signs of its recovery, but it is likely that the Goliath will remain a protected animal for many years to come.

We encourage our readers to educate themselves about the fishes they encounter. Most fishes have defense mechanisms meant to protect them from other fishes. Many fishes will defend themselves if threatened. A venomous fish would rather be left alone than use its precious venom. We also encourage our readers to:

· Handle fishes gently.
· Please respect seasonal and catch limits.
· Obtain proper permits and licenses, as many states use sales revenues to fund conservation and enforcement.
· Do not release non-native fishes into any open body of water.
· Whenever possible, practice catch-and-release and respirate the fish before releasing it back into the water.
· Release native live bait back into the water.
· Dispose of waste properly.
· Anchor in designated areas.
· Participate in fisheries management by reporting tagged fishes.
· Use circle hooks to help prevent the fish from swallowing the hook.

In short, if we care for fishes and their environment, they will be here in the future for us to enjoy.

FAMILIES

Order MYXINIFORMES - Hagfishes

Family Myxinidae - Hagfishes

Hagfishes are finless, scaleless, and eel-shaped. They possess a single nostril above the mouth. True jaws are absent. The mouth is circled by a series of barbels and bears rows of keratinized, rasp-like teeth. Eyes are rudimentary and lack lenses and irises. Hagfishes are bottom-dwelling scavengers that occur worldwide in marine and estuarine waters. One shallow water species in the area. Page 62.

Order PETROMYZONTIFORMES - Lampreys

Family Petromyzontidae - Northern lampreys

Lampreys are long, scaleless, and eel-shaped. There is a single nostril above the disk-shaped, jawless mouth. The mouth consists of a series of horny teeth arranged in a circular pattern. A series of gill pouches connect separately to a row of external openings. Dorsal fins are paired on posterior half of body. Lampreys are anadromous or land-locked in northern temperate waters. Adults are parasitic or non-parasitic. One species in the area. Page 62.

Order CHIMAERIFORMES - Chimaeras

Family Rhinochimaeridae - Longnose chimaeras

Longnose chimaeras have a long, pointed, and fleshy snout. Lateral-line canals form grooved lines over the snout and head and along the sides of the body. First dorsal fin is erect with a spine that is toxic in some. The tail tapers to a pointed tip. Pectoral fins are broad. Longnose chimaeras are bottom-dwelling over deep oceanic continental slopes and shelves of the Atlantic, Pacific, and Indian oceans. The snout is used to rout prey from muddy bottoms. Females lay paired egg cases. One species in the area. Page 62.

Family Chimaeridae - Shortnose chimaeras

Shortnose chimaeras have a large head with a blunt snout. Lateral-line canals form grooved lines over the snout and head and along the body. First dorsal fin is erect with a spine that is toxic in some. The tail tapers to a pointed tip. Shortnose chimaeras occur on or near deep muddy bottoms of the Atlantic and Pacific oceans. They swim by flapping broad, wing-like pectoral fins. Females lay paired egg cases. One relatively shallow water species in the area. Page 62.

Order ORECTOLOBIFORMES - Carpet sharks

Family Ginglymostomatidae - Nurse sharks

Nurse sharks are almost cylindrical in shape with a long, low caudal fin. The snout is short and rounded. Nostrils bear obvious barbels. Gill slits are small and spiracles are smaller than eyes. Dorsal fins are similar in size and shape. Nurse sharks are bottom-dwelling and occur worldwide in tropical to subtropical seas. One species in the area. Page 62.

24

Order ORECTOLOBIFORMES - Carpet sharks, *cont.*

Family Rhincodontidae - Whale shark

Whale sharks are the largest fish on Earth. The mouth is wide, and when open, exposes five rows of very long gill plates. The eyes are small and positioned just behind the mouth. Spiracles are similar in size to the eyes and are set just behind the eyes. Three dorsal ridges are present, with the lowest ridge becoming a strong keel at the caudal peduncle. The first dorsal fin is larger than the second, and the caudal fin is very tall. Whale sharks occur worldwide in warm seas. Filterfeed mostly on plankton. One species in this family. Page 64.

Order LAMNIFORMES - Mackerel sharks

Family Odontaspididae - Sand tiger sharks

Sand tiger sharks are stout with a conical, depressed snout. The large mouth extends beyond the eyes and contains protruding teeth. Gill slits are low on the body and anterior to the pectoral fins. Dorsal, pelvic and anal fins are similar in size. Pectoral fins are relatively small. They occur worldwide in warm marine waters. Sand tiger shark embryos feed on lesser developed embryos, fertilized eggs, and unfertilized eggs before birth. Two species in the area. Page 64.

Family Alopiidae - Thresher sharks

Thresher sharks have a round, streamlined body, and a very long, asymmetrical caudal fin. The head is short and the snout is pointed. The forth and fifth gill slits are above the pectoral-fin base. Pectoral fins are as long or longer than the head. Second dorsal fin and anal fin are very small. Thresher sharks occur wordwide in tropical to cold seas. The elongated tail is used to disrupt schooling fishes and to stun prey. Two species in the area. Page 64.

Family Cetorhinidae - Basking shark

Basking sharks are very large and stout. The snout is pointed and may be long. Eyes are relatively small. The large mouth opens to reveal five long gill plates. Gill slits nearly circle the throat and are anterior to pectoral fins. The caudal peduncle possesses a distinct keel. Basking sharks live worldwide in temperate seas and are usually found near the surface. They swim slowly through the water with their mouths agape. Plankton is captured in the numerous gill rakers. One species in this family. Page 66.

Family Lamnidae - Mackerel sharks

Mackerel sharks have a round body and a conical snout. The large mouth has triangular-shaped teeth. Gill slits are located anterior to pectoral fins. Internal gills lack gill rakers. The first dorsal fin is angular and erect. Second dorsal and anal fins are small. Caudal peduncle is heavily keeled. Caudal-fin lobes are similarly sized. Mackerel sharks are found worldwide in tropical to temperate seas. They have specialized circulatory systems that keep them warmer than the ambient water temperature. Ancestors of this family reached a length of 65 ft. Four species in the area. Page 66.

Family Scyliorhinidae - Catsharks

Catsharks are typically small and cigar-shaped with a low, asymmetrical caudal fin. The snout is short and depressed. The mouth extends behind the posterior margin of the small, oval to slit-like eyes. The first dorsal fin is small and originates above or posterior to the origin of the pelvic fins. Catsharks are bottom-dwelling in circumglobal temperate to tropical marine waters from intertidal areas to about 6,500 ft. There are 65 known species in the catshark family. Two shallow water species in the area. Page 68.

Family Triakidae - Hound sharks

Hound sharks are slender with a low, asymmetrical caudal fin, a depressed snout, and slit-like eyes. Gill slits are small and spiracles are close to the eyes. Dorsal and ventral fins are similarly shaped. Hound sharks are primarily demersal and occur worldwide in warm temperate and tropical seas, rarely in fresh water. They occur from near shore to the outer continental shelf and feed on invertebrates and bony fishes. Three species in the area. Page 68.

Family Carcharhinidae - Requiem sharks

The requiem shark family is diverse and consists of about 50 species with several common traits. The head is neither flattened nor laterally expanded. The eyes are circular or oval with well developed nictitating membranes. Spiracles are absent or very small. The mouth is usually large and extends well beyond the eyes. The caudal fin is long, strongly asymmetrical, and with rippled or undulating dorsal margin. Requiem sharks are circumglobal in a wide variety of marine to freshwater habitats. They are strong and active. Seventeen species in the area. Page 68.

Family Sphyrnidae - Hammerhead sharks

Hammerhead sharks are moderately slender with a greatly flattened and laterally expanded head. The eyes and nostrils are on the outer margins of the modified head. Spiracles are absent. First dorsal fin is tall. Hammerhead sharks occur worldwide in warm marine to brackish waters. The flattened head creates more surface area for increased electrochemical perception. The position of the eyes enhances vision. Most hammerhead sharks are shy and difficult to approach. Four species in the area. Page 76.

Order HEXANCHIFORMES - Sixgill sharks

Family Hexanchidae - Cow sharks

Cow sharks are slender to stout in shape and are distinguished by the presence of six or seven long gill slits. Most other shark families have five gill slits. The mouth is very long. Cow sharks have only one dorsal fin which is located close to the caudal fin. The caudal fin is long and low. Cow sharks are circumglobal in shallow to deep marine waters from bays to submarine canyons. Females bear live young. Two shallow water species in the area. Page 76.

Order ECHINORHINIFORMES - Bramble sharks

Family Echinorhinidae - Bramble sharks

Bramble sharks are relatively large and stout. Body and fins are covered with irregular, thorn-like denticles. The dorsal fins are small and similar in size. Anal fin is absent. Caudal fin is low and pointed. Bramble sharks are bottom-dwelling over continental shelves and slopes. Feed on other sharks, fishes, and crabs. Litters contain 15 to 24 young. One species in the area. Page 78.

Order SQUALIFORMES - Dogfish sharks

Family Squalidae - Dogfish sharks

Dogfish sharks are moderately slender with a conical or depressed head. The spiracles are relatively large and close to the eyes. Gill slits are small and low on the body. Dorsal fins are small, with concave margins, and have a small or prominent grooveless spine. The anal fin is absent. The caudal fin is low and asymmetrical. Dogfishes occur worldwide in arctic to tropical seas. They have strong jaws and prey mainly on fishes and invertebrates. Some form large schools. Four species in the area. Page 78.

Family Centrophoridae - Gulper sharks

Gulper sharks are small to moderate in size and cylindrical in shape. The eyes are large. Spiracles are relatively large and just behind eyes. Gill slits are low on the body. Dorsal fins have a strong, grooved spine which may be short to long. Anal fin is absent. Caudal fin is asymmetrical and notched. Gulper sharks are demersal in worldwide tropical to warm temperate seas. They prey on a wide variety of benthic organisms. One shallow water species in the area. Page 78.

Family Etmopteridae - Lantern sharks

Lantern sharks are small and cylindrical in shape. The snout is short to moderately long. Eyes are large and well developed. Spiracles are large and set just behind eyes. The snout is short. Gill slits are small and low on the body or near body midline. Dorsal fins possess a strong, grooved spine. First dorsal fin is usually smaller than second. Anal fin is absent. Caudal fin is low and notched. Lantern sharks are darkly colored and usually have luminescent organs on the abdomen, over the pelvic fin, and on the caudal peduncle and base. They occur worldwide in deep water over continental shelves and slopes. Some form large schools. Four shallow water species in the area. Page 80.

Family Somniosidae - Sleeper sharks

The sleeper shark family includes small to giant species. Spiracles are large and set just behind the eyes. The gill slits are low on the body. The dorsal fins usually lack small, grooved spines. When present, spines occur on both dorsal fins. Anal fin is absent. Caudal fin is strongly asymmetrical and notched. Most possess luminous organs. Sleeper sharks are circumglobal, primarily close to the bottom over continental shelves and slopes. Some are oceanic or nearshore. Prey is diverse. Females bear live young. Two shallow water species in the area. Page 80.

Family Dalatiidae - Kitefin sharks

Kitefin sharks are very small to moderately large. The head is short, the eyes are large. Some have modified jaws to take crater-like bites out of prey. The gill slits are small and anterior to the pectoral fins. Dorsal fins either lack spines, or the first dorsal fin may have a small, grooved spine. Luminous organs are present on ventral surface. Kitefin sharks are typically demersal in deep water over circumglobal continental slopes. One shallow water species in area. Page 80.

Order SQUATINIFORMES - Angel sharks

Family Squatinidae - Angel sharks

Angel sharks have flattened heads, bodies, and fins. The eyes and spiracles are on top of the head. The nostrils and mouth are at the front of a rounded snout. Pectoral fins are separate and triangular. Dorsal fins are near the end of the tail. Angel sharks occur worldwide at or near bottom in tropical to temperate seas, from shore to about 4,200 ft. One species in the area. Page 82.

Order TORPEDINIFORMES - Electric rays

Family Torpedinidae - Torpedo electric rays

Torpedo electric rays are flattened with head, body, and pectoral fins forming an almost circular disc. The eyes and spiracles are on top of the head. Pelvic fins are rounded. First dorsal fin is larger than second. Caudal fin is triangular. Kidney-shaped electric organs are visible from above. They are circumglobal in tropical to temperate seas. Can discharge up to 45 volts to stun prey and defend against predators. One species in the area. Page 82.

Family Narcinidae - Numbfishes

Numbfishes have flattened bodies and pectoral fins that form an oblong disc. The eyes and spiracles are on top of the head. Pelvic fins have an almost straight margin. Dorsal fins are large and almost equal in size. Caudal fin is triangular. A pair of electric organs is visible from above. Numbfishes are demersal and circumglobal in tropical to warm temperate seas. Electric charge is used as defense and to stun prey. One shallow water species in area. Page 82.

Order PRISTIFORMES - Sawfishes

Family Pristidae - Sawfishes

Sawfishes are long and flattened. The eyes and spiracles are on top of the head. The skin is rough. The pectoral, pelvic, and dorsal fins are triangular in shape. The rostrum is modified into a long, narrow, flattened 'saw.' The saw is armed on each side with many sharp teeth and is used to flush benthic prey or to slash and disable schooling fishes. Sawfishes are demersal in worldwide tropical to subtropical coastal waters. Two species in area. Page 82.

Family Rhinobatidae - Guitarfishes

Guitarfishes are moderately flattened with a wedge-shaped snout. The pectoral fins are moderately broad and rounded. Eyes and spiracles are on top of the head. Dorsal fins are located on top of an elongated caudal peduncle. The skin is rough and covered in denticles. Their small, rounded teeth are used to crush bottom-dwelling crustaceans and mollusks. Guitarfishes are demersal over shallow sandy and muddy bottoms of coastal areas in tropical to warm temperate seas. One species in the area. Page 84.

Family Rajidae - Skates

Skates are flattened with broadly expanded pectoral fins. The snout may be elongate or blunt. The pelvic fins are typically bilobed and the tail is moderately slender. Eyes and spiracles are on top of the head. Most have denticles and thorns dorsally. Skates are bottom-dwelling in worldwide polar to tropical seas. Female skates deposit large fertilized eggs in rectangular, leathery cases that are often called 'mermaids' purses.' Swim by undulating the pectoral fins. Over 240 species in the family. Fourteen shallow water species in the area. Page 84.

Order MYLIOBATIFORMES - Stingrays

Family Urotrygonidae - American round stingrays

American round stingrays are small to moderate in size. The flattened body and pectoral fins form an oval to almost round disc. Eyes and spiracles are on top of the head. The tail is slender, almost as long as the body, and has one or more venomous spines. Dorsal fins are absent. American round stingrays occur in warm, shallow waters of the Atlantic and eastern Pacific. They conceal themselves in sand or mud. One species in the area. Page 88.

Family Dasyatidae - Whiptail stingrays

Whiptail stingrays are moderate to very large in size. The body is flattened, expanded, and rhomboid or oval in shape. The pectoral fins are very broad and extend to the tip of the snout. The snout may be pointed or blunt. Eyes and spiracles are on top of head. Pelvic fins are single-lobed. Caudal fin is absent. The tail is long and whiplike and possesses one or more serrated, venomous spines. The spine is used in self defense. Whiptail stingrays occur worldwide in tropical to warm temperate seas. Most are demersal. They prey on fishes and invertebrates. Five species in the area. Page 90.

Family Gymnuridae - Butterfly rays

Butterfly rays are very flattened and laterally expanded. The pectoral fins and body form a diamond-shaped disc. Eyes and spiracles are on top of the head. The tail is short and pointed and may have a serrated spine. Butterfly rays occur worldwide in tropical to temperate seas. Demersal over shallow coastal sandy and muddy bottoms, also in estuaries and river mouths. Feed on benthic invertebrates and fishes. Two species in the area. Page 92.

Order MYLIOBATIFORMES - Stingrays, *cont.*

Family Myliobatidae - Eagle rays

 Eagle rays are moderate to large in size, flattened, and rhomboid to diamond-shaped. The pectoral fins are broad and pointed and attached to the elevated head. Eyes and spiracles are on the sides of the head. Tail may be long and whiplike and may have a serrated spine. Caudal fin absent. Eagle rays occur worldwide in tropical to temperate seas. Some feed on mollusks and crustaceans, others on plankton. Eagle rays swim in a flapping motion. May leap from water in pursuit of prey. Six species in the area. Page 92.

Order ACIPENSERIFORMES - Sturgeons

Family Acipenseridae - Sturgeons

Sturgeons are elongate and stout with five rows of keeled, bony scutes on body. The head is covered in bony plates. The mouth is inferior and preceded by four barbels. Sturgeons occur in Northern Hemisphere fresh water or are anadromous. Demersal. They are modern descendants of primitive fishes. Females scatter thousands of eggs. Two species in the area. Page 94.

Order LEPISOSTEIFORMES - Gars

Family Lepisosteidae - Gars

Gars are elongate and cylindrical with a body covered in diagonal, rhomboid scales. The jaws are long and forceps-like. Single dorsal fin is close to the caudal fin which is attached to the tail at an angle. Gars occur in fresh and brackish North American waters. They have an air bladder that allows them to gulp air. Eggs are reported to be toxic. Three species in the area. Page 96.

Order ELOPIFORMES - Tenpounders

Family Elopidae - Tenpounders

Tenpounders are elongate and cylindrical in shape. The upper jaw extends past the eyes. The single dorsal fin has a concave margin. The caudal fin is deeply forked. Tenpounders occur primarily in coastal waters of tropical and subtropical oceans. Some enter brackish or fresh water. Tenpounder larvae are transparent and ribbon-like. One species in the area. Page 96.

Family Megalopidae - Tarpons

Tarpons are elongate and moderately compressed with a single dorsal fin and a deeply forked caudal fin. The mouth is large and upturned, with the upper jaw extending past the eyes. Anal fin is long-based. Scales are large. The tarpon family consists of two species: one in the Indo-Pacific, the second in the Atlantic Ocean. They are primarily marine but may enter fresh water. Juveniles are often in estuaries and around mangroves. Tarpons are able to gulp air from the water surface. One species in the area. Page 96.

Order ALBULIFORMES - Bonefishes

Family Albulidae - Bonefishes

Bonefishes are elongate and cylindrical in shape with a single dorsal fin and a deeply forked caudal fin. The snout is conical and the mouth is subterminal. The body is translucent. Bonefishes occur near shore in worldwide tropical to warm temperate seas. They are active fishes that forage over sandy and muddy bottoms for invertebrates and small fishes. Bonefish larvae are transparent and ribbon-like. A thick coating of slime covers the skin. One described and at least two possibly undescribed species in the area. Page 96.

Order ANGUILLIFORMES - Eels

Family Anguillidae - Freshwater eels

Freshwater eels are long and round in cross-section. The dorsal and anal fins merge with the caudal fin. The snout is short and acute to rounded. Lower jaw protrudes. Teeth are small. Pectoral fins are well developed. The body is scaled. Freshwater eels occur in tropical to temperate eastern Atlantic, Indian, and western Pacific oceans. They are usually catadromous, living in fresh water and spawning at sea. Larvae remain at sea for up to two years, move back to shore, mature, and migrate into fresh water. One species in the area. Page 98.

Family Muraenidae - Moray eels

Moray eels are elongate and somewhat laterally compressed. The dorsal and anal fins merge with the caudal fin. The snout is short to relatively long. Teeth are well-developed. Pectoral fins are absent. The body is scaleless. Moray eels occur worldwide in primarily tropical to subtropical seas. Most are marine, inhabiting rocky or coralline holes and crevices. Some live around mangroves and tidal creeks. They are predators and scavengers. The flesh of some is toxic. Nineteen species in the area. Page 98.

Family Ophichthidae - Snake eels and Worm eels

Snake and Worm eels are elongate and serpentine or worm-like. The caudal fin is usually absent with the tail ending as a hardened tip. When present, dorsal, anal, and caudal fins are confluent. Snout is short to moderately elongate. Body is scaleless. Pectoral fins are resent or absent. They occur in a variety of worldwide tropical to warm temperate marine and brackish habitats. Thirty-three species in the area. Page 102.

Family Congridae - Conger eels

Conger eels are elongate, serpentine, and medium to very large in size. Dorsal, anal, and caudal fins are confluent. The snout is short to long and usually longer than the lower jaw. Upper lips are usually grooved. Eyes are large and well developed. Pectoral fins are usually present. They occur at or near bottom in worldwide tropical to warm temperate seas. Some species form large or small colonies. Many burrow or seek shelter during the day and forage at night. Nineteen shallow water species in the area. Page 104.

Order CLUPEIFORMES - Herrings

Family Engraulidae - Anchovies

Anchovies are relatively small, with a blunt, rounded snout and a single dorsal fin. The snout extends beyond the jaws. Jaws are long and slender and extend almost to end of gill cover. Eyes are large. Lateral line is absent. All have a thin to wide silvery stripe on each side. Scales are delicate and easily shed. Anchovies occur worldwide in tropical to temperate marine and brackish waters. Most are schooling and filter-feed on plankton. Eight species in the area. Page 106.

Family Clupeidae - Herrings

Herrings are cylindrical in shape or laterally compressed. The body is typically silvery. The mouth is usually upturned and may have a notch at the upper jaw tip. A row of scutes is usually present along the abdomen. The dorsal fin is single, the caudal fin is deeply forked. Herrings occur worldwide in tropical to polar seas. They are usually marine, coastal, and schooling. Some tolerate low salinities, others are anadromous. Most are plankton filter-feeders with numerous gill rakers. Twenty-three species in the area. Page 106.

Order SILURIFORMES - Catfishes

Family Ariidae - Sea catfishes

Sea catfishes are moderately elongate with long barbels around a broad mouth. The head is depressed and has a bony shield. Dorsal and pectoral fins possess serrated spines. An adipose fin is always present. The skin is scaleless. Sea catfishes occur worldwide in tropical to warm temperate marine, brackish, and fresh water. Some form schools. In most species, the male mouth-broods eggs. Two species in the area. Page 114.

Order ARGENTINIFORMES - Marine smelts

Family Argentinidae - Argentines or Herring smelts

Argentines are elongate with a small mouth and large eyes. The single dorsal fin is located at the midbody line. An adipose fin is always present. Scales are easily shed. They are pelagic or demersal over outer shelves and upper slopes of worldwide tropical to warm temperate seas. Prey on planktonic invertebrates and small fishes. Three shallow water species in the area. Page 116.

Order OSMERIFORMES - Freshwater smelts

Family Osmeridae - Smelts

Smelts are small and elongate with a large mouth and large eyes. The single dorsal fin originates posterior to the mid-body line. An adipose fin is always present. Smelts occur in the Northern Hemisphere, in coastal marine and fresh waters. They are anadromous and schooling. Smelts are an important commercial and environmental food source. Two species in the area. Page 116.

Order SALMONIFORMES - Trouts

Family Salmonidae - Salmonids

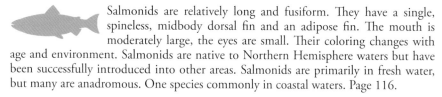

Salmonids are relatively long and fusiform. They have a single, spineless, midbody dorsal fin and an adipose fin. The mouth is moderately large, the eyes are small. Their coloring changes with age and environment. Salmonids are native to Northern Hemisphere waters but have been successfully introduced into other areas. Salmonids are primarily in fresh water, but many are anadromous. One species commonly in coastal waters. Page 116.

Order AULOPIFORMES - Lizardfishes

Family Aulopidae - Flagfins

Flagfins are elongate and oval in cross-section. The mouth is wide and toothy. The single dorsal fin is expanded and originates on the anterior one-third of the body. An adipose fin is present. Flagfins are demersal over continental shelves and slopes of worldwide tropical to warm temperate seas. One species in the area. Page 116.

Family Synodontidae - Lizardfishes

Lizardfishes are elongate and cylindrical in shape. The mouth is wide and toothy. The single dorsal fin is located over the midbody line. An adipose fin is present. Lizardfishes are demersal over a variety of bottoms in the Atlantic, Pacific, and Indian oceans. Some occur in brackish waters. They are swift, voracious predators. Eight species in the area. Page 118.

Family Chlorophthalmidae - Greeneyes

Greeneyes are small and slender. The mouth is large. Eyes are large, pupil is teardrop-shaped. Dorsal fin inserts on anterior third of body. Adipose fin is present. They are demersal in deep tropical to temperate Atlantic, Pacific, and Indian oceans. Two species in the area. Page 120.

Family Alepisauridae - Lancetfishes

Lancetfishes are elongate, slender, and laterally compressed. The mouth is large and toothy and extends past the large eyes. The dorsal fin is long-based and sail-like. A small adipose fin is present. The caudal peduncle has a lateral keel. Lancetfishes are found worldwide in warm temperate to tropical seas from surface to about 3,300 ft. Two species in the area. Page 120.

Order MYCTOPHIFORMES - Lanternfishes

Family Myctophidae - Lanternfishes

Lanternfishes are small, dark, and slender to somewhat deep-bodied, with a large head and eyes. The snout is short and pointed to blunt. The mouth is very large. An adipose fin is present. Most possess a series of photophores along abdomen and on sides. Lanternfishes occur worldwide from tropical to polar seas. Family account only.

Family Lampridae - Opahs

Opahs are deep-bodied, compressed, and oval in shape. The mouth is small and protrusible. The pectoral fins are vertically oriented. Dorsal and anal fins are long-based. Scales are minute and reflective. Opahs are pelagic and found worldwide in tropical to temperate seas. They may wander north during summer. They swim by flapping their pectoral fins. One species in the area. Page 120.

Family Lophotidae - Crestfishes

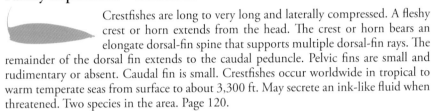

Crestfishes are long to very long and laterally compressed. A fleshy crest or horn extends from the head. The crest or horn bears an elongate dorsal-fin spine that supports multiple dorsal-fin rays. The remainder of the dorsal fin extends to the caudal peduncle. Pelvic fins are small and rudimentary or absent. Caudal fin is small. Crestfishes occur worldwide in tropical to warm temperate seas from surface to about 3,300 ft. May secrete an ink-like fluid when threatened. Two species in the area. Page 120.

Family Trachipteridae - Ribbonfishes

Ribbonfishes are elongate and laterally compressed. The body is deep anteriorly and tapers to a very narrow caudal peduncle. The eyes are large and the mouth is protrusible. The dorsal fin extends from the top of the head to the tail, and the first several rays are elongate in some species. Caudal fin is upturned and fan-like in some. Anal fin is absent. Pelvic fins are well-developed to rudimentary. Ribbonfishes occur worldwide in warm temperate to tropical seas from surface to about 4,000 ft. Three species in the area. Page 120.

Family Regalecidae - Oarfishes

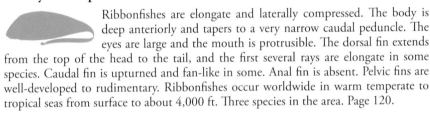

Oarfishes are very long and laterally compressed. The head is angular, the eyes are large, and the mouth is highly protrusible. The dorsal fin runs from the top of the head to the tail with anterior rays that are long and trailing. The caudal fin is usually absent in large specimens. Anal fin is always absent. Pelvic fins are long and trailing and consist of a single ray with paddle-like projections. Oarfishes occur worldwide in warm temperate to tropical seas from surface to about 3,300 ft. Swim vertically in water column by undulating the dorsal fin. One species in the area. Page 122.

Family Polymixiidae - Beardfishes

Beardfishes are relatively deep-bodied and compressed. The snout is short and rounded, the eyes are large. Lower jaw has a pair of long barbels. Single dorsal fin is long-based. The caudal fin is forked. Beardfishes live near bottom over shelves and upper slopes of worldwide tropical to temperate seas. They use barbels to locate benthic prey. Two species in the area. Page 122.

Family Bregmacerotidae - Codlets

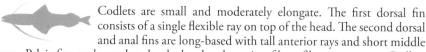 Codlets are small and moderately elongate. The first dorsal fin consists of a single flexible ray on top of the head. The second dorsal and anal fins are long-based with tall anterior rays and short middle rays. Pelvic fins are located under the head and consist of long filamentous rays. Codlets occur worldwide in tropical to subtropical seas. Most are pelagic from near surface to mid-depths of about 6,500 ft. Some occur in coastal waters and estuaries. Four species in the area. Family account only.

Family Macrouridae - Grenadiers

Grenadiers have a large head, a pointed to blunt snout, and a body that tapers to a pointed tail. The eyes are very small to large. A small chin barbel is usually present. The first dorsal fin is short-based and relatively tall. The second dorsal fin is long-based, low, and continuous with the anal fin. There is a gap between the first and second dorsal fins. Caudal fin is absent. Some possess a light organ. They occur worldwide in tropical to polar seas. Most live near bottom from about 330 to 13,000 ft. Eight shallow water species in the area. Page 122.

Family Moridae - Deepsea cods

Deepsea cods are relatively elongate with one, two, or rarely, three dorsal fins. They have one or two anal fins. The pelvic fins are small or filamentous. Caudal peduncle is narrow and caudal fin is small. Chin barbel is present or absent. Some possess a light organ. Deepsea cods are found near bottom of worldwide, deep continental shelves and slopes to about 8,200 ft. Six relatively shallow water species in the area. Family account only.

Family Merlucciidae - Merlucciid hakes

Merluccid hakes are relatively elongate and laterally compressed posteriorly. The snout is long and depressed. The jaws are large and have strong, pointed teeth. There is a V-shaped ridge on the head. The first dorsal fin is short-based, the second is long-based and notched at midlength. Chin barbel is absent. Merluccid hakes occur in the northern and southern Atlantic, eastern Pacific, and western South Pacific oceans. Most are found near bottom over continental shelves and upper slopes. Some occur inshore. Some are of commercial importance. Two species in the area. Page 122.

Family Phycidae - Phycid hakes

Phycid hakes are relatively elongate, soft-bodied, and rounded in cross-section anteriorly. The mouth is large. The snout is rounded to moderately long. Chin barbel is present. Two dorsal fins are usually present. Rarely, the first dorsal fin is a single ray followed by a series of short rays and a long-based third fin. Anal fin is long-based. Pelvic fins are typically long and slender. Caudal fin is well-developed. Phycid hakes occur in the Atlantic, western Pacific and western Indian oceans. They are demersal primarily over soft bottoms from near shore to upper continental slopes. Eight species in the area. Page 124.

Order GADIFORMES - Cods, *cont.*

Family Gadidae - Cods

Cods are small to large with moderately elongate bodies and one, two, or three dorsal fins. The mouth is large and a chin barbel is usually present. The caudal fin is well-developed and symmetrical. V-shaped ridge on head is absent. The swim bladder is not connected to the skull. Cods are typically cold water marine fishes found in the Northern Hemisphere. Most are demersal. Seasonal migrations are common. Some are able to detect food with taste buds located on the barbels and pectoral fins. Cods are commercially important but have been overfished. Five species in the area. Page 126.

Order OPHIDIIFORMES - Cusk-eels

Family Carapidae - Pearlfishes

Pearlfishes have elongate, scaleless bodies that taper to a pointed tail. The anal fin originates anteriorly on the body and its rays are taller than the dorsal-fin rays. Chin barbel and pelvic fins are absent. Pearlfishes occur worldwide in tropical to warm temperate seas. They are found from near shore to about 5,200 ft. Some are free swimming. Others live commensally in sea cucumbers, starfishes, clams, or tunicates. Two shallow water species in the area. Page 128.

Family Ophidiidae - Cusk-eels

Cusk-eels are elongate and scaled. The anal fin originates posteriorly to the dorsal-fin origin. Dorsal and anal fins are continuous with caudal fin and have rays of similar height. Chin barbel and opercular spine may be present or absent. Pelvic fins are usually present and are located on the throat. Cusk-eels are bottom-dwelling in shallow to deep waters of worldwide tropical to temperate seas. Sixteen shallow water species in the area. Page 128.

Family Bythitidae - Viviparous brotulas

Viviparous brotulas are moderately elongate. Dorsal and anal fins are either continuous with or free of caudal fin. Scales are present in most. Chin barbel is absent. Opercle has a well-developed spine. Pelvic fins, when present, have one or two rays. Viviparous brotulas are bottom-dwelling in worldwide tropical to temperate seas. Some occur in freshwater caves. Females give birth to live young. Four species in the area. Page 132.

Order BATRACHOIDIFORMES - Toadfishes

Family Batrachoididae - Toadfishes

Toadfishes are small to medium in size. The head is broad and flattened, with eyes on top. Barbels or fleshy tabs are sometimes around mouth and head. First dorsal fin has two or three spines. Pelvic fins are on the throat. Toadfishes are demersal and occur worldwide in tropical to temperate waters. Most are coastal, some live in freshwater or on continental shelves. They are sluggish but voracious predators. Four species in the area. Page 134.

Order LOPHIIFORMES - Anglerfishes

Family Lophiidae - Goosefishes

 Goosefishes are moderately to greatly flattened with a broad head and body that tapers to the tail. The mouth is broad, toothy, and usually bordered by fleshy tabs. Gill openings are behind the pectoral fins. The first dorsal-fin spine may be isolated on the snout and act as a lure. Goosefishes are bottom-dwelling in worldwide tropical to temperate seas. Three shallow water species in the area. Page 134.

Family Antennariidae - Frogfishes

 Frogfishes are very small to medium-sized with rounded bodies. The mouth is large, oblique, and toothy. Gill openings are located behind the pectoral fin. The first dorsal-fin spine is separate and modified and bears a lure. Pectoral fins are elongate and leg-like. Frogfishes occur worldwide in tropical to subtropical seas. They use their pectoral fins to cling to or 'walk' across substrate. They are usually well camouflaged, often blending with surroundings. Six species in the area. Page 136.

Family Ogcocephalidae - Batfishes

 Batfishes are flattened with head and body forming a circular to triangular disk. The rostrum may be short or long. A cavity under the snout contains a short lure. The mouth is small. Gill openings are behind the limb-like pectoral fins. The body is covered in tubercles and/or bucklers. Batfishes are bottom-dwelling in worldwide tropical to subtropical seas from shore to about 3,000 ft. Use pelvic and pectoral fins to 'walk' across the bottom. Ten species in the area. Page 138.

Order MUGILIFORMES - Mullets

Family Mugilidae - Mullets

 Mullets are medium to large in size. The head is typically broad and flattened. The eyes are usually partly covered by adipose lids. The snout is short and the mouth is small or moderate in size. First dorsal fin has four spines. Pectoral fins are high on the body. Mullets occur worldwide in tropical to temperate salt, brackish, and fresh water. Most tolerate varying salinities. They feed by filtering detritus or plankton. Five species in the area. Page 142.

Order ATHERINIFORMES - Silversides

Family Atherinopsidae - New World silversides

New World silversides are typically small, translucent fishes with a silvery stripe on each side. The upper jaw is protractile. They have two well-separated dorsal fins, the first with two to nine spines. Pectoral fins are high on the body. New World silversides occur in marine to fresh water of North, Central, and South America. They form small to large schools near the surface. Usually omnivorous planktivores. Six species in the area. Page 144.

Family Atherinidae - Old World silversides

 Old World silversides are small translucent fishes with a silvery stripe on each side. The stripe usually has a dark upper border. The upper jaw is not protractile. They have two well-separated dorsal fins, the first with two to five spines. Pectoral fins are high on the body. Old World silversides mainly occur in warm marine waters of the Atlantic and Indo-West Pacific oceans. Some are found in estuaries or fresh water. They form large schools and feed on a variety of planktonic organisms. Two species in the area. Page 144.

Order BELONIFORMES - Needlefishes

Family Exocoetidae - Flyingfishes

 Flyingfishes are small to medium-sized with very long pectoral fins that are set high on the body. The pectoral fins almost always extend past the dorsal-fin origin. Pelvic fins are usually expanded. The caudal fin is deeply forked with a long lower lobe. The lateral line is along the lower margin of the body. Flyingfishes live at the surface from inshore to well offshore in all tropical to subtropical oceans. Known for leaping and gliding for long distances. Feed on zooplankton. Eleven nearshore species in the area. Page 146.

Family Hemiramphidae - Halfbeaks

Halfbeaks are elongate and slender. The upper jaw is short; the lower jaw is usually very long with a fleshy tip. Single dorsal and anal fins are near the tail. The pectoral fins are short to long and set high on the body. Scales are easily shed. Lateral line runs near lower margin of body. Halfbeaks occur near surface of all tropical to warm temperate seas. They are typically omnivorous and feed on a variety of grasses, invertebrates, and small fishes. Some are actively sought as baitfish for game fishing. Seven species in the area. Page 150.

Family Belonidae - Needlefishes

Needlefishes are long and slender with long, pointed jaws. The mouth has many sharp teeth. Single dorsal and anal fins are near the tail. Caudal fin may be emarginate or asymmetrical. Body is either oval or round in cross-section. Needlefishes occur worldwide in tropical to warm temperate seas. Some enter brackish and fresh water. They are usually found near the surface and feed primarily on small, schooling fishes. Seven species in the area. Page 152.

Family Scomberesocidae - Sauries

Sauries are elongate and slender with long, pointed jaws. Dorsal and anal fins are near the tail and are followed by five to six finlets. Caudal fin is deeply forked and symmetrical. The body is covered in small scales. Sauries are found in open waters of tropical to temperate seas. They are schooling and feed on a wide variety of small prey items. Sauries are an important prey item for many marine animals, including squids, billfishes, sharks, tunas, marine mammals, and sea birds. One species in the area. Page 154.

Order CYPRINODONTIFORMES - Killifishes

Family Rivulidae - New World rivulines

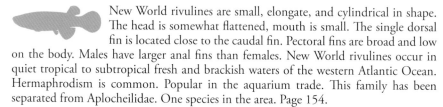

New World rivulines are small, elongate, and cylindrical in shape. The head is somewhat flattened, mouth is small. The single dorsal fin is located close to the caudal fin. Pectoral fins are broad and low on the body. Males have larger anal fins than females. New World rivulines occur in quiet tropical to subtropical fresh and brackish waters of the western Atlantic Ocean. Hermaphrodism is common. Popular in the aquarium trade. This family has been separated from Aplocheilidae. One species in the area. Page 154.

Family Fundulidae - Topminnows

Topminnows are small with elongate to moderately deep bodies. The head is usually flattened. The snout is short and the mouth is protrusible. The lower jaw protudes beyond the upper. The single dorsal fin is located posterior to midlength. Males have larger anal fins than females. Females are larger than males. Topminnows occur in tropical to temperate fresh to coastal marine waters of the western Atlantic. Many adapt to a wide range of salinities and temperatures. Ten species in the area. Page 154.

Family Cyprinodontidae - Pupfishes

Pupfishes are small, long, and robust to short and deep-bodied. The single dorsal fin is located posteriorly. Pectoral fins are set low on the body. Caudal peduncle and caudal fin are broad. Females are larger than males. Pupfishes occur in tropical to temperate fresh, brackish, and marine waters of the Americas and West Indies. Most occur in shallow water. Pupfishes adapt to varying conditions. Two species in the area. Page 158.

Family Poeciliidae - Livebearers

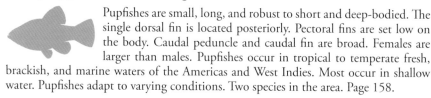

Livebearers are small with elongate to moderately deep bodies. The head is flattened. The single dorsal fin is located posteriorly and may be fan-like. Caudal peduncle is elongate, caudal fin is broad. Livebearers occur in quiet fresh, brackish, and marine coastal waters of the Americas. Most live near the surface. Most males have a modified anal fin used for internal fertilization of females. Females may carry more than one brood and give birth to live young. Some livebearers hybridize and produce only females. Popular in aquarium trade. Seven species in the area. Page 158.

Order BERYCIFORMES - Alfonso squirrelfishes

Family Trachichthyidae - Roughies

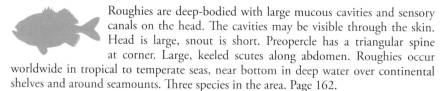

Roughies are deep-bodied with large mucous cavities and sensory canals on the head. The cavities may be visible through the skin. Head is large, snout is short. Preopercle has a triangular spine at corner. Large, keeled scutes along abdomen. Roughies occur worldwide in tropical to temperate seas, near bottom in deep water over continental shelves and around seamounts. Three species in the area. Page 162.

Family Berycidae - Alfonsinos

Alfonsinos are moderate-sized with moderately to deeply compressed bodies. The mouth is large and oblique. Eyes are very large. Preopercle is spineless. Dorsal fin is single and short-based. Alfonsinos occur in the Atlantic, Indian, and Pacific oceans on or near the bottom of continental shelves and slopes. They are also found over seamounts. Sought commercially in many locations. Two species in the area. Page 162.

Family Holocentridae - Squirrelfishes

Squirrelfishes are small to medium-sized with oval to moderately elongate bodies. The eyes are large. The head has ridges and mucous channels. The gill covers are serrated or spiny. Some are reported to have venomous spines on the preopercle. Squirrelfishes are found in the tropical Atlantic, Pacific, and Indian oceans. Most occur around hard bottoms and reefs. They are typically nocturnal. Ten species in the area. Page 162.

Order ZEIFORMES - Dories

Family Grammicolepididae - Tinselfishes

Tinselfishes are deep-bodied and laterally compressed. The head is relatively small. The first dorsal fin is short-based, the second is long-based. First dorsal- and anal-fin spines may be elongate and shorten with age. Scales are vertically elongate. Tinselfishes are scattered in warm, deep Atlantic and Pacific marine waters. Little is known of their habits. Two species in the area. Page 166.

Family Zeidae - Dories

Dories are deep-bodied, laterally compressed, and oval in shape. The mouth is large and protrusible. Body scales are minute or absent. May have large buckler scales along base of dorsal and anal fins. Scutes may be present along abdomen. Pelvic fins may be greatly expanded. Dories occur in relatively deep marine waters over continental shelves and upper slopes of the Atlantic, Indian and Pacific oceans. Commercially important. Two species in the area. Page 168.

Order GASTEROSTEIFORMES - Sticklebacks

Family Gasterosteidae - Sticklebacks

Sticklebacks are small, typically elongate, and scaleless. Some have bony plates on sides. The dorsal-fin spines are isolated and usually stout and sharp. Pelvic fins usually consist of a stout spine and one or two soft spines. Sticklebacks are found in the Northern Hemisphere in marine, brackish, and fresh water. Males build nests, court females, and defend young. Sticklebacks have a wide range of salinity tolerance. Some are entirely marine, one is entirely freshwater, others are anadromous. Four species in the area. Page 168.

Family Syngnathidae - Pipefishes and Seahorses

Pipefishes and seahorses are included in the same family despite the outward difference in appearance. All are elongate with a body covered in a series of bony rings. The mouth is small, the snout is tube-like. Dorsal fin is single. Anal fin, when present, is very small. Pelvic fins are absent in all. Pipefishes and seahorses are found worldwide in warm temperate to tropical marine and brackish waters. Some are found in fresh water. Females deposit eggs into males' pouch or attach eggs to males' abdomen. Males then care for eggs until eggs hatch. Twenty-one species in the area. Page 170.

Family Aulostomidae - Trumpetfishes

Trumpetfishes are elongate with a small mouth and a tube-like snout. The chin bears a small barbel. First dorsal fin is series of isolated spines. Caudal fin is rounded. Trumpetfishes occur in tropical marine waters of the Atlantic and Indo-Pacific. Usually occur around coral reefs. May hover facing downward. Often hunt alongside other fishes. One species in the area. Page 174.

Family Fistulariidae - Cornetfishes

Cornetfishes are slender, elongate, and slightly depressed in shape. The mouth is small and the snout is tube-like. First dorsal fin is absent. Second dorsal and anal fins are similar in shape. Caudal fin is forked, with a long filament trailing from the middle rays. Cornetfishes are found worldwide in tropical to subtropical seas. Occur along coastal areas, over seagrass beds, rubble, and soft bottoms, and around reefs. They may change color to match surroundings. Feed on fishes and shrimps. Two species in the area. Page 176.

Family Macroramphosidae - Snipefishes

Snipefishes are small, laterally compressed, and moderately deep-bodied with a long tube-like snout. Teeth are absent. First dorsal fin has five to eight spines. The second spine is enlarged and serrated along the rear margin. Body has bony plates above pectoral fins and/or along ventral midline. Snipefishes occur worldwide in tropical to subtropical seas. Juveniles are pelagic, adults live near bottom. Two species in the area. Page 176.

Order SCORPAENIFORMES - Mail-cheeked fishes

Family Dactylopteridae - Flying gurnards

Flying gurnards are elongate and squarish in cross-section with a bony head, a blunt snout, and expanded pectoral fins. The top and sides of the head are covered in bony plates. Keeled spines extend from nape. The preopercle bears a long spine. Flying gurnards are demersal over soft bottoms in temperate to tropical waters of the Atlantic and Indo-West Pacific oceans. They use pelvic fin rays to 'walk' across the bottom and to locate prey. Pectoral fins are spread when the fish is alarmed. Able to change color to blend with surroundings. One species in the area. Page 176.

Family Scorpaenidae - Scorpionfishes

 Scorpionfishes are small to moderate in size with a relatively large head bearing spines and ridges. The pectoral fins are rounded to fan-like. The dorsal fin is continuous and notched. Venom glands are usually present at base of dorsal, pelvic, and anal fin spines. Venom may be slightly to highly toxic. They are demersal or pelagic and occur worldwide in tropical to temperate seas. Most are well camouflaged and are adept at ambushing prey. Twenty-two shallow water species in the area. Page 178.

Family Triglidae - Searobins

 Searobins are moderately elongate with a large, bony head that is armored with plates, ridges, and spines. Venom glands and chin barbels are absent. Pectoral fins may be small or very broad. First three pectoral-fin rays are free and fleshy. Searobins are demersal and occur worldwide from near shore to continental shelves of tropical to warm-temperate brackish and marine waters. The free pectoral-fin rays are used for locomotion and to detect benthic prey items. Searobins use the swim bladder for sound production, particularly during spawning. Fifteen species in the area. Page 184.

Family Peristediidae - Armored searobins

 Armored searobins are moderately elongate. The head is large and armored with plates, ridges, and spines. Rows of spiny scutes cover the body. Flattened projections extend from snout and sides of head. Lip and chin barbels are usually present. First two pectoral-fin rays are free and fleshy. Armored searobins are demersal and occur worldwide on continental and insular slopes in deep tropical to temperate seas. Barbels and free pectoral-fin rays are used for locomotion and to locate prey. Five shallow water species in the area. Page 188.

Family Cottidae - Sculpins

Sculpins are moderately elongate with fan-like pectoral fins and large eyes set high on the head. The head is broad with prominent spines on the sides. The body may be scaleless or covered with prickles or plate-like scales. Adults lack a swim bladder. Most occur in cold, northern marine waters, some occur in freshwater. All are bottom-dwelling. Sculpins are well camouflaged and some may inflate themselves with air or water when threatened. Capable of sound production. Six species in the area. Page 188.

Family Hemitripteridae - Searavens

 Searavens are moderately elongate to robust. The head is knobby. The preopercle has blunt, skin-covered spines. Fins are typically broad. The body is covered in tiny prickles. Swim bladder is absent. Most searavens occur in the North Pacific. One species occurs in the northwestern Atlantic. All live on or near bottom. Searavens prey primarily on invertebrates. This is a small family of fishes which is closely related to Poachers. One species in the area. Page 190.

Family Agonidae - Poachers

Poachers are typically elongate with overlapping, bony plates on the body. Most have two dorsal fins, some have one. Swim bladder is absent. Poachers occur in shallow to deep northern waters of the Atlantic and Pacific and also off southern South America. They are sluggish, and females scatter their eggs. Poachers are closely related to Sculpins and Searavens despite the differences in outward appearance. One species in the area. Page 192.

Family Psychrolutidae - Fathead sculpins

Fathead sculpins have a large head and forebody which tapers to a narrow caudal peduncle. Bony arches on the head may bear spines. Dorsal fins are usually continuous. Body is scaleless or has prickly plates. Females are larger than males. Fathead sculpins occur in the Atlantic, Indian and Pacific oceans from shallow inshore waters to about 9,000 ft. Most are demersal. One species in the area. Page 192.

Family Cyclopteridae - Lumpfishes

Lumpfishes have rounded bodies that are usually covered with tubercles. Two short dorsal fins are usually present, the first sometimes covered with skin. Pelvic fins, when present, form a suction disk used to adhere to hard surfaces. Pectoral fins extend ventrally and in front of suction disk. Lumpfishes occur in the cold, northern waters of the Pacific and Atlantic oceans. Most are demersal, some are pelagic. Some migrate hundreds of miles to spawn. After spawning, males defend egg clusters. Some are sought for their eggs, which are sold as caviar. Two species in the area. Page 192.

Family Liparidae - Snailfishes

Snailfishes are small, tadpole-shaped, and soft-bodied. Dorsal and anal fins are confluent, or nearly so, with the caudal fin. Pectoral fins are usually bilobed. Pelvic fins, when present, form a suction disk used to adhere to hard surfaces. Snailfishes are a large family. They are widely distributed and are found in all oceans from the intertidal zone to about 23,000 ft. Some are commensal. Three species in the area. Page 192.

Order PERCIFORMES - Perches

Family Centropomidae - Snooks

Snooks are moderately elongate and laterally compressed. The mouth is large with a protruding lower jaw. Preopercles have a serrated lower margin. Head profile is almost straight to concave. Lateral line is well developed and extends onto caudal fin, which is rounded, truncate, or forked. Dorsal fins are separate. Anal fin has three strong spines, the second being the stoutest. Snooks occur worldwide in shallow, coastal tropical to subtropical marine, brackish, and sometimes fresh water. They are voracious predators sought for sport and food. Four species in the area. Page 194.

Family Moronidae - Temperate basses

Temperate basses are moderately elongate and somewhat laterally compressed. The opercle has two rounded to pointed spines on the posterior margin. The two dorsal fins are completely to nearly separate. The anal fin has three strong spines and the lateral line extends onto the caudal fin. Temperate basses are native in fresh, brackish, and inshore marine waters of North America, Europe, and North Africa. They have been successfully introduced to other areas. Species are either tolerant or intolerant of changes in salinity. Some are important game and food fish. Two species in the area. Page 196.

Family Acropomatidae - Lanternbellies

Lanternbellies are oblong and moderately compressed. The snout is short and rounded. Upper jaw is protrusible. Eyes are large. Opercle has one to two spines on rear margin. Dorsal fins are nearly to completely separate. Three species have light organs. Lanternbellies occur in the water column over continental shelves and slopes of the Atlantic, Indian, and Pacific oceans. Previously called Temperate ocean-basses. Family account only.

Family Polyprionidae - Wreckfishes

Wreckfishes are large, oblong, and moderately compressed. The head has a bony knob between the eyes and one at the nape. The opercle has two spines, with the lower spine at the end of a long, horizontal ridge. Wreckfishes occur in the Atlantic, Indian, and Pacific oceans and in the Mediterranean. Adults are found over continental shelves and seamounts. Juveniles are pelagic. One species in the area. Page 196.

Family Serranidae - Sea basses

Seabasses include a large variety of fishes that share several characteristics. The body is usually robust and the mouth is large. The opercle has three spines; the middle spine is usually the most conspicuous. The dorsal fin is usually continuous, and the anal fin has three spines. Lateral line extends only to end of caudal peduncle. Seabasses occur worldwide from near shore to near bottom of shallow to moderately deep tropical to temperate seas. A few live in fresh water. The majority are hermaphrodites that mature first as females and then change into males. Some function simultaneously as both sexes. Seabasses include important game and food fishes as well as fishes sought for the aquarium trade. Sixty-seven species in the area. Page 198.

Family Grammatidae - Basslets

Basslets are small and moderately elongate. The snout is short and the eyes are large. Opercle has up to two small spines. Dorsal fin is continuous. Pelvic fins are moderately long to filamentous. Lateral line is interrupted or absent. Basslets occur primarily over coral reefs, rocky ledges, and drop-offs of the tropical western Atlantic. They feed on plankton and are often oriented upside down on substrate. Three species in the area. Page 222.

Family Opistognathidae - Jawfishes

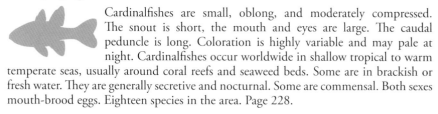

Jawfishes are small and oblong to moderately elongate. The head is rounded. The eyes are large and the jaws are very large. The dorsal fin is continuous and the pelvic fins are located anterior to the pectoral fins. Jawfishes occur worldwide in tropical to subtropical seas on muddy, sandy, and rubble bottoms. Most are found near coral reefs. All live in vertical burrows that each constructs and maintains by moving sediment with its mouth. Males mouth-brood eggs until hatching. Nine species in the area. Page 224.

Family Priacanthidae - Bigeyes

Bigeyes are moderately deep-bodied and laterally compressed with oblique jaws and very large, reflective eyes. The preopercle has a single spine at the lower corner. The dorsal fin is continuous. Pelvic fins are small to greatly expanded with inner rays attached to the abdomen by a membrane. Color of head, body, and irises are shades of red. Bigeyes occur worldwide in tropical to warm temperate seas. They are generally found near bottom around reefs and rock formations, occasionally in open water. Most Bigeyes are nocturnal. Four species in the area. Page 226.

Family Apogonidae - Cardinalfishes

Cardinalfishes are small, oblong, and moderately compressed. The snout is short, the mouth and eyes are large. The caudal peduncle is long. Coloration is highly variable and may pale at night. Cardinalfishes occur worldwide in shallow tropical to warm temperate seas, usually around coral reefs and seaweed beds. Some are in brackish or fresh water. They are generally secretive and nocturnal. Some are commensal. Both sexes mouth-brood eggs. Eighteen species in the area. Page 228.

Family Malacanthidae - Tilefishes

Tilefishes are oblong to elongate with a single, low dorsal fin. The head profile is gently to steeply sloping. The mouth is moderately large and fleshy. A prominent to reduced predorsal ridge is present in most. The opercle has a single blunt or notched spine. Tilefishes are demersal and occur worldwide in tropical to warm temperate seas from shore to about 1,600 ft. They inhabit caves and crevices or may construct mounds or burrows. Some authors divide Tilefishes into two families. Six species in the area. Page 236.

Family Pomatomidae - Bluefish

Bluefish are moderately elongate and compressed with separate dorsal fins and a forked caudal fin. The jaws have prominent, sharp teeth. The lower jaw protrudes slightly and the opercle has a single broad, flat spine. Bluefish occur worldwide in eight populations in subtropical to warm temperate seas from shore to over continental shelves. They are schooling, swift, and ravenous and follow schools of squids and small fishes. Females carry up to one million eggs. Important sportfish. There is one species in this family. Page 238.

Family Coryphaenidae - Dolphinfishes

Dolphinfishes are elongate and laterally compressed. The head profile is rounded in females, steeply sloping in adult males. The spineless dorsal fin originates on the head and reaches to the caudal peduncle. The caudal fin is deeply forked. Dolphinfishes occur worldwide in tropical to warm temperate seas. They live in open ocean waters over continental shelves and slopes from surface to about 660 ft. They are often associated with *Sargassum* seaweed, flotsam, and oil rigs. Sought as food fish. Two species in this family. Page 238.

Family Rachycentridae - Cobia

Cobia are elongate and round in cross-section with a broad mouth and a flattened head. Six to nine short spines precede the second dorsal fin. The pectoral fins are long and pointed, the caudal fin is forked. Cobia occur in tropical to warm temperate waters of the Atlantic and Indo-Pacific over a variety of habitats including shallow coral reefs, rocky shores, and continental shelves to about 165 ft. Cobia are fast-growing carnivores of crabs, benthic invertebrates, and fishes. There is one species in this family. Page 238.

Family Echeneidae - Remoras

Remoras are elongate and round in cross-section. The head is broad and flattened. The first dorsal fin is modified into an oval-shaped cephalic disk that the remora uses to attach itself to a host. Remoras occur worldwide in tropical to warm temperate seas. They are associated with a variety of hosts, including sharks, rays, tarpons, seabasses, cobias, jacks, parrotfishes, billfishes, molas, dolphins, and whales. Some are host-specific, some attach to a range of hosts, and others are free-swimming. Eight species in the area. Page 240.

Family Carangidae - Jacks and Pompanos

Jacks and Pompanos are small to large fishes with body shapes varying from deep and compressed to elongate and fusiform. The eyes have an adipose lid that is either poorly or strongly developed. First dorsal fin may be well-developed or a series of spines. Most have a forked caudal fin. First one or two anal-fin spines are separate and may be embedded. Many have bony scutes along lateral line. Body shape and coloring change dramatically with age. Jacks and Pompanos occur worldwide in tropical to warm temperate brackish to oceanic habitats. Most are schooling and pursue pelagic prey. Twenty-eight species in the area. Page 242.

Family Bramidae - Pomfrets

Pomfrets are medium to large, laterally compressed, and round to teardrop-shaped. The eyes are large, the lower jaw protrudes. Dorsal and anal fins are long-based and may be either low, lobed, or fan-like. Scales covering head and body are large and often keeled. Pomfrets occur worldwide in temperate to warm-temperate seas. Most are pelagic and schooling. Five or more species in the area. Page 252.

Family Emmelichthyidae - Rovers

Rovers are moderately elongate with dorsal and ventral profiles similarly shaped. The mouth is highly protrusible, with the lower jaw projecting beyond the upper. Teeth, when present, are conical and located anteriorly. Most of the head and maxilla are covered in scales. The dorsal fin is either deeply notched or divided. Color is usually reddish dorsally, silvery below. Rovers occur worldwide in tropical to warm temperate seas near bottom over continental shelves and upper slopes. One species commonly in the area. Page 254.

Family Lutjanidae - Snappers

Snappers are elongate and moderately deep-bodied. The upper jaw is moderately protrusible. Well-developed canine-like teeth are usually present. The cheeks are scaled, the snout is scaleless. Maxilla are usually scaleless. Dorsal fin is unnotched to slightly notched. Color is often reddish but may be purplish to grayish with stripes or bars. Snappers occur worldwide in tropical to warm temperate seas from shore to about 1,800 ft. Most are demersal, some live in freshwater. Most are nocturnal. Females spawn several times per season. Seventeen species in the area. Page 256.

Family Lobotidae - Tripletails

Tripletails are oval to oblong, deep-bodied, and laterally compressed. The head profile is steeply sloping. The snout is short and blunt. Second dorsal, anal, and caudal fins are broad and rounded. Tripletails occur worldwide in tropical to warm temperate seas from coastal waters to well offshore. Some live in brackish or fresh water. They are sluggish fishes, often floating on their sides with flotsam or *Sargassum* seaweed. Juveniles are believed to mimic mangrove leaves. One species in the area. Page 262.

Family Gerreidae - Mojarras

Mojarras are moderately slender to deep-bodied and laterally compressed. The upper jaw is highly protrusible. The dorsal fin is continuous with a slight to deep notch between the spiny and soft portions. The bases of the dorsal and anal fins are scaled. Caudal fin is deeply forked. Mojarras occur worldwide in tropical to warm temperate seas. Most are found along the coast, many enter brackish water, some live in freshwater. Eleven species in the area. Page 262.

Family Haemulidae - Grunts

Grunts are oblong, moderately deep-bodied, and compressed. The dorsal head profile is almost straight to convex. The snout is moderately short to long. Mouth is small to moderate with thick lips. Dorsal fin is continuous and either notched or unnotched. Scales are absent on snout and lips. Grunts occur worldwide in tropical to warm temperate seas in shallow coastal, brackish, and occasionally, fresh water. All are capable of sound production. Sixteen species in the area. Page 266.

Family Inermiidae - Bonnetmouths

Bonnetmouths are relatively small, elongate, and fusiform. The upper jaw is highly protrusible. Dorsal fins are either widely separated or divided by a deep notch. The caudal fin is forked. The anal fin is similar in shape to, and located below, second dorsal fin. Bonnetmouths occur in the tropical western Atlantic. They are pelagic in open water and around islands and coral heads to about 980 ft. Bonnetmouths feed on zooplankton and small fishes. Two species in the family; both occur in the area. Page 272.

Family Sparidae - Porgies

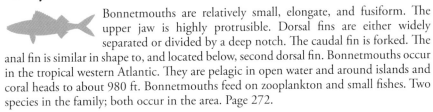

Porgies are small to medium in size and oblong to oval in profile. The body is usually deep and moderately to deeply compressed. The dorsal head profile is usually steep. The mouth is small with a slightly protrusible upper jaw. Porgies in the area have conical or insisor-like front teeth. Teeth in sides of jaws are molar-like. Dorsal fin is continuous and weakly to slightly notched. Porgies are found worldwide in tropical to warm temperate seas. Most occur over continental shelves, some enter estuaries or live in freshwater. Some are hermaphroditic. Fifteen species in the area. Page 272.

Family Polynemidae - Threadfins

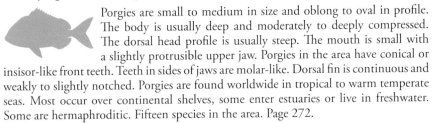

Threadfins are moderately elongate with widely separated dorsal fins and a deeply forked caudal fin. The snout is short and projects beyond the large, inferior mouth. The pectoral fin is low on the body with lower rays that are separate and filamentous. Threadfins occur worldwide in shallow tropical to warm temperate seas. Some enter estuaries and rivers. Some are hermaphroditic. Three species in the area. Page 278.

Family Sciaenidae - Drums

Drums are small to large in size with a gently to steeply sloping dorsal head profile. The body may be short and deep to moderately elongate. All have pores on the snout and chin, a bony flap above the gill opening, and lateral-line scales that extend onto the caudal fin. The first dorsal fin is short-based and usually continuous with the long-based second dorsal fin. A gas bladder is always present and is used to produce drumming sounds. Drums are found in a wide range of habitats in worldwide tropical to warm temperate seas. Some live in estuaries or fresh water. Many are migratory and use estuaries as nursery grounds. Many sought as sport and food fish. Twenty-three species in the area. Page 280.

Family Mullidae - Goatfishes

Goatfishes are moderately elongate with a convex dorsal profile and nearly straight ventral profile. The dorsal fins are separate, the caudal fin is forked. Two well-developed barbels extend from chin. The upper jaw is slightly protrusible. Goatfishes are found near bottom over continental shelves and slopes of worldwide tropical to warm temperate seas. Some enter estuaries. Use barbels to locate prey. Four species in the area. Page 288.

Family Pempheridae - Sweepers

Sweepers are small, compressed, and deep-bodied. The mouth and eyes are large. The dorsal fin is short-based, the anal fin is long-based. Posterior profile of back and caudal peduncle is nearly horizontal. A few possess luminous organs. Sweepers occur over inner continental shelves in tropical to subtropical marine and brackish waters of the western Atlantic, Indian, and Pacific Oceans. They are common over coral reefs and in caves and crevices. Nocturnal and schooling. One species in the area. Page 290.

Family Kyphosidae - Sea chubs

Sea chubs are oval to oblong in profile and moderately compressed. The head is short, the snout is blunt. The mouth is small and horizontal. The dorsal fin is continuous, with the spiny portion depressible into a groove. Sea chubs are found worldwide in tropical to subtropical seas. They are associated with coral reefs, rocky bottoms, hard structures, and *Sargassum* seaweed. Two species in the area. Page 290.

Family Chaetodontidae - Butterflyfishes

Butterflyfishes are oval, round, or rhomboid in profile, deep-bodied, and compressed. The mouth is small and the snout is blunt to long and pointed. Preopercular margin may be serrated, but a prolonged spine is absent at the lower angle. Dorsal fin is continuous to slightly notched. Spiny portion may be tall and deeply incised. Butterflyfishes occur worldwide in shallow tropical to subtropical seas. Most are associated with coral reefs. Some are found in brackish or fresh water. They graze on small invertebrates and plankton. Butterflyfishes usually occur singly or in pairs. Popular in the aquarium trade. Six species in the area. Page 292.

Family Pomacanthidae - Angelfishes

Angelfishes are oval to round in profile, deep-bodied, and compressed. The mouth is small and the snout is blunt. Preopercular margin is often serrated and a prolonged spine is always present at the lower angle. Dorsal fin is continuous. Second dorsal fin may have long, trailing middle rays. Angelfishes occur worldwide in shallow tropical to subtropical seas. They are generally associated with coral reefs, where they graze on sponges, invertebrates, and algae. Often seen singly or in pairs. Popular in the aquarium trade. Six species in the area. Page 294.

Family Cirrhitidae - Hawkfishes

Hawkfishes are small and oval to oblong in shape. The snout is blunt to pointed. Dorsal fin is continuous with cirri projecting from tips of first dorsal-fin spines. The lower pectoral fin is incised with long, unbranched rays. Hawkfishes occur worldwide in shallow tropical seas. Most are found in the Indo-Pacific. They use the thickened lower pectoral-fin rays to move over substrate. One species in the area. Page 298.

Family Pomacentridae - Damselfishes

Damselfishes are small, oval to oblong, and laterally compressed. The mouth is small and oblique. The lateral line is either incomplete or interrupted. The dorsal fin is continuous, with the spiny portion having a longer base than the soft portion. Caudal fin is shallowly to deeply forked. Damselfishes are found worldwide in tropical to warm temperate seas. Most occur in the Indo-Pacific. They occur over reefs, sandy, rubble, and seagrass bottoms. Most are territorial and will defend eggs and feeding areas. Feed on small invertebrates and algae. Fourteen species in the area. Page 298.

Family Labridae - Wrasses

Wrasses are oblong in shape and slightly to strongly compressed. The upper head profile may be gently to steeply sloping. Upper jaw is protrusible, lips are often fleshy. Teeth are separate and canine-like. Dorsal fin is continuous. Second dorsal fin may have elongated middle rays. Scales are relatively large. Wrasses occur worldwide in tropical to warm temperate seas from shore to about 330 ft. Many are reef associated, others are found over seagrass beds and sandy to rocky bottoms. Most wrasses are hermaphroditic. Color and pattern often change with sex. Most bury themselves in sand at night. Second largest family of marine fishes. Twenty-two species in the area. Page 304.

Family Scaridae - Parrotfishes

Parrotfishes are oblong and somewhat compressed. The upper jaw is not protrusible. The snout is usually bluntly rounded, and the front teeth are usually partially fused. Pectoral-fin base is oblique. Dorsal fin is continuous. Scales are large. Parrotfishes occur worldwide in shallow tropical to warm temperate seas. Many are reef associated. Some are found in lagoons, creeks, and over seagrasss beds. Most parrotfishes are hermaphroditic. Color and pattern often change with sex. Some species secrete a cocoon of mucus in which they rest at night. Fourteen species in the area. Page 312.

Family Zoarcidae - Eelpouts

Eelpouts are elongate, rounded to oval in cross-section anteriorly, and compressed posteriorly. Dorsal and anal fins are long-based and confluent with caudal fin. Pectoral fins are large to small. Pelvic fins, when present, are very small and located anteriorly to pectoral fins. Eelpouts are demersal to bottom-dwelling in worldwide seas. Three shallow water species in the area. Page 318.

Family Stichaeidae - Pricklebacks

Pricklebacks are elongate, rounded to oval in cross-section anteriorly, and compressed posteriorly. The dorsal fin is long-based and entirely spiny in most species. Caudal fin is separate. Pectoral fins are small to large. Pelvic fins, when present, are small and located anteriorly to pectoral fins. Pricklebacks are found in cold northern waters of the Atlantic and Pacific from shore to about 820 ft. Four species in the area. Page 320.

Family Cryptacanthodidae - Wrymouths

Wrymouths are slender and eel-like. The mouth is large and strongly oblique. Dorsal fin is entirely spiny and confluent with caudal and anal fins. Pelvic fins are absent. Wrymouths occur in shallow inshore waters of the northwest Atlantic and northern Pacific oceans. Some are found in mud burrows. This is a small family of fishes that is closely related to Pricklebacks. One species in the area. Page 320.

Family Pholidae - Gunnels

Gunnels are small and elongate and laterally compressed. The mouth is relatively small and oblique. Dorsal fin is entirely spiny. Anal fin is about half as long as dorsal fin. Pectoral fins are small or rudimentary. Pelvic fins are very small in most, absent in others. Body is covered in minute embedded scales. Gunnels occur in northern Atlantic and Pacific coastal waters. Often found around rocks or in tide pools. One species in the area. Page 320.

Family Anarhichadidae - Wolffishes

Wolffishes are large, robust, and blenny-like in shape. The jaws are large and strong with large conical teeth anteriorly and molar-like teeth laterally. Dorsal fin is tall, flexible, and entirely spiny. Caudal fin is separate from dorsal and anal fins. Pectoral fins are large and fan-like. Pelvic fins are absent. Body is naked or with minute scales. Lateral line is faint. Wolffishes are found near bottom in coastal northern waters of the Atlantic and Pacific oceans. Females produce large eggs that are believed to be internally fertilized. Some species exhibit brooding behavior. Wolffishes feed on a variety of bottom-dwelling crustaceans, urchins, and starfish. Two species in the area. Page 322.

Family Percophidae - Duckbills

Duckbills are elongate with a flattened head and snout. The mouth is very large. The eyes are large and closely set on top of head. Dorsal fins are separate. First dorsal fin (when present) is short-based, second dorsal fin is long-based. Duckbills are bottom-dwelling over outer continental shelves and upper slopes of the Atlantic, Indo-West Pacific, and southeast Pacific oceans. They feed on shrimps and small fishes. Some authors refer to this family as Flatheads. Two shallow water species in the area. Page 322.

Family Ammodytidae - Sand lances

Sand lances are slender and elongate. The head is small, the snout is pointed. The lower jaw protrudes beyond the upper jaw. Teeth are absent. Dorsal and anal fins lack spines. Caudal fin is separate and forked. Pelvic fins are usually absent. Lateral line is high on the body. Sand lances occur from shore to offshore in the cold to tropical marine waters of the Atlantic, Arctic, Pacific, and Indian oceans. Sand lances swim in an eel-like motion. Some burrow, rather quickly, into bottom sediment. Sand lances are an important prey for larger fishes, marine mammals, and birds. Sought commercially. Two species in the area. Page 322.

Family Uranoscopidae - Stargazers

Stargazers have a large head and a robust body. The mouth is large and oblique to nearly vertical. Lips are fringed. Eyes are either on or near top of head. First dorsal fin is present or absent. Pectoral fins are fan-like. A blunt or sharp spine is present over the pectoral fins. These spines are venomous in most species. The body is naked or covered in small, smooth scales. Stargazers are demersal and occur worldwide in tropical to temperate seas from shore to about 3,000 ft. Some species have a lure-like filament in the mouth. One genus has an electric organ used to stun prey. Four species in the area. Page 324.

Family Tripterygiidae - Triplefin blennies

Triplefins are small, blenny-like fishes. The snout is short, the eyes are large. Nostrils may bear a tentacle or cirri. Eyes may have a tentacle. Nape always lacks cirri. There are three distinct dorsal fins: the first two dorsal fins are spiny, the third is soft. Head and abdomen scaleless in most specimens. Triplefins are demersal over coral reefs and rocky areas from shore to about 1,800 ft. in worldwide tropical seas. Three species in the area. Page 326.

Family Dactyloscopidae - Sand stargazers

Sand stargazers are small and moderately elongate. Eyes are set on top of the head, are very small to somewhat large, and are either stalked or unstalked. Snout is very short. Mouth is oblique to almost vertical and may have fringed lips. Upper edge of opercle is often fringed. The lower portion of the opercle is expanded. The dorsal fin may be continuous, incised anteriorly, or notched. Sand stargazers are demersal in tropical to warm temperate western Atlantic and eastern Pacific oceans. Eight species in the area. Page 326.

Family Blenniidae - Combtooth blennies

Combtooth blennies are small and scaleless. The head is usually very blunt. Nostrils sometimes with fleshy flaps. Eyes often with cirri, fleshy flaps, or tentacles. Mouth is usually small. Comb-like teeth are closely packed in a single row and are fixed or moveable. Some have canine teeth. Pectoral-fin rays are unbranched. Caudal-fin rays are branched or unbranched. Combtooth blennies occur worldwide in tropical to warm temperate seas. All are demersal, most are in shallow water. Eighteen species in the area. Page 330.

Family Labrisomidae - Labrisomid blennies

Labrisomid blennies are small with a body that is usually scaled. The head is blunt to pointed. Cirri are often present on nostrils and eyes. Nape may have a single cirrus, multiple cirri, or fleshy tabs. Mouth is small to moderate with fleshy lips. Teeth are variable in size and arrangement. All fin rays, including caudal, are unbranched. Labrisomid blennies are demersal in the tropical to warm temperate western Atlantic and eastern Pacific oceans. They are generally reef associated in shallow water. Males and females often colored and patterned differently. Nineteen species in the area. Page 336.

Family Chaenopsidae - Tube blennies

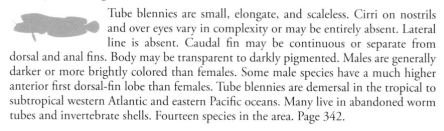

Tube blennies are small, elongate, and scaleless. Cirri on nostrils and over eyes vary in complexity or may be entirely absent. Lateral line is absent. Caudal fin may be continuous or separate from dorsal and anal fins. Body may be transparent to darkly pigmented. Males are generally darker or more brightly colored than females. Some male species have a much higher anterior first dorsal-fin lobe than females. Tube blennies are demersal in the tropical to subtropical western Atlantic and eastern Pacific oceans. Many live in abandoned worm tubes and invertebrate shells. Fourteen species in the area. Page 342.

Family Gobiesocidae - Clingfishes

Clingfishes are small and scaleless. The head and anterior portion of the body are generally rounded and flattened. Eyes are on top of the head. Dorsal fin is single. Pelvic fins are modified into an adhesive sucking disc. Clingfishes occur worldwide in shallow tropical to warm temperate seas and in brackish and fresh water. Most are bottom-dwelling. Use adhesive disc to attach to hard substrates and plants. Three species in the area. Page 348.

Family Callionymidae - Dragonets

Dragonets are small, elongate, and variably flattened anteriorly. The head may be moderately to greatly expanded. Gill openings are small pores behind upper head. Preopercle has a strong, variably shaped spine. First dorsal fin is often tall and sometimes filamentous. Pelvic fins are attached by a membrane to the pectoral-fin base. Color may be bright to cryptic. Dragonets are circumglobal in shallow to deep tropical to temperate seas. A few enter fresh water. Four species in the area. Page 348.

Family Eleotridae - Sleepers

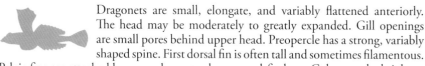

Sleepers are small to moderately sized and elongate to somewhat stout. The head is usually flattened on top. Eyes widely separated. Two dorsal fins present. Second dorsal-fin base is shorter than the distance between its rear origin and the caudal-fin origin. Pelvic fin bases close together or united, but fins are separate. Sleepers are found in most shallow tropical to subtropical fresh, brackish, and marine waters. Four species in the area. Page 350.

Family Gobiidae - Gobies

Gobies are small to very small and elongate. The head is short and broad. First dorsal fin, when present, is separate from second dorsal fin. Second dorsal-fin base is usually longer than the distance between its rear origin and caudal-fin origin. Pelvic fins separate, somewhat connected, or completely united to form a disc. Lateral line is absent. Body is scaleless, partially scaled, or entirely scaled. Gobies occur worldwide in a wide variety of tropical to warm temperate waters. They are most diverse over reefs. Most are marine and demersal. Some are free-swimming. Gobies are the largest family of marine fishes, with more than 1,500 species. Sixty or more species in the area. Page 352.

Family Microdesmidae - Wormfishes

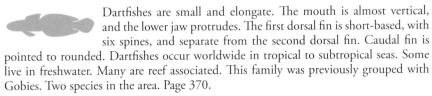

Wormfishes are small, elongate to eel-like, and laterally compressed. Th snout is blunt and the lower jaw protrudes. Dorsal fin is single and long-based. Caudal fin is separate or continuous with dorsal and anal fins. Pelvic fins are small and separate. Lateral line is absent. Wormfishes occur worldwide in shallow, nearshore, tropical to warm temperate seas. They are found around coral reefs, over muddy bottoms, and in tidepools. Wormfishes often burrow into soft muddy and sandy bottoms. Three species in the area. Page 370.

Family Ptereleotridae - Dartfishes

Dartfishes are small and elongate. The mouth is almost vertical, and the lower jaw protrudes. The first dorsal fin is short-based, with six spines, and separate from the second dorsal fin. Caudal fin is pointed to rounded. Dartfishes occur worldwide in tropical to subtropical seas. Some live in freshwater. Many are reef associated. This family was previously grouped with Gobies. Two species in the area. Page 370.

Family Ephippidae - Spadefishes

Spadefishes are deep-bodied and circular to oblong in profile. The head is short and the body is deeply compressed. Jaws have slender, movable, brush-like teeth. First dorsal fin is distinct from second and is usually notched in most. Preopercle is smooth to serrate and lacks a prominent spine at corner. Spadefishes occur worldwide in tropical to warm temperate seas, and rarely in brackish waters. They form schools over coral and artificial reefs and rocky areas. Juveniles are pelagic; some mimic floating leaves or debris. One species in the area. Page 370.

Family Luvaridae - Louvar

Louvar are oblong in profile with a slender, keeled, caudal peduncle. The snout is blunt, the forehead is compressed and keel-like. Dorsal fin originates just posterior to eyes in juveniles and posterior to body midpoint in adults. First dorsal- and anal-fin spines are long in juveniles and shorten with age. Louvar are oceanic and pelagic in relatively deep water in worldwide tropical to temperate seas. Females can produce up to 47 million eggs. There is one species in this family. Page 372.

Family Acanthuridae - Surgeonfishes

Surgeonfishes are deep-bodied, laterally compressed, and round to oval in profile. Head profile is steep. The eyes are relatively small. Mouth is small and not protrusible. Dorsal fin is continuous. The lateral caudal peduncle has one or more spines, or keeled, bony plates. Surgeonfishes occur worldwide in tropical to subtropical seas. They are usually associated with shallow coral and rocky reefs. They are herbivores that feed during the day. The caudal peduncle spine is used in defense and in aggression against competitors. Three species in the area. Page 372.

Family Scombrolabracidae - Longfin escolar

Longfin escolar are moderately elongate and laterally compressed. The head, mouth, and eyes are large. The pectoral fins are very long. The lateral line is close to the dorsal profile. Occur over relatively deep continental shelves and slopes of worldwide seas. There is one species in this family. Page 374.

Family Sphyraenidae - Barracudas

Barracudas are elongate and small to moderately large. The head is long, the snout is pointed, and the lower jaw protrudes. The large jaws and roof of the mouth have numerous sharp conical or flattened teeth. Dorsal fins are short-based and widely separated. Barracudas occur worldwide in tropical to warm temperate seas. They occur in a wide variety of coastal habitats from surface to near bottom. Small species and juveniles of large species form schools. Barracudas are swift, voracious predators. Three species in the area. Page 374.

Family Gempylidae - Snake mackerels

Snake mackerels are very elongate to moderately deep-bodied. The lower jaw protrudes. Teeth are strong, some are fang-like. First dorsal fin is long-based. Finlets follow second dorsal fin in some species. Pelvic fins are small, rudimentary, or absent. Lateral line is single or double. They occur worldwide in tropical to subtropical seas. Seven shallow water species in the area. Page 374.

Family Trichiuridae - Cutlassfishes

Cutlassfishes are elongate and strongly compressed. The lower jaw protrudes. Fang-like teeth are usually present. Dorsal fin is extremely long-based and is notched in some. Pelvic fins are very small or absent. Caudal fin is small or absent. They occur worldwide in tropical to warm temperate seas from surface to about 6,500 ft. One shallow water species in the area. Page 378.

Family Scombridae - Mackerels and Tunas

Mackerels and Tunas are medium to large and elongate to very robust. The body is somewhat compressed to rounded in cross-section. Dorsal fins are short-based and well-separated and insert into grooves. A series of finlets follow second dorsal and anal fins. Two oblique keels at caudal-fin base are present in all. Many also with a lateral keel on the caudal peduncle. Occur worldwide in tropical to temperate seas. Most are marine and pelagic, some are coastal. Some are highly migratory. Important foodfish. Sixteen species in the area. Page 378.

Family Xiphiidae - Swordfish

Swordfish are large, robust anteriorly, and rounded in cross-section. The upper jaw forms a flattened, sword-like bill. First dorsal fin is short-based and widely separated from second dorsal fin. The caudal peduncle is deeply notched at the caudal-fin base and has strong lateral keels. Pelvic fins absent. Oceanic in worldwide offshore tropical to temperate seas. Highly migratory. Important foodfish. There is one species in this family. Page 384.

Family Istiophoridae - Billfishes

 Billfishes are elongate and moderately compressed. The body tapers from head to tail. The upper jaw forms a spear-like bill that is round in cross-section. First dorsal fin is long-based in all and sail-like in one species. The first dorsal and first anal fins insert into grooves. Pelvic fins are long and narrow. The caudal peduncle is shallowly notched in front of the caudal-fin base. Caudal fin with two keels at base. Occur worldwide in tropical to warm temperate seas. They are usually offshore in upper water column. Swift predators. Use bill to injure and stun prey. Important sport fishes. Five species in the area. Page 384.

Family Centrolophidae - Medusafishes

 Medusafishes are medium to large and elongate to robust. The mouth is relatively large. The jaws have a single row of small, conical teeth. The dorsal fin is continuous and the caudal peduncle is thick. Pelvic fins insert into a groove. Medusafishes occur worldwide in tropical to temperate seas. Absent from eastern Pacific. Adults are found over deep shelves, canyons, and submerged islands. Juveniles are associated with jellyfishes and flotsam. Four species in the area. Page 386.

Family Nomeidae - Driftfishes

 Driftfishes are moderate to large, slender to deep-bodied, and laterally compressed. The mouth is small, the snout is moderately blunt to very blunt. Dorsal fins are separate. First dorsal fin and pelvic fins insert into a groove. Second dorsal and anal fins are about the same shape and are scaled at the base. Caudal keels are absent. Driftfishes occur worldwide in tropical to subtropical oceanic waters. Adults inhabit midwaters or are demersal over continental slopes. Juveniles occur near surface and are associated with jellyfishes. Seven species in the area. Page 388.

Family Ariommatidae - Ariommatids

Ariommatids are small, slender to moderately deep-bodied, and laterally compressed to rounded in cross-section. The mouth is small, the snout is short and blunt. Dorsal fins are separate. First dorsal fin and pelvic fins insert into a groove. Second dorsal and anal fins are about the same shape. Two low, fleshy keels on caudal-fin base. Ariommatids occur worldwide in tropical to subtropical oceanic waters. Absent from eastern Pacific. Adults are offshore in deep water. Juveniles occur near surface. Three species in the area. Page 390.

Family Tretragonuridae - Squaretails

Squaretails are medium-sized, elongate, and rounded in cross-section. The snout is blunt. The lower jaw fits completely within the upper jaw when closed. First dorsal fin is long-based and inserts into a groove. Second dorsal and anal fins are similar in shape. Modified scales form two low keels on caudal-fin base. Body scales are keeled and geometrically arranged. Squaretails occur in deep water of tropical to subtropical oceanic seas. One species in the area. Page 390.

Family Stromateidae - Butterfishes

Butterfishes are deep-bodied and laterally compressed. Snout is short, mouth is small. Eyes are covered by adipose eyelids. Dorsal and anal fins are single, long-based, scaled, and similar in shape. Anterior fin lobes may be elongate. Pelvic fins are absent. Butterfishes are found worldwide in tropical to warm temperate seas. Absent from southwestern Pacific and Indian Oceans. They are pelagic along the coast, over continental shelves, and sometimes in estuaries. Three species in the area. Page 392.

Family Caproidae - Boarfishes

Boarfishes are moderately to extremely deep-bodied and laterally compressed. The snout is relatively short. The mouth is small with a protrusible upper jaw. Head profile above eyes is concave. Dorsal and anal fins have strong anterior spines and are notched and long-based. Pelvic fins are present. Boarfishes occur worldwide from tropical to temperate seas. Absent from eastern Pacific. Adults live near bottom. Juveniles are pelagic, schooling. Two species in the area. Page 392.

Order PLEURONECTIFORMES - Flatfishes

Family Scophthalmidae - Turbots

Turbots are deep-bodied, rhomboid in shape, and extremely compressed. The mouth is large. The eyes are comparatively large and left-facing. Anterior dorsal-fin rays are long and mostly free of membrane. Pelvic-fin base is long, extending from opercular margin to very close to anal-fin origin. Most fin rays are branched. Turbots occur in the western and northeastern Atlantic. Demersal over soft bottoms in relatively shallow waters. One species in the area. Page 394.

Family Paralichthyidae - Sand flounders

Sand flounders are deep-bodied and laterally compressed. The mouth is large and protrusible. Eyes are large to relatively small and usually left-facing. Anterior dorsal-fin rays may be long and mostly free of membrane. Pelvic fins are symmetrical in shape and short-based. Sand flounders occur worldwide in tropical to temperate seas. Some in fresh water. All are demersal and usually over soft bottoms. Usually burrow into sediment to await prey. Many are quick color-changers. Twenty-three species in the area. Page 394.

Family Pleuronectidae - Righteye flounders

Righteye flounders are deep-bodied and laterally compressed. The mouth is very small to large. Eyes almost always face right. Pelvic fins are symmetrical in shape and short-based. Lateral line is well developed on both sides of body. Righteye flounders primarily occur in temperate to cold marine waters. Some in deep, tropical, or brackish waters. Demersal but may leave bottom to find prey. Seven species in the area. Page 402.

Family Bothidae - Lefteye flounders

 Lefteye flounders are somewhat to very deep-bodied and laterally compressed. The mouth is moderate to large and protrusible. Eyes are left-facing (rarely right-facing) and are close-set or widely separated. Pelvic fin on eyed side is larger and with longer base than blind-side fin. Lefteye flounders occur worldwide in tropical to warm temperate seas. Demersal, usually over soft bottoms. In many species males have separated eyes and elongate pectoral fin. Nine species in the area. Page 404.

Family Achiridae - American soles

 American soles are deep-bodied and laterally compressed. The mouth is small. The eyes are small to minute and right-facing (rarely left-facing). The preopercular margin is concealed or appears as a groove. Pectoral fins are small to absent. Eyed-side pelvic fin may be free or joined to anal fin. American soles occur in the western Atlantic along North, Central and South America. Most are found near shore over soft bottoms. Five species in the area. Page 408.

Family Cynoglossidae - Tonguefishes

 Tonguefishes are moderately deep-bodied and laterally compressed. The body is lance- or tongue-shaped and tapers to a blunt or pointed tail. The mouth is small. Eyes are small and left-facing. Dorsal and anal fins are confluent with caudal fin. Pectoral fins are absent. Eyed-side pelvic fin, when present, is confluent with anal fin. Lateral line is absent. Tonguefishes are demersal over a variety of bottoms from shore to about 5,000 ft. in worldwide tropical to temperate seas. Thirteen species in the area. Page 410.

Order TETRAODONTIFORMES - Plectognaths

Family Triacanthodidae - Spikefishes

 Spikefishes are small, deep-bodied, and laterally compressed. The mouth is small. The snout is short to elongate and tubular in some. The eyes are large. Gill openings are short and slit-like. First dorsal fin with six strong spines. Pelvic fins with one long spine and one to two short rays. Scales are prickly. Spikefishes occur worldwide in tropical to warm temperate seas. Demersal. Two species in the area. Page 412.

Family Balistidae - Triggerfishes

 Triggerfishes are deep-bodied and moderately compressed. The mouth is small with a single row of strong teeth. Gill openings are short and slit-like. A patch of enlarged scales above the pectoral fins is usually present. First dorsal fin with three stout spines; second locks first upright. Second dorsal- and anal-fin rays branched. Scales plate-like and geometric. Triggerfishes occur worldwide in tropical to warm temperate seas from surface to bottom in a variety of habitats. Six species in the area. Page 412.

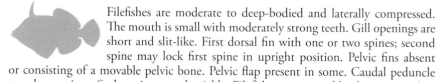

Family Monacanthidae - Filefishes

Filefishes are moderate to deep-bodied and laterally compressed. The mouth is small with moderately strong teeth. Gill openings are short and slit-like. First dorsal fin with one or two spines; second spine may lock first spine in upright position. Pelvic fins absent or consisting of a movable pelvic bone. Pelvic flap present in some. Caudal peduncle may have spines. Scales minute and prickly. Filefishes occur worldwide in tropical to temperate seas from shore to about 650 ft. Ten species in the area. Page 414.

Family Ostraciidae - Boxfishes

Boxfishes are triangular and square to pentangular in cross-section. The mouth is small with fleshy lips and moderately strong teeth. Gill openings are slit-like. Spiny dorsal fin and pelvic skeleton are absent. The body is encased in close-set hexagonal to polygonal plates. Plates are either visible or concealed by skin. Plates may form 'horns' over eyes in some. Boxfishes occur worldwide in tropical to warm temperate seas. Demersal over rocky and coral reefs, sandy and seagrass bottoms. Five species in the area. Page 418.

Family Tetraodontidae - Puffers

Puffers are slender to somewhat robust, becoming rounded when inflated. The mouth is small with four fused teeth forming a 'beak.' Gill openings are slit-like. Spiny dorsal fin and pelvic skeleton are absent. Body is covered in tough, elastic skin. Most with small, spine-like prickles on abdomen, sides, and back. Puffers occur worldwide in tropical to warm temperate seas. Most in shallow, nearshore waters. Some in brackish or fresh water. Many have highly toxic organs and flesh. Eleven species in the area. Page 420.

Family Diodontidae - Porcupinefishes

Porcupinefishes are somewhat robust, becoming round when inflated. The mouth is small, with fused teeth forming a strong 'beak.' Gill openings are slit-like. Spiny dorsal fin and pelvic skeleton are absent. Body is covered in tough, elastic skin, and short to long spine-like scales. Some with spines always erect, others with spines erect only when inflated. Porcupinefishes occur worldwide in tropical to warm temperate seas. Most are demersal and solitary; one is pelagic, offshore, and schooling. Seven species in the area. Page 424.

Family Molidae - Molas

Molas are large, deep-bodied, and laterally compressed. Body profile is rounded or oval to oblong. Mouth and gill openings are small. Dorsal and anal fins are set back on the body and are similar in shape. A caudal peduncle and true caudal fin are absent. Caudal fin is replaced by a leathery structure that is supported by dorsal- and anal-fin rays. Pelvic fins are absent. Molas are pelagic in worldwide tropical to warm temperate seas. Often seen swimming on their sides at the surface. Juveniles are schooling. Three species in the area. Page 426.

SPECIES

Myxinidae - Hagfishes

Atlantic hagfish - *Myxine glutinosa* Linnaeus, 1758

FEATURES: Reddish brown to grayish dorsally. Paler below. There is one gill opening per side. Fins are absent, tail is a skin-flap. There are six barbels around the mouth. Eyes are rudimentary. HABITAT: Baffin Island, Canada, and Greenland to FL. Over muddy bottoms to about 3,000 ft. BIOLOGY: Atlantic hagfish scavenge on dead or dying fishes. They locate prey by smell, then rasp through the flesh. Mucus is secreted through the series of pores along the lower sides.

Petromyzontidae - Northern lampreys

Sea lamprey - *Petromyzon marinus* Linnaeus, 1758

FEATURES: Usually olive brown above, sides mottled yellow brown. Rounded mouth, single nostril on top of head. Seven gill openings. Small eyes. Two separate dorsal fins. HABITAT: Greenland to N FL. Rarely NW FL. In streams, estuaries and offshore to about 3,000 ft. BIOLOGY: Sea lamprey are anadromous and parasitic. Larvae live in freshwater, metamorphose into juveniles, and return to sea.

Rhinochimaeridae - Longnose chimaeras

Atlantic spearnose chimaera - *Rhinochimaera atlantica* Holt & Byrne, 1909

FEATURES: Body tan brown; fins dark. Snout is elongate, broadly compressed. Pectoral fins are broad. Anal fin absent. Caudal fin long, tapered. HABITAT: Nova Scotia to New England. Rarely W Gulf of Mexico. Bottom dwelling, 2,500 to 5,900 ft. BIOLOGY: Atlantic spearnose chimaera have electrochemical receptors around head and down the snout. The eyes are well developed. Feed on bottom invertebrates.

Chimaeridae - Shortnose chimaeras

Deepwater chimaera - *Hydrolagus affinis* (de Brito Capello, 1868)

FEATURES: Overall tan brown, dark gray, or dark brown. Snout is blunt, slightly upturned. First dorsal fin is erect. Pectoral fins are broad. Body tapers to short, pointed tail. HABITAT: Nova Scotia to NY. Bottom dwelling, 980 to 7,800 ft. BIOLOGY: Deepwater chimaera have well-developed elecrochemical receptors on head. Eyes adapted to low light. Dorsal-fin spine is venomous. Females lay egg cases.

Ginglymostomatidae - Nurse sharks

Nurse shark - *Ginglymostoma cirratum* (Bonnaterre, 1788)

FEATURES: Brown, gray, or yellowish; paler below. Two nasal barbels. Snout short, rounded. Dorsal fins closely spaced. HABITAT: RI (rare) to Brazil, Gulf of Mexico, and Caribbean Sea. Over sand flats, around mangroves, channels, and coral reefs. BIOLOGY: Nurse shark are social and nocturnal. They return to favored caves/crevices to rest. Feed on variety of invertebrates and fishes. Litters contain up to 28 young.

Hagfishes

Atlantic hagfish
to 3 ft.

Northern lampreys

Sea lamprey
to 3 ft.

Longnose chimaeras

Atlantic spearnose chimaera
to 5 ft.

Shortnose chimaeras

Deepwater chimaera
to 4 ft.

Nurse sharks

Nurse shark
to 10.5 ft.

Rhincodontidae - Whale shark

Whale shark - *Rhincodon typus* Smith, 1828

FEATURES: Dark gray, greenish gray, or reddish dorsally. Pale below. Pattern of white or yellow spots and bars covers upper body and fins. Snout short; mouth broad. Largest fish in the world. HABITAT: Circumglobal in tropical and warm temperate seas. NY to Brazil and Gulf of Mexico. Pelagic, at or near surface, near shore or in the open sea. BIOLOGY: Whale shark are harmless filter feeders. Large quantities of planktonic crustaceans, squids, and fishes are captured as water flows over large, fibrous gill plates. Often seen near the surface. Placid and approachable by divers. Females carry up to 300 young. Currently listed as Vulnerable.

Odontaspididae - Sand tiger sharks

Sand tiger - *Carcharias taurus* Rafinesque, 1810

FEATURES: Grayish to gray brown dorsally. Pale below. Irregular dark spots and blotches on upper body and fins. Snout pointed. Narrow, pointed teeth protrude. Dorsal fins nearly equal in size. HABITAT: Worldwide in shallow, inshore waters on or near bottoms from surf zone to continental shelf. Gulf of Maine to TX. BIOLOGY: Sand tiger feed on a variety of fishes, sharks, rays, and invertebrates. They occur singly or in pairs. Females gestate for about nine months. Litters contain only one to two young. Sand tiger sharks are popular public aquaria specimens. May be subject to exploitation. Protected in the US. Currently listed as Vulnerable.

ALSO IN THE AREA: Ragged-tooth shark - *Odontaspis ferox*, p. 431.

Alopiidae - Thresher sharks

Bigeye thresher - *Alopias superciliosus* (Lowe, 1841)

FEATURES: Bluish brown dorsally. Pale below. Skin has iridescent highlights. Forehead prominently indented over very large eyes. Tail almost as long as body. HABITAT: Worldwide in tropical and warm temperate seas. NY to FL, Gulf of Mexico, Bahamas, Cuba, Venezuela, and S Brazil. Found coastally over the continental shelf to deep water. BIOLOGY: Bigeye thresher feed on pelagic and bottom fishes and squids. Use tail to stun prey. Litters usually contain only two young. Commercially fished for flesh and oil. Currently listed as Vulnerable.

Thresher shark - *Alopias vulpinus* (Bonnaterre, 1788)

FEATURES: Brown, gray, blackish, or blue gray dorsally and under snout. Two distinct white patches ventrally. Tail almost as long as body. HABITAT: Worldwide in tropical to temperate seas. Newfoundland to FL and Gulf of Mexico; Venezuela to Argentina. Coastal and pelagic from surface to about 1,200 ft. BIOLOGY: Thresher shark are active and strong swimming, often leaping out of the water. Feed on small schooling fishes, squids, octopods, and crustaceans. They herd prey into tight groups. The tail is then used to stun prey. Sometimes hunt in groups. Prized for its flesh. Taken by longlines and nets. Currently listed as Vulnerable.

Whale shark

Whale shark
to 40 ft.

Sand tiger sharks

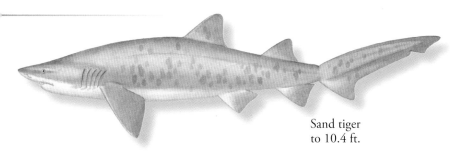

Sand tiger
to 10.4 ft.

Thresher sharks

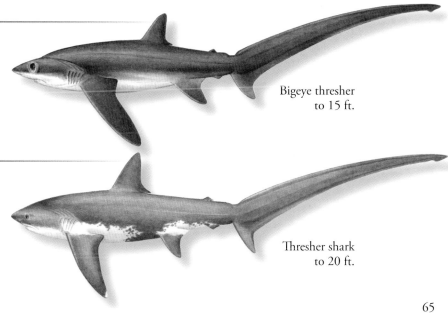

Bigeye thresher
to 15 ft.

Thresher shark
to 20 ft.

Cetorhinidae - Basking shark

Basking shark - *Cetorhinus maximus* (Gunnerus, 1765)

FEATURES: Brownish, grayish, or blackish; often with obscure, irregular stripes. Mouth large; snout long and fleshy. Gill slits nearly circle throat. A very large shark. HABITAT: Worldwide in cold to warm temperate seas. Newfoundland to FL, rarely NE Gulf of Mexico. Pelagic. Near shore and offshore. BIOLOGY: Basking shark are sluggish and nonagressive. Gill rakers possess numerous bristle-like combs that filter large quantities of plankton. Rakers are shed in the winter and then regrown. Basking shark are often seen feeding at the surface. They migrate long distances. Occur singly or in groups up to one hundred. Currently listed as Vulnerable.

Lamnidae - Mackerel sharks

White shark - *Carcharodon carcharias* (Linnaeus, 1758)

FEATURES: Pale to dark gray, blue gray, or blackish dorsally. White ventrally. Snout blunt, pointed. Prominent caudal keels. Large-bodied. HABITAT: Newfoundland to Gulf of Mexico. Worldwide in tropical to temperate seas, less common in tropical waters. Nearshore. BIOLOGY: White shark are strong swimmers and predators. Inquisitive. They feed on a wide variety of marine life, from sea birds to seals. Slow-growing. Females produce small litters. Well documented but often maligned. Threatened by overfishing and exploitation. Currently listed as Vulnerable.

Shortfin mako - *Isurus oxyrinchus* Rafinesque, 1810

FEATURES: Gray blue, purplish, or dark blue dorsally. White ventrally. Snout pointed. Pectoral fins shorter than head. Prominent caudal keels. HABITAT: Worldwide in tropical to temperate seas. Nova Scotia to FL, Gulf of Mexico to Argentina. Pelagic. BIOLOGY: Shortfin mako are active and strong-swimming. Acrobatic and bold. They feed on schooling fishes and other sharks. Populations are threatened by overfishing.

Longfin mako - *Isurus paucus* Guitart Manday, 1966

FEATURES: Slate blue or gray black dorsally and on sides. Dark chin strap. Snout pointed. Pectoral fins broad, longer than head. Broad caudal keels. HABITAT: Worldwide in tropical to warm temperate seas. Georges Bank to Cuba and Gulf of Mexico. Offshore. BIOLOGY: To date, longfin mako are not well-researched. Litters contain two to eight young. Not known to be bold or aggressive. Poorly known. Taken by longlines, nets, and as bycatch. Currently listed as Vulnerable.

Porbeagle - *Lamna nasus* (Bonnaterre, 1788)

FEATURES: Bluish gray dorsally. White ventrally. Lower posterior portion of dorsal fin white. Snout pointed. Prominent caudal keels. Stout, round-bodied. HABITAT: Primarily in North Atlantic. Also in southern oceans. Newfoundland to NC. Pelagic. A cold water species. BIOLOGY: Porbeagle are active and strong. They sometimes travel in schools and feed in aggregations. Able to maintain higher body temperature than surrounding water. Severely depleted by overfishing. Currently listed as Vulnerable.

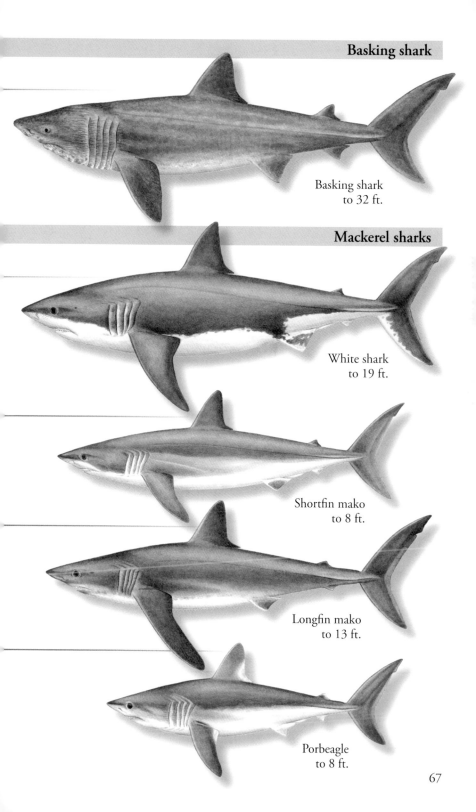

Basking shark
to 32 ft.

White shark
to 19 ft.

Shortfin mako
to 8 ft.

Longfin mako
to 13 ft.

Porbeagle
to 8 ft.

Scyliorhinidae - Catsharks

Marbled catshark - *Galeus arae* (Nichols, 1927)

FEATURES: Yellowish brown to brown with rows of dark brown spots and saddles dorsally and on sides. Dorsal and caudal fins tipped white. HABITAT: NC to FL, northeastern Gulf of Mexico. Bottom-dwelling, deep-water, on upper continental shelf in depths of about 95 to 2,400 ft. BIOLOGY: Feed on deep-water shrimps. May school in large numbers. Similar in appearance to the Antilles catshark.

Chain dogfish - *Scyliorhinus retifer* (Garman, 1881)

FEATURES: Dusky pale brown with dark 'chain' pattern over body. Alternating dark saddles within chain pattern. Dorsal fins set posteriorly. HABITAT: MA to FL and northern Gulf of Mexico to Nicaragua. Bottom-dwelling on the outer continental slope, in depths of about 240 to 2,000 ft. BIOLOGY: Chain dogfish are sluggish. Feed on bottom fishes and invertebrates. Females lay eggs that attach to bottom materials via long, filamentous tendrils. Young incubate for about seven months.

Triakidae - Hound sharks

Smooth dogfish - *Mustelus canis* (Mitchill, 1815)

FEATURES: Uniformly slate gray or gray dorsally. May pale or darken. White or yellowish ventrally. Fins edged white. Moderately sized. Pectoral fins broad, caudal fin deeply notched. Teeth lack cusps. HABITAT: Bay of Fundy to FL, Gulf of Mexico, Caribbean Sea to Argentina. Coastal. On muddy bottoms to about 300 ft. BIOLOGY: Smooth dogfish are active and predatory. Feed on variety of invertebrates. Often kept in public aquaria. Threatened by heavy fishing. SIMILAR SPECIES: Gulf smoothhound, *Mustelus sinusmexicanus*, p. 431.

Florida smoothhound - *Mustelus norrisi* Springer, 1939

FEATURES: Uniformly gray or grayish brown dorsally. Pale ventrally. Head compressed. Moderately sized. Pectoral fins relatively narrow. Caudal fin deeply notched. HABITAT: S FL, northern Gulf of Mexico to Venezuela. Found over mud and sandy bottoms, near shore to about 250 ft. BIOLOGY: Florida smoothhound are common and migratory; move inshore in winter. Found offshore otherwise. Prey include shrimps, crabs, and small fishes.

Carcharhinidae - Requiem sharks

Blacknose shark - *Carcharhinus acronotus* (Poey, 1860)

FEATURES: Yellowish, greenish-gray, yellowish-brown, or bronze dorsally. White or pale ventrally. Tip of snout has dusky blotch. Blotch paler on adults. Tip of caudal fin dusky. HABITAT: NC to FL, Gulf of Mexico, Bahamas, Caribbean Sea to southern Brazil. Inshore to continental slopes and on sandy or coralline bottoms. BIOLOGY: Blacknose shark feed on small fishes. Females give birth to 3–6 young. Fished heavily by longlines and gill nets. May be vulnerable to decline.

Catsharks

Marbled catshark
to 13 in.

Chain dogfish
to 23 in.

Hound sharks

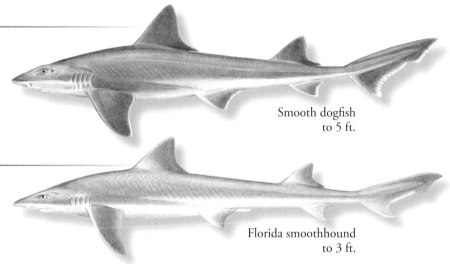

Smooth dogfish
to 5 ft.

Florida smoothhound
to 3 ft.

Requiem sharks

Blacknose shark
to 4.6 ft.

Bignose shark - *Carcharhinus altimus* (Springer, 1950)

FEATURES: Gray to bronze dorsally. White ventrally. All fins with dusky tips except pelvic fins. Inner corners of pectoral fins blackish. Snout large, bluntly rounded. Body stout. Interdorsal ridge present. HABITAT: FL to Bahamas and Cuba; northern Gulf of Mexico to Venezuela. Spotty distribution elsewhere in all warm temperate and tropical seas. Bottom-dwelling offshore from about 75 to 1,300 ft. BIOLOGY: Bignose shark prey on small fishes, rays, sharks, and cephalopods. Litters contain 3–15 young. Protected in US waters.

Spinner shark - *Carcharhinus brevipinna* (Müller & Henle, 1839)

FEATURES: Gray dorsally. White ventrally. Dorsal, pectoral, anal, and lower lobe of caudal fins tipped dusky or black. Snout long and pointed. Upper and lower teeth similar, narrow, and pointed. HABITAT: NC to FL, Gulf of Mexico, Bahamas, and Cuba. Wide ranging in warm coastal waters to about 250 ft. BIOLOGY: Spinner shark feed on small schooling fishes, sharks, rays, and squids. Often leap and spin out of the water while feeding. Active, schooling, migratory.

Silky shark - *Carcharhinus falciformis* (Müller & Henle, 1839)

FEATURES: Dark gray, gray brown, or blackish dorsally. White ventrally. Fins with inconspicuous, dusky tips. Snout moderately long and rounded. Slender body. Long, narrow pectoral fins. HABITAT: Worldwide in tropical seas. Occur along coast from MA to S Brazil. Coastal and oceanic. BIOLOGY: Silky shark feed on fishes, tunas, octopods, squids, and some invertebrates. They are bold, active, inquisitive, and sometimes aggressive. Sought commercially for their flesh, skins, and livers. Vulnerable to overfishing and exploitation.

Finetooth shark - *Carcharhinus isodon* (Müller & Henle, 1839)

FEATURES: Bluish gray dorsally. White ventrally. No distinct markings on body or fins. Snout moderately long and pointed. Relatively long gill slits. Small pectoral fins. Teeth very small. HABITAT: Rarely NY to NC. Usually from SC to FL and Gulf of Mexico; Trinidad, Guyana, and southern Brazil. Occur in warm temperate to tropical waters from intertidal zone to about 65 ft. Coastal. BIOLOGY: Finetooth shark are highly active. Schooling and migratory with changing water temperatures. Feed on fishes and shrimp.

Bull shark - *Carcharhinus leucas* (Müller & Henle, 1839)

FEATURES: Gray dorsally. White ventrally. Fins with dusky tips. Eyes well forward on head. Snout very short, wide, and blunt. Body stout. Broad triangular to falcate dorsal fins. HABITAT: Worldwide in warm temperate to topical seas. MA to FL, Gulf of Mexico, Bahamas, Caribbean Sea to southern Brazil. Generally near shore. May enter bays, estuaries, and rivers. BIOLOGY: Bull shark are large and powerful and may be aggressive. They have the uncommon ability to thrive in fresh water. Their diet is wide and varied.

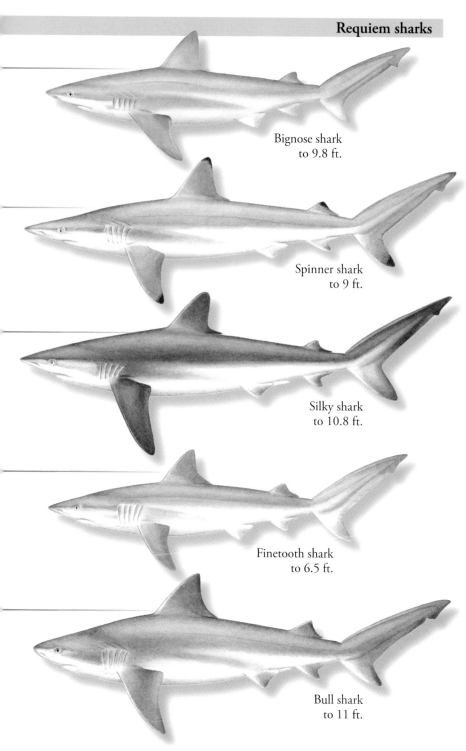

Bignose shark
to 9.8 ft.

Spinner shark
to 9 ft.

Silky shark
to 10.8 ft.

Finetooth shark
to 6.5 ft.

Bull shark
to 11 ft.

Blacktip shark - *Carcharhinus limbatus* (Müller & Henle, 1839)

FEATURES: Gray, brown gray, or blue gray dorsally. Pale ventrally. Dorsal fins, pectoral fins, and ventral lobe of caudal fin tipped black. Tips of pelvic and anal fins sometimes blackish. HABITAT: Occur worldwide in warm, shallow seas. MA to FL, Gulf of Mexico, Bahamas, Caribbean Sea to southern Brazil. BIOLOGY: Blacktip shark are active, fast, and social. Adults form feeding schools. May leap from the water while feeding. Females are migratory. Vulnerable to overfishing.

Oceanic whitetip shark - *Carcharhinus longimanus* (Poey, 1861)

FEATURES: Gray or brown dorsally. White below. Tips of first dorsal, pectoral, and caudal fins mottled white. Black blotches present on fin tips of immature specimens. Dorsal fin very tall and rounded at top. Pectoral fins large and paddle-like. Body large and stout. HABITAT: Circumglobal from edges of continental shelves to mid-ocean. ME to FL, Gulf of Mexico to Argentina. BIOLOGY: Oceanic whitetip shark feed primarily upon fishes, but also invertebrates, sea birds, and carrion. Inquisitive and competitive. Currently listed as Vulnerable.

Dusky shark - *Carcharhinus obscurus* (Lesueur, 1818)

FEATURES: Shades of gray to bronze dorsally. White ventrally. Most fins dusky-tipped. Pectoral fins long and pointed. Low interdorsal ridge present. A large-bodied shark. HABITAT: Worldwide in tropical to warm temperate seas. MA to FL, Gulf of Mexico, Bahamas, Cuba to Nicaragua. Occur from shoreline to outer continental shelves in warm waters. Worldwide. BIOLOGY: Dusky shark are active and migratory. They are slow to mature and have small litters. Prey include fishes, other sharks, rays, carrion, and a variety of invertebrates. Protected in US. Currently listed as Vulnerable.

Reef shark - *Carcharhinus perezii* (Poey, 1876)

FEATURES: Gray brown, dark gray, or olive gray dorsally. White to yellow ventrally. No distinct markings on body or fins. Snout short, bluntly rounded. Interdorsal ridge present. Pectoral fins and lower caudal-fin lobe pointed. HABITAT: E FL, Bermuda, Bahamas, Cuba, Caribbean Sea to Venezuela. Also Flower Garden Banks. Inhabit shallow coastal waters. BIOLOGY: Reef shark are the most common shark on Caribbean reefs and are popular tourist attractions. They may lie motionless on the bottom while pumping water over their gills. Litters are small. Protected in US.

Sandbar shark - *Carcharhinus plumbeus* (Nardo, 1827)

FEATURES: Grayish dorsally. White ventrally. Snout is broad, rounded, and short. Eyes small. Interdorsal ridge present. Body medium-sized and stout. First dorsal fin tall, pectoral fins long. HABITAT: Worldwide in tropical to warm temperate seas. MA to FL, Gulf of Mexico, Bahamas, Cuba to Brazil. Occur over sandy or muddy bottoms, mostly near shore. BIOLOGY: Sandbar shark are slow-growing and late-maturing and have small litters. Kept in public aquaria. They have been heavily fished for meat, skins, and fins. US protection has helped recovery. Currently listed as Vulnerable.

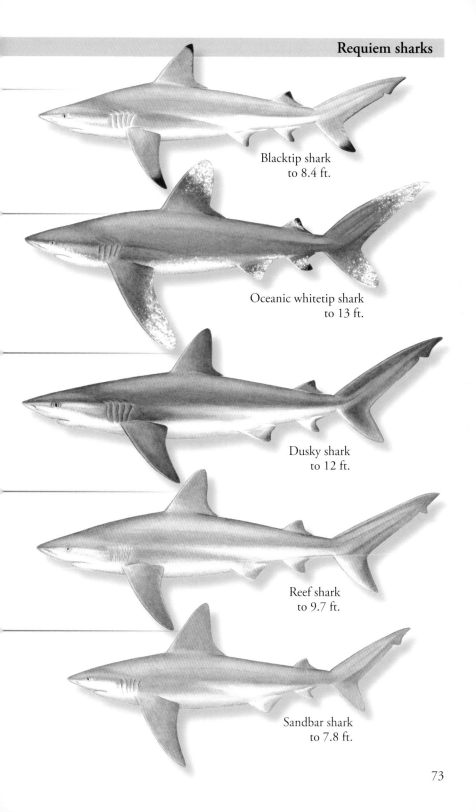

Blacktip shark
to 8.4 ft.

Oceanic whitetip shark
to 13 ft.

Dusky shark
to 12 ft.

Reef shark
to 9.7 ft.

Sandbar shark
to 7.8 ft.

Smalltail shark - *Carcharhinus porosus* (Ranzani, 1839)

FEATURES: Bluish gray or gray dorsally. Pale to white ventrally. Snout long, pointed. Eyes relatively large. Gill slits short. Body slender, small. Pectoral fins small. Pectoral, dorsal, and caudal fins dusky edged or unmarked. HABITAT: N Gulf of Mexico to southern Brazil. Also eastern Pacific. In warm water along coasts, in estuaries, and over muddy bottoms to about 100 ft. BIOLOGY: Smalltail shark feed upon small fishes, invertebrates, and other sharks. Litters are small and produce only 2–7 young.

Tiger shark - *Galeocerdo cuvier* (Péron & Lesueur, 1822)

FEATURES: Shades of gray with darker bars and spots dorsally. Markings muted in older sharks. Snout short, blunt. Body large and stout. Trailing fin tips pointed. Small caudal keel present. HABITAT: Circumglobal in tropical to temperate seas. MA to FL, Gulf of Mexico and Caribbean Sea to Uruguay. Coastal and offshore, at surface or bottom. BIOLOGY: Tiger shark are voracious predators with an indiscriminate and varied diet. Prey include: invertebrates, fishes, sharks, marine reptiles, birds, and mammals. Litters with 10–82 young.

Lemon shark - *Negaprion brevirostris* (Poey, 1868)

FEATURES: Pale yellow brown. Snout short. Body large and stocky. Dorsal fins almost equal in size. Pectoral fins broad and short. Caudal fin low. HABITAT: NJ to FL, Gulf of Mexico and Caribbean Sea to southern Brazil. Also in eastern Pacific. Around coral, mangroves, docks, bays, or river mouths. Inshore and coastal to about 300 ft. BIOLOGY: Lemon shark can adapt to low oxygen levels. They may occasionally enter river mouths in search of prey. They occur singly or in groups of about 20. They are most active at dusk and dawn. Popular as tourist attractions.

Blue shark - *Prionace glauca* (Linnaeus, 1758)

FEATURES: Shades of blue dorsally. White ventrally. Snout is pointed. Body is slender. Pectoral fins long; caudal fin long and tall. HABITAT: Circumglobal in all tropical to temperate seas. Newfoundland to Argentina; rare in Gulf of Mexico. Widely distributed. Oceanic, pelagic. BIOLOGY: Blue shark are inquisitive and brazen. They are most active in early evening and night. Migrate long distances with prey and during reproductive seasons, sometimes crossing the Atlantic. A heavily fished shark. Listed as Near Threatened.

Atlantic sharpnose shark - *Rhizoprionodon terraenovae* (Richardson, 1836)

FEATURES: Gray or brownish gray dorsally. White ventrally. Dorsal fins with dusky tips. Adults with scattered white spots dorsally and white pectoral-fin margins. Eyes large. Body small, slender. Pectoral fins short. HABITAT: New Brunswick to FL, Gulf of Mexico to Honduras. Common in bays, estuaries, sounds, river mouths, and surf zone of sandy beaches. Typically intertidal to about 35 ft. BIOLOGY: Migratory; move inshore in summer. Feed on small fishes, crustaceans, and mollusks.

ALSO IN THE AREA: Night shark - *Carcharhinus signatus*, p. 431.

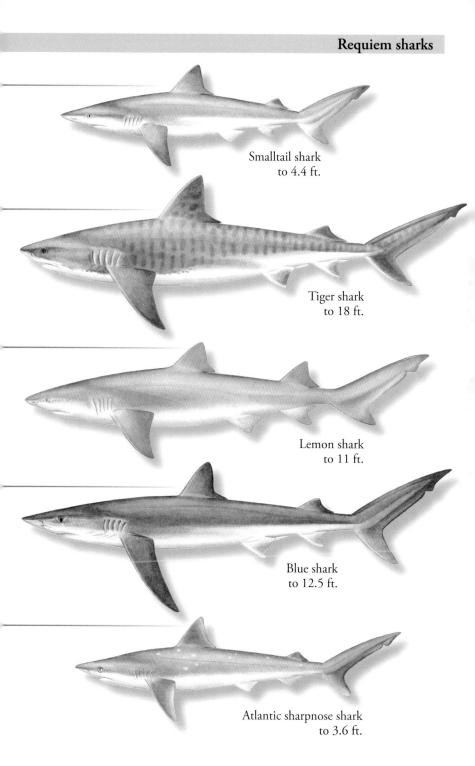

Smalltail shark
to 4.4 ft.

Tiger shark
to 18 ft.

Lemon shark
to 11 ft.

Blue shark
to 12.5 ft.

Atlantic sharpnose shark
to 3.6 ft.

Sphyrnidae - Hammerhead sharks

Scalloped hammerhead - *Sphyrna lewini* (Griffith & Smith, 1834)

FEATURES: Brownish gray or gray dorsally. White ventrally. Head flattened; anterior edge arched, with three notches near middle. Fins usually tipped dusky or black. HABITAT: Worldwide in warm seas over continental shelves. NJ to Brazil, Gulf of Mexico, and Caribbean Sea. BIOLOGY: Scalloped hammerhead occur singly, in pairs, or in huge congregations. Migratory. Prey on fishes, crustaceans, and squids. Adults placid and popular with tourists. Vulnerable to overfishing. Listed as Endangered.

Great hammerhead - *Sphyrna mokarran* (Rüppell, 1837)

FEATURES: Gray or gray brown dorsally. Pale ventrally. Body, fins unmarked. Head flattened, broad, T-shaped; large specimens with a notch at center. Dorsal fin very tall. All fins pointed, with concave rear margins. HABITAT: Worldwide in warm seas from near shore to over continental shelves. NC to Brazil, Gulf of Mexico, and Caribbean Sea. BIOLOGY: Great hammerhead are powerful, migratory, and nomadic. Feed on variety of fishes, squids, and crustaceans. Litters contain 13–42 young. Common target or bycatch of fisheries. Currently listed as Endangered.

Bonnethead - *Sphyrna tiburo* (Linnaeus, 1758)

FEATURES: Gray or grayish brown dorsally; may have small dark spots. Pale to white ventrally. Head flattened, convex; front edge lacks notches. Second dorsal and anal fins similar in size. HABITAT: RI to FL, Gulf of Mexico, Bahamas, all Antilles to southern Brazil. Also eastern Pacific. Occur in warm, shallow, coastal waters over sand and mud, in river mouths, and around reefs. BIOLOGY: Bonnethead are sluggish. Preyed upon by other, larger sharks. They prey primarily on crabs and shrimps. Common bycatch of shrimp fisheries. Popular in public aquaria.

Smooth hammerhead - *Sphyrna zygaena* (Linnaeus, 1758)

FEATURES: Gray or brownish gray dorsally. Pale or white ventrally. Fins unmarked or tipped dusky or black. Head flattened, arched, with no notch at center. HABITAT: Worldwide in tropical to temperate seas. Nova Scotia to FL. Occur close inshore to offshore. Near surface in deep water. BIOLOGY: Smooth hammerhead are strong and seasonally migratory. Young form very large schools. Prey include fishes, other sharks, rays, crustaceans, and squids. Listed as Vulnerable.

Hexanchidae - Cow sharks

Sharpnose sevengill shark - *Heptranchias perlo* (Bonnaterre, 1788)

FEATURES: Brownish gray dorsally. Pale ventrally. Head narrow, pointed. Mouth long, narrow, arched. Eyes large. Seven long gill slits. Tips of dorsal and caudal fins black in young, fading with age. HABITAT: Circumglobal in deep tropical to temperate seas. NC, N Gulf of Mexico to Cuba; Venezuela to Argentina. Demersal. BIOLOGY: Sharpnose sevengill shark prey on bony fishes, crustaceans, and squids. Litters contain 9–20 young. SIMILAR SPECIES: Bluntnose sixgill shark, *Hexanchus griseus*, p. 431.

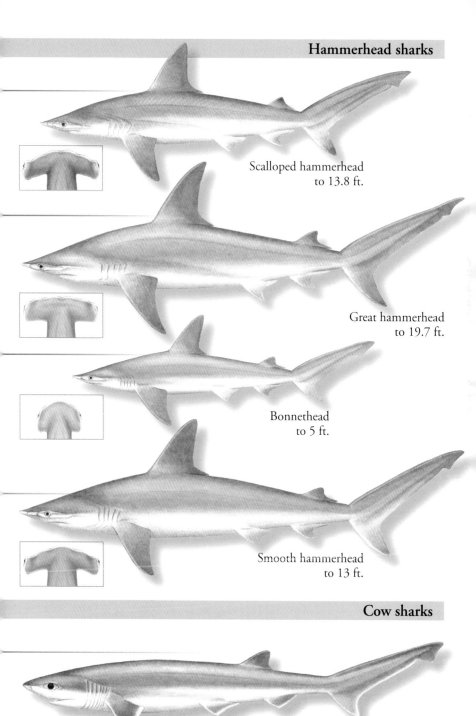

Scalloped hammerhead
to 13.8 ft.

Great hammerhead
to 19.7 ft.

Bonnethead
to 5 ft.

Smooth hammerhead
to 13 ft.

Cow sharks

Sharpnose sevengill shark
to 4.5 ft.

77

Echinorhinidae - Bramble sharks

Bramble shark - *Echinorhinus brucus* (Bonnaterre, 1788)

FEATURES: Brown, dark gray, or blackish dorsally. Paler below. Small, thorn-shaped denticles pepper body, fins, and underside of snout. Denticles may fuse to form small plates. Dorsal fins almost identical. Caudal fin wedge-shaped. HABITAT: Scattered worldwide in warm and temperate seas, to about 2,900 ft. Occur off VA coast and NE Gulf of Mexico. BIOLOGY: Bramble shark are not currently well studied. Caught on longlines near bottom. Prey on small fishes, sharks, and crustaceans.

Squalidae - Dogfish sharks

Roughskin dogfish - *Cirrhigaleus asper* (Merrett, 1973)

FEATURES: Gray, light gray, or brownish gray dorsally. Pale ventrally. Margins of all fins pale to white. Snout and head broad, blunt. Dorsal fins similarly-sized with tall, round spines. HABITAT: Scattered in warm temperate and tropical seas, near bottom from about 210 to 1,800 ft. NC to FL and northern Gulf of Mexico. BIOLOGY: Roughskin dogfish feed on fishes and squids. Litters with 20–21 young.

Spiny dogfish - *Squalus acanthias* Linnaeus, 1758

FEATURES: Gray or bluish gray dorsally. Pale ventrally. White spots usually on upper sides. Pectoral fins with pale margins. Snout pointed. Body slender. Dorsal fins with round spines. HABITAT: Greenland to FL, Cuba, and Bahamas. Found in cold to warm water from intertidal zone to continental shelves. BIOLOGY: Unusually slow-growing and long-lived. Seasonally migratory or residential; schooling. May form large feeding packs. Once very abundant. Populations declining from overfishing. Vulnerable.

Cuban dogfish - *Squalus cubensis* Howell Rivero, 1936

FEATURES: Gray dorsally. Pale to white ventrally. Dorsal and caudal fins with dark inner margins, white outer margins. Pectoral-fin margin white. Snout blunt. Body slender. HABITAT: NC to FL, Gulf of Mexico, Cuba to northern South America. Bottom-dwelling from about 195 to 1,200 ft. BIOLOGY: Cuban dogfish form large, dense schools. Litters contain about 10 young.

ALSO IN THE AREA: Shortspine dogfish, *Squalus mitsukurii,* p. 431.

Centrophoridae - Gulper sharks

Gulper shark - *Centrophorus granulosus* (Bloch & Schneider, 1801)

FEATURES: Light brown dorsally. Paler ventrally. Eyes large. Gill slits relatively long. Dorsal fins have long, grooved spines. Pectoral fins have long posterior lobes. Skin moderately smooth. HABITAT: Worldwide in tropical to temperate seas. Northern Gulf of Mexico to Brazil. On or near bottom to about 4,000 ft. BIOLOGY: Gulper shark are solitary and prey on fishes, squids, and crustaceans. Litters are small with only 1 or 2 young. Currently listed as Vulnerable.

Bramble sharks

Bramble shark
to 10 ft.

Dogfish sharks

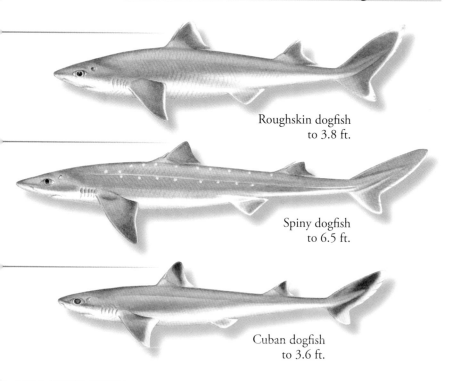

Roughskin dogfish
to 3.8 ft.

Spiny dogfish
to 6.5 ft.

Cuban dogfish
to 3.6 ft.

Gulper sharks

Gulper shark
to 3.2 ft.

Etmopteridae - Lantern sharks

Black dogfish - *Centroscyllium fabricii* (Reinhardt, 1825)

FEATURES: Uniform blackish brown. Snout short. Eyes, spiracles large. Gill slits low, small. Dorsal fins with prominent, grooved spines. Second dorsal fin larger than first. Pectoral fins small, paddle-shaped. Skin contains luminescent organs. HABITAT: Greenland to VA. Possibly Gulf of Mexico. Also eastern Atlantic. Over outer continental slopes. BIOLOGY: Black dogfish are schooling, migratory.

Broadband lantern shark - *Etmopterus gracilispinis* Krefft, 1968

FEATURES: Dark blackish brown. Abdomen black, lower caudal peduncle with black markings. Snout blunt. Eyes large. Gill slits low and small. Body small, stout. Dorsal fins with grooved spines. Pectoral fins small. HABITAT: VA to FL, Uruguay, Argentina, and S Africa. Occur over continental shelves and slopes. BIOLOGY: Broadband lantern shark feed on bony fishes, octopi, squids, and shrimp.

ALSO IN THE AREA: Blurred lantern shark, *Etmopterus bigelowi*, p. 431.

Somniosidae - Sleeper sharks

Portuguese shark - *Centroscymnus coelolepis* Barboza du Bocage & de Brito Capello, 1864

FEATURES: Uniform blackish brown. Snout short, blunt. Eyes large. Gill slits low and short. Dorsal fins small, almost equal in size, with a small spine. Pectoral fins small. HABITAT: Newfoundland to FL. Scattered circumglobally. Bottom-dwelling over outer continental slopes. BIOLOGY: Portuguese shark are nocturnal feeders. Jaws are adapted to bite chunks out of large prey. Currently listed as Near Threatened.

Greenland shark - *Somniosus microcephalus* (Bloch & Schneider, 1801)

FEATURES: Uniform dark gray, blackish, brownish, or ashy. Snout short, blunt. Eyes and spiracles small. Gill slits low, short. Body very large. Dorsal and pelvic fins small, similar in size. Pectoral fins small. HABITAT: North Atlantic and Arctic waters to MA. Seasonally migratory. BIOLOGY: Feed on invertebrates, sea birds, fishes, seals, and carrion. They are sluggish and gregarious. Growth rate very slow. Fished for liver oil, skin, and meat. Listed as Near Threatened.

Dalatiidae - Kitefin sharks

Kitefin shark - *Dalatias licha* (Bonnaterre, 1788)

FEATURES: Uniform dark brownish or blackish. Short, blunt snout. Eyes and spiracles large. Gill slits small and low. Dorsal fins lack spines. Posterior margins of fins may be translucent. HABITAT: Scattered circumglobally. Georges Bank to N Gulf of Mexico. On or near bottom in deep water over continental shelves. BIOLOGY: Kitefin shark have large, buoyant livers that allow them to hover in the water column. Litters with 10–16 young. Prey on fishes and invertebrates.

Lantern sharks

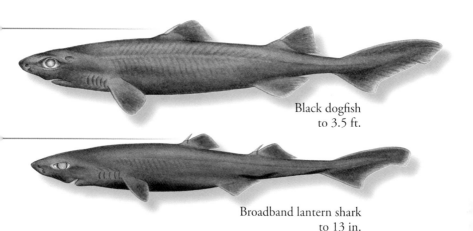

Black dogfish
to 3.5 ft.

Broadband lantern shark
to 13 in.

Sleeper sharks

Portuguese shark
to 4 ft.

Greenland shark
to 21 ft.

Kitefin sharks

Kitefin shark
to 5.2 ft.

Squatinidae - Angel sharks

Atlantic angel shark - *Squatina dumeril* Lesueur, 1818

FEATURES: Gray to brownish dorsally; may have darker spots. White ventrally. Body and fins flattened. Thorny denticles along midline. Pectoral fins broad and squared at tips. Dorsal fins small. Unequal caudal fin at end of long tail. HABITAT: MA to FL, Gulf of Mexico, Cuba, Jamaica, Nicaragua, Venezuela, Argentina. Bottom-dwelling, close inshore to 4,000 ft. BIOLOGY: Atlantic angel shark feed on bottom-dwelling fishes and invertebrates. Litters contain up to 25 young.

Torpedinidae - Torpedo electric rays

Atlantic torpedo - *Torpedo nobiliana* Bonaparte, 1835

FEATURES: Uniform shades of dark gray or brown dorsally; with or without darker spots. Ventral area pale. Head, body, and fins form round disc. First dorsal fin larger than second. Pectoral fins rounded. Caudal fin triangular. HABITAT: Nova Scotia to FL and northern Gulf of Mexico. Also eastern Atlantic. Bottom dwelling from near shore to about 2,000 ft. BIOLOGY: Atlantic torpedo bury in sediment, await prey, then stun them with electrical charge. Prey mostly on fishes.

Narcinidae - Numbfishes

Lesser electric ray - *Narcine bancroftii* (Griffith & Smith, 1834)

FEATURES: Shades of brown dorsally with darker, irregular, ocellated blotches. Whitish ventrally. Snout always with dark blotches. Head, body, and fins form oval shape. Pelvic fin margins almost straight. Caudal fin triangular. HABITAT: NC to FL and Gulf of Mexico. Scattered throughout Caribbean Sea. Demersal between surf and about 120 ft. BIOLOGY: Lesser electric ray prey on invertebrates and fishes. Females carry up to 18 young of various developmental stages. Seasonally migratory. Critically Endangered.

Pristidae - Sawfishes

Smalltooth sawfish - *Pristis pectinata* Latham, 1794

FEATURES: Dark brownish gray to blackish dorsally. Pale to white ventrally. Rostrum elongate with 24–32 pairs of teeth. Body and fins flattened. First dorsal fin almost directly above pelvic fins. HABITAT: NY to FL, Gulf of Mexico, Bahamas, Bermuda, Caribbean Sea to Argentina. Near shore and in bays, estuaries, and rivers. BIOLOGY: Feed on invertebrates and fishes. Saw used to dislodge, injure prey. Critically Endangered.

Largetooth sawfish - *Pristis pristis* (Linnaeus, 1758)

FEATURES: Dark gray to yellowish brown dorsally. Grayish ventrally. Rostrum elongate with 16–20 pairs of teeth. Body and fins flattened. First dorsal fin anterior to pelvic fins. HABITAT: Worldwide in tropical to warm termperate seas. S FL, Gulf of Mexico to Brazil. Inshore. Occur in bays, estuaries, and rivers. BIOLOGY: Prey on bottom-dwelling invertebrates and fishes. Use saw to rout and injure prey. Critically Endangered.

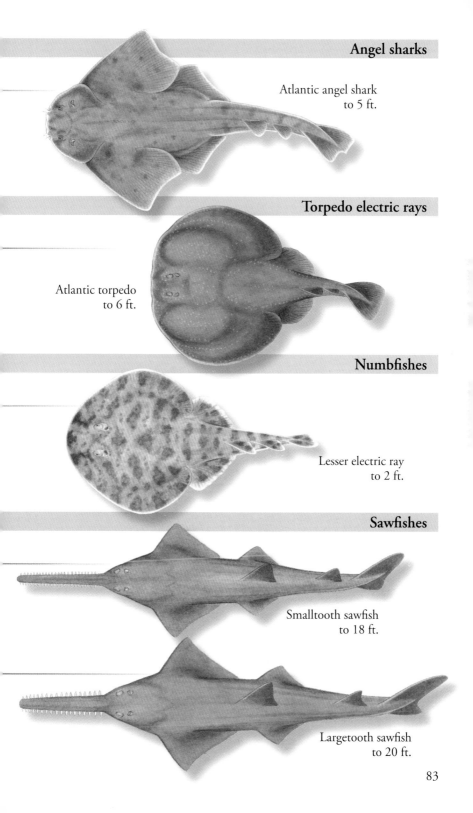

Angel sharks

Atlantic angel shark
to 5 ft.

Torpedo electric rays

Atlantic torpedo
to 6 ft.

Numbfishes

Lesser electric ray
to 2 ft.

Sawfishes

Smalltooth sawfish
to 18 ft.

Largetooth sawfish
to 20 ft.

Rhinobatidae - Guitarfishes

Atlantic guitarfish - *Rhinobatos lentiginosus* Garman, 1880

FEATURES: Color and pattern varies. Plain grayish or brownish dorsally; generally with many small, pale spots. Ventral area pale. Snout pointed. Body and fins flattened. Body and fins form rhomboid shape. Dorsal fins widely spaced on long tail. HABITAT: NC to Yucatán. Coastal and bottom-dwellling over sandy, weedy, or muddy shallow waters. BIOLOGY: Atlantic guitarfish bury themselves in bottom sediment. They prey on crustaceans and mollusks. Sluggish. Harmless.

Rajidae - Skates

Thorny skate - *Amblyraja radiata* (Donovan, 1808)

FEATURES: Uniformly brown dorsally, or brown with irregular dark and pale spots. Ventral area whitish with dark blotches. Large, distinct thorns behind spiracles, at shoulders, along midline, on pectoral fins, and down tail. Denticles on snout, around eyes, and on pectoral fins. HABITAT: Greenland to SC. Also eastern South Atlantic. Occur over variety of bottom matter from 60 to about 3,000 ft. Southern specimens found in deep water. BIOLOGY: Thorny skate are sedentary. They prey on a variety of bottom-dwelling fishes and invertebrates. Fished commercially. Listed as Vulnerable.

Clark's fingerskate - *Dactylobatus clarkii* (Bigalow & Schroeder, 1958)

FEATURES: Pale brown dorsally with dark markings and symmetrically arranged, white, ocellated spots. Ventral area white with irregular grayish band along posterior margins. Blunt snout set between concave anterior margins. Disk heart-shaped. Band of thornlets from snout to pectoral tips. Thorns on snout, around eyes, at shoulder, and down tail. HABITAT: E FL; scattered in Gulf of Mexico and Caribbean Sea. Demersal on continental slopes to 3,000 ft. BIOLOGY: Clark's fingerskate feed on bony fishes. Lay eggs on or near bottom. Other name: Hookskate.

Barndoor skate - *Dipturus laevis* (Mitchill, 1818)

FEATURES: Brown dorsally with vaguely ocellated, irregular, dark spots. Pair of eye-spots on each pectoral fin. Ventral area pale or grayish. Snout long, pointed. Trailing tips of pointed pectoral fins slightly concave. Denticles on snout, around eyes, and along tail. HABITAT: Newfoundland to NC. Found on soft muddy, sandy, or gravelly bottoms, from shore to about 2,400 ft. BIOLOGY: Barndoor skate are one of the largest skates. They prey on a variety of fishes and invertebrates. Females lay a single, rectangular egg case. Fished commercially. Currently listed as Endangered.

Spreadfin skate - *Dipturus olseni* (Bigelow & Schroeder, 1951)

FEATURES: Dark brown to olive brown dorsally with many small, dark, obscure spots. Ventral area gray to black. Thorns around eyes and along tail midline. Dorsal fins distinctly separate. Caudal fin is bilobed. HABITAT: Northern Gulf of Mexico from FL to TX. Demersal in waters to 1,250 ft. BIOLOGY: Spreadfin skate lay their eggs on or near bottom. Egg cases have horn-like projections.

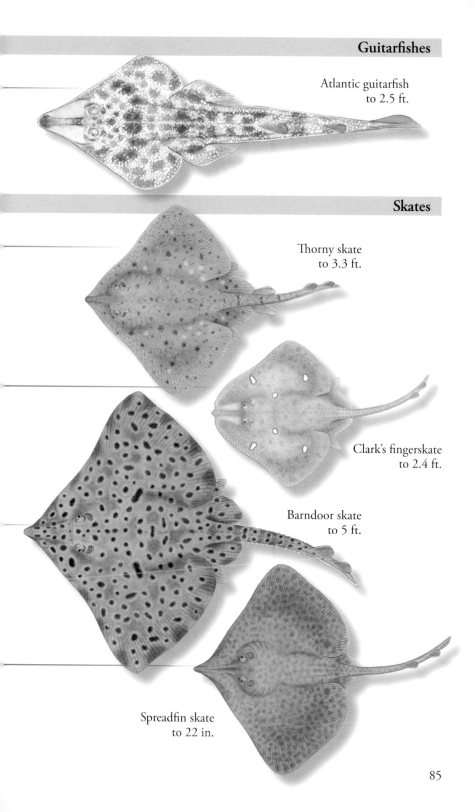

Guitarfishes

Atlantic guitarfish
to 2.5 ft.

Skates

Thorny skate
to 3.3 ft.

Clark's fingerskate
to 2.4 ft.

Barndoor skate
to 5 ft.

Spreadfin skate
to 22 in.

Rajidae - Skates, *cont.*

Prickly brown ray - *Dipturus teevani* (Bigelow & Schroeder, 1951)

FEATURES: Pale brown dorsally. Creamy to dusky ventrally. Dorsal and caudal fins black. Snout very long, pointed. Middle anterior margins of pectoral fins concave. Trailing tips of pectoral fins slightly concave. Dorsal fins connected at base. Few thorns around eyes. Single row of thorns on tail. HABITAT: NC to FL Keys, N Gulf of Mexico, Bahamas, Lesser Antilles; coasts of Honduras, Nicaragua, and Colombia. Demersal along continental slope to about 2,400 ft. BIOLOGY: Prickly brown ray egg cases have horn-like projections.

Underworld windowskate - *Fenestraja plutonia* (Garman, 1881)

FEATURES: Yellowish brown, grayish brown, or purplish brown dorsally with dark spots and blotches. Ventral area yellowish white. Long tail has dark bands. Snout short, blunt. Disk heart-shaped. Thorns around eyes, between spiracles, at shoulder, and along midline. Tail covered in rows of thorns and denticles. HABITAT: NC to FL Keys, northern Gulf of Mexico. Also off Cuba, Costa Rica, and northern South America. Demersal along continental slope to about 3,000 ft. BIOLOGY: Egg cases have horn-like projections. Other name: Pluto skate.

Little skate - *Leucoraja erinacea* (Mitchill, 1825)

FEATURES: Brown to grayish dorsally with small, irregular scattered spots. Snout short, blunt. Disk heart-shaped. Denticles on snout, around eyes, on pectoral fins, and down sides of tail. Lacks row of denticles along midline. HABITAT: SE Newfoundland to NC. Bottom-dwelling. Occur on sandy, gravelly, and muddy bottoms from shore to about 1,200 ft. BIOLOGY: Little skate are seasonally migratory. They feed on a variety of bottom-dwelling fishes and invertebrates. Females are reproductive year-round and lay a single, rectangular egg case.

Rosette skate - *Leucoraja garmani* (Whitley, 1939)

FEATURES: Tan to brown dorsally with small, dark and pale spots that form rosettes around a dark center. Spots form bands on tail. Ventral area whitish. Disk heart-shaped. Patch of thorns on shoulder. Thorns and denticles along tail. HABITAT: MA to FL Keys. Demersal on soft bottoms of outer continental shelves and upper slopes. BIOLOGY: Rosette skate have two similar subspecies: *Leucoraja garmani virginica*, which occur from MA to NC; and *Leucoraja garmani garmani*, which occur from NC to FL. Classification ongoing.

Freckled skate - *Leucoraja lentiginosa* (Bigelow & Schroeder, 1951)

FEATURES: Tan to brown dorsally with scattered, dark freckles and indistinct, pale spots. Ventral area whitish. Heart-shaped disk. Patch of thorns on shoulder. Thorns and denticles along tail. HABITAT: NW FL to Yucatán. Demersal on soft bottoms along outer continental shelves and upper slopes. BIOLOGY: Freckled skate likely feed on bottom-dwelling invertebrates and fishes. Formerly considered a subspecies of *Leucoraja garmani*.

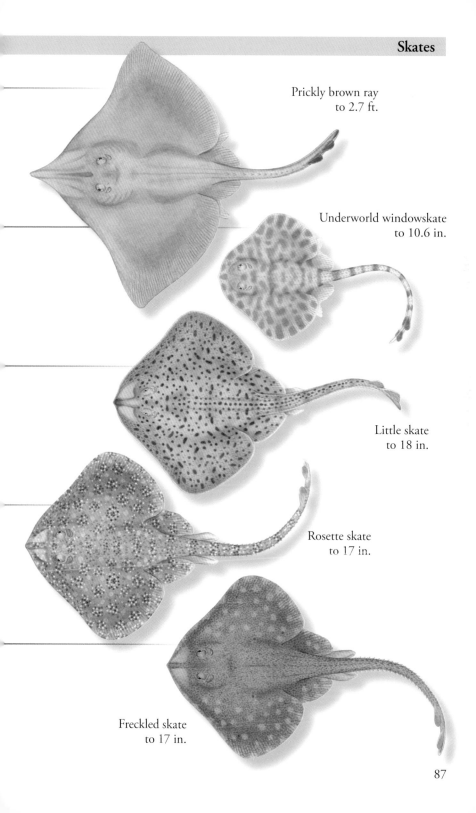

Prickly brown ray
to 2.7 ft.

Underworld windowskate
to 10.6 in.

Little skate
to 18 in.

Rosette skate
to 17 in.

Freckled skate
to 17 in.

Rajidae - Skates, *cont.*

Winter skate - *Leucoraja ocellata* (Mitchill, 1815)

FEATURES: Light brown dorsally with dark, closely scattered spots. May also have larger dark blotches. White, ocellated eye-spots usually on pectoral fins. Disk is heart-shaped. Rows of three or more thorns on mid-disk and on top of tail. HABITAT: S Newfoundland to SC. Found over sandy to gravelly bottoms from shoreline to about 1,200 ft. More abundant in water shallower than 360 ft. BIOLOGY: Winter skate feed on bottom-dwelling fishes and invertebrates. Seasonally migratory. Endangered.

Smooth skate - *Malacoraja senta* (Garman, 1885)

FEATURES: Pale brown dorsally with darker, obscure spots. Ventral area either plain white or with dark blotches. Snout tip acutely pointed. Disk heart-shaped. Small thorns around snout, eyes, and at shoulder. A row of thorns along midline. HABITAT: Gulf of St. Lawrence and southern Grand Bank to Gulf of Maine and Georges Bank. Found on a variety of bottom substrates. BIOLOGY: Smooth skate feed on a wide variety of invertebrates and fishes. Females lay single egg case. Currently listed as Endangered.

Clearnose skate - *Raja eglanteria* Bosc, 1800

FEATURES: Shades of brown to gray dorsally with numerous darker spots and lines. May also have scattered pale spots and blotches. Ventral area whitish to yellowish. Small thorns around eyes and spiracles. Rows of thorns down back and tail. HABITAT: MA to FL and northern Gulf of Mexico to TX. Occur on soft bottoms from shore to about 1,000 ft. BIOLOGY: Clearnose skate are seasonally migratory.

Roundel skate - *Raja texana* Chandler, 1921

FEATURES: Brown dorsally with ocellated eye spots on pectoral fins. Each spot has a brown to black center with a bright yellow ring. Ventral area white. Denticles on snout tip, around eyes, down back and tail. HABITAT: SE FL to Yucatán. Demersal over soft bottoms of continental shelves from about 50 to 360 ft. BIOLOGY: Roundel skate feed primarily on bottom-dwelling crustaceans. Caught in trawls and marketed.

ALSO IN THE AREA: Lozenge skate, *Dipturus bullisi*; Gulf skate, *Fenestraja sinusmexicanus*, p. 431.

Urotrygonidae - American round stingrays

Yellow stingray - *Urobatis jamaicensis* (Cuvier, 1816)

FEATURES: Mottled tan to brown dorsally and covered in small, irregular gold to white spots. Spots cluster to form larger blotches. Body and fins flattened into oval disk shape. Single spine on tail. Caudal fin rounded. HABITAT: NC to FL, Gulf of Mexico, Bahamas, Antilles; Caribbean coast of Central America and northern South America. Demersal. Nearshore. BIOLOGY: Yellow stingray feed on crustaceans and fishes. Females gather on eel grass beds to give birth to live young. Young are born tail first, with barbs safely sheathed.

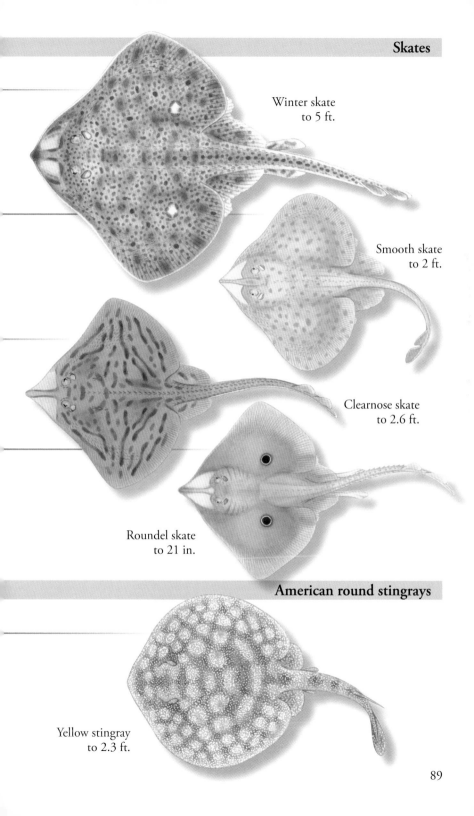

Winter skate
to 5 ft.

Smooth skate
to 2 ft.

Clearnose skate
to 2.6 ft.

Roundel skate
to 21 in.

American round stingrays

Yellow stingray
to 2.3 ft.

Dasyatidae - Whiptail stingrays

Southern stingray - *Dasyatis americana* Hildebrand & Schroeder, 1928

FEATURES: Color varies with bottom substrate. May be gray, brown, or olive dorsally. Small, pale spot between eyes. Ventral area white with darker margins. Snout continuous with almost-straight anterior fin margins. Anterior pectoral margins almost straight. Series of denticles on shoulder. Row of denticles down back to base of tail. Tail has ventral skin fold. HABITAT: NJ to FL, Gulf of Mexico and Caribbean Sea to Brazil. Near shore in shallow water. BIOLOGY: Southern stingray bury themselves in sandy bottoms. Feed mainly on bivalves and worms. Migrate northward during summer. Use serrated barb in self-defense.

Roughtail stingray - *Dasyatis centroura* (Mitchill, 1815)

FEATURES: Dark brown to olive dorsally. Whitish ventrally. Snout projects slightly from barely convex fin margins. Denticles on snout, behind spiracles. Tubercles scattered on shoulder, at pectoral-fin base, along back, and down tail. Tail with ventral fold. HABITAT: MA to FL, northern Gulf of Mexico, and Bahamas. Demersal along intercontinental shelves to about 300 ft. BIOLOGY: Roughtail stingray may be the largest known stingray. They prey on a variety of invertebrates and bony fishes. Litters contain 2–6 young.

Atlantic stingray - *Dasyatis sabina* (Lesueur, 1824)

FEATURES: Brown to yellowish brown dorsally; occasional dark stripe on midline. Ventral area whitish; may have darker margins. Snout projects from concave anterior fin margins. Pectoral-fin tips and trailing edges rounded. Tubercles from midline to serrated barb on tail. Dorsal and ventral folds on tail. HABITAT: VA to FL, Gulf of Mexico to Yucatán. Coastal and in estuaries. BIOLOGY: Atlantic stingray feed on bottom-dwelling invertebrates and fishes. Litters contain 1–3 young.

Bluntnose stingray - *Dasyatis say* (Lesueur, 1817)

FEATURES: Grayish brown, olive brown, reddish brown to dusky green dorsally. Ventral area whitish; may have blotches or dark margin. Snout projects slightly from slightly convex anterior fin margins. Pectoral-fin tips slightly rounded. Rows of denticles or thorns at shoulder and down tail. Dorsal and ventral folds on tail. HABITAT: Rarely MA and NJ, more commonly VA to FL, Gulf of Mexico, Greater and Lesser Antilles. From shore to about 65 ft. BIOLOGY: Bluntnose stingray feed on invertebrates and fishes. Litters contain 2–4 young.

Pelagic stingray - *Pteroplatytrygon violacea* (Bonaparte, 1832)

FEATURES: Dark bluish gray to purplish black dorsally. Ventral area is a paler shade of dorsal color. Snout short, barely protrudes from rounded anterior fin margins. Tail thick at base, narrows at barb. Narrow ventral tail fold present, dorsal fold absent. HABITAT: Newfoundland to NC, northern Gulf of Mexico, and Lesser Antilles. Pelagic over continental shelves to oceanic waters. BIOLOGY: Pelagic stingray feed on animals associated with *Sargassum* seaweed. Litters contain 3–9 young.

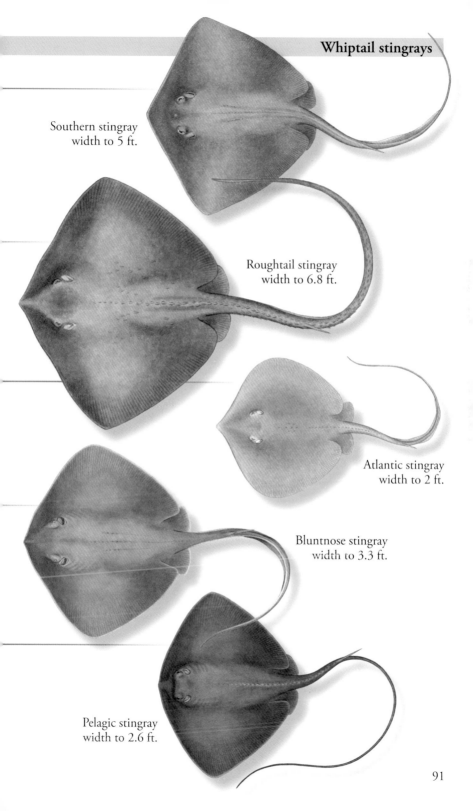

Southern stingray
width to 5 ft.

Roughtail stingray
width to 6.8 ft.

Atlantic stingray
width to 2 ft.

Bluntnose stingray
width to 3.3 ft.

Pelagic stingray
width to 2.6 ft.

Gymnuridae - Butterfly rays

Spiny butterfly ray - *Gymnura altavela* (Linnaeus, 1758)

FEATURES: Shades of brown with small, pale spots and ocelli scattered on dorsal surface. Darker spots form irregular blotches. Ventral area creamy white. Small 'tentacles' at inner margins of spiracles. Pectoral-fin tips rounded and trailing. Tail has one or more barbs. HABITAT: MA to FL, rarely Gulf of Mexico. At or near bottom from shoreline to about 180 ft. BIOLOGY: Spiny butterfly ray feed on crustaceans, mollusks, and plankton. Females gestate for about 6 months. Currently listed as Vulnerable.

Smooth butterfly ray - *Gymnura micrura* (Bloch & Schneider, 1801)

FEATURES: Brown, gray, olive, or purplish dorsally with darker and lighter irregular spots and blotches. Ventral area whitish with darker margins. Tail banded. Middle anterior disk margins weakly concave. Pectoral-fin tips blunt. Tail barb absent. HABITAT: VA to FL, Gulf of Mexico to Brazil. Demersal over sandy bottoms, close to shore, and in estuaries. BIOLOGY: Smooth butterfly ray are harmless. They feed on bivalves, crustaceans, and fishes.

Myliobatidae - Eagle rays

Spotted eagle ray - *Aetobatus narinari* (Euphrasen, 1790)

FEATURES: Dorsal surface blackish, gray, olive gray, or brownish with numerous pale spots, crescents, or circles. Ventral area whitish. Snout protruding. Pectoral-fin tips pointed, trailing edges slightly concave. Dorsal fin located between pelvic fins. HABITAT: Worldwide in temperate to tropical seas. NC to FL, Gulf of Mexico, Bahamas, and Antilles to Brazil. Coastal. BIOLOGY: Spotted eagle ray are active, acrobatic, and migratory. Often leap from the water during spawning or in pursuit of prey. May congregate in large numbers for spawning. Prey include mollusks, squids, shrimps, and fishes. Species name may change in future.

Bullnose ray - *Myliobatis freminvillei* Lesueur, 1824

FEATURES: Reddish brown, dark brown, or grayish dorsally with numerous irregular pale spots. Ventral area white. Snout broad, protruding. Pectoral-fin tips pointed, trailing edges slightly concave. Dorsal fin located close to pelvic fins. HABITAT: MA (rare) to FL, northern Gulf of Mexico, northern coast of South America. Coastal waters and estuaries to about 33 ft. BIOLOGY: Bullnose ray are migratory, may travel long distances, and may leap from surface. Feed on invertebrates.

Southern eagle ray - *Myliobatis goodei* Garman, 1885

FEATURES: Uniform dark brown to grayish brown dorsally. Ventral area brownish white with dusky margins. Snout slightly protruding, rounded. Pectoral-fin tips pointed, trailing edges slightly concave. Small dorsal fin located away from pelvic fins. HABITAT: NC to S FL. Also from E Yucatán to Argentina. Occur at bottom to near surface of coastal waters. BIOLOGY: Southern eagle ray feed on crustaceans and bivalves. They are capable of traveling long distances.

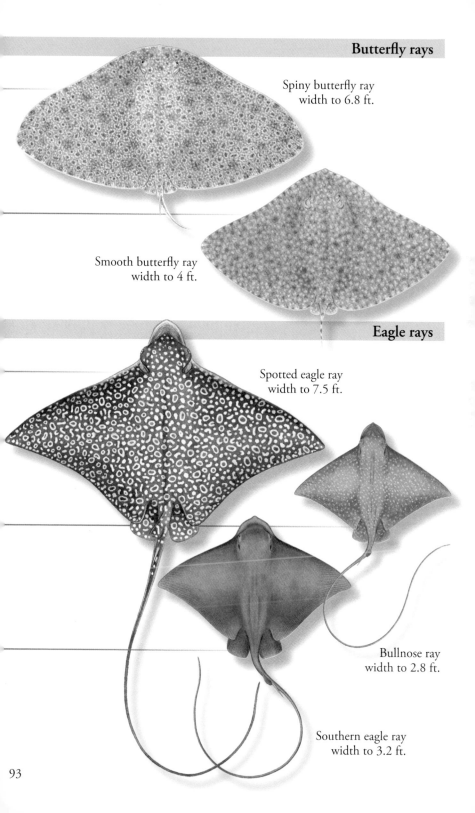

Spiny butterfly ray
width to 6.8 ft.

Smooth butterfly ray
width to 4 ft.

Spotted eagle ray
width to 7.5 ft.

Bullnose ray
width to 2.8 ft.

Southern eagle ray
width to 3.2 ft.

Myliobatidae - Eagle rays, *cont.*

Cownose ray - *Rhinoptera bonasus* (Mitchill, 1815)

FEATURES: Uniform brown to reddish brown dorsally. Ventral surface white to yellowish. Snout bilobed, indented. Pectoral fins pointed, trailing edges concave. Dorsal fin located between pelvic fins. HABITAT: MA to FL, Gulf of Mexico to Argentina. Also Cuba. Found at bottom or near surface from shore to over continental shelves. May enter estuaries. BIOLOGY: Cownose ray congregate in salty bays and inshore shelves during summer. Prey on mollusks and crustaceans.

Giant manta - *Manta birostris* (Walbaum, 1792)

FEATURES: Color varies. Blackish, reddish, olivaceous to brown dorsally. Dorsal surface either uniform or with pale patches, spots, or chevrons. Ventral area either white or blotched. Head very broad with mouth at front. Pectoral fins wide, arched, and pointed. Anterior margin of pectoral fins modified into cephalic fins. HABITAT: Worldwide in tropical to warm temperate seas over continental shelves. MA (rare), NC to Brazil, Gulf of Mexico, Caribbean Sea to Brazil. BIOLOGY: Giant manta use cephalic fins to direct flow of water into mouth. Use specialized gill plates to strain food. Migratory. May be seen leaping from the water. Listed as Near Threatened.

Devil ray - *Mobula hypostoma* (Bancroft, 1831)

FEATURES: Uniform dark bluish black to dark brownish black dorsally. Ventral area whitish or yellowish. Head broad with mouth on underside. Pectoral fins wide, arched, and pointed. Anterior margin of pectoral fins modified into cephalic fins. HABITAT: NC to FL. Gulf of Mexico and Caribbean Sea to Argentina. Absent from Bahamas. Pelagic. Found near surface of warm, tropical waters over continental shelves. BIOLOGY: Devil ray may travel singly, in pairs, or in schools. Use cephalic fins to direct plankton over gills. May leap into air.

Acipenseridae - Sturgeons

Shortnose sturgeon - *Acipenser brevirostrum* Lesueur, 1818

FEATURES: Grayish, brownish, to greenish dorsally. Pale ventrally. Snout short with four barbels. Five rows of keeled, bony plates on back, sides, and abdomen. Fins may be tinged pinkish. HABITAT: St. John's River, New Brunswick, to St. John's River, FL. Demersal. In rivers and low-salinity estuaries and occasionally near coast. BIOLOGY: Adult shortnose sturgeon are anadromous. More active at night. Subject to pollution, habitat destruction, and overfishing. Currently listed as Vulnerable.

Atlantic sturgeon - *Acipenser oxyrinchus* Mitchill, 1815

FEATURES: Brownish to bluish black dorsally. Ventral area white. Fins have pale margins. Snout long, conical, upturned. Four barbels under snout. Body armored with five rows of keeled, bony plates. HABITAT: Labrador to St. Johns River, FL. Suwannee River, FL, to the Mississippi River. Demersal. Inshore, in rivers and tidal tributaries. BIOLOGY: Atlantic sturgeon are anadromous. Slow-growing, long-lived. Populations severely diminished due to overfishing. Currently listed as Near Threatened.

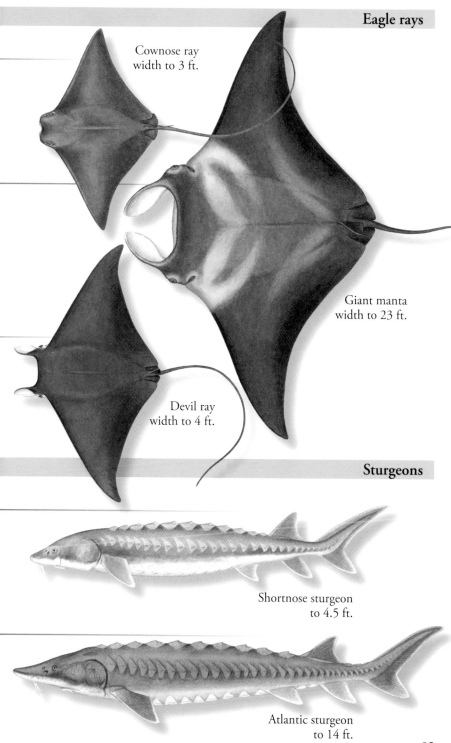

Cownose ray
width to 3 ft.

Giant manta
width to 23 ft.

Devil ray
width to 4 ft.

Shortnose sturgeon
to 4.5 ft.

Atlantic sturgeon
to 14 ft.

95

Lepisosteidae - Gars

Alligator gar - *Atractosteus spatula* (Lacepède, 1803)

FEATURES: Greenish gray above, pale below. Occasional spots on fins. Snout long. Upper jaw overhangs lower. Body elongate and covered in diagonal, rhomboid scales. Dorsal and anal fins set close to tail. HABITAT: N to W Gulf of Mexico. Inshore in fresh, brackish, and marine waters. BIOLOGY: Alligator gar are sluggish but voracious predators. Spawn in fresh water.

Longnose gar - *Lepisosteus osseus* (Linnaeus, 1758)

FEATURES: Color varies. Commonly olive gray dorsally, pale to silver ventrally. Fins and body may be spotted. Snout very long. Body long, cylindrical. Dorsal and anal fins near tail. Anal fin forward of dorsal. HABITAT: NJ to S TX. Adults occur in brackish or coastal marine waters, especially in winter. BIOLOGY: Longnose gar spawn in fresh water and are active predators. Tolerant of low oxygen levels.

Elopidae - Tenpounders

Ladyfish - *Elops saurus* Linnaeus, 1766

FEATURES: Greenish gray to bluish dorsally. Silvery below. Snout short, pointed. Mouth and eyes large. Body elongate. Dorsal-fin origin over pelvic-fin origin. Caudal fin deeply forked. HABITAT: MA (rare) to FL, Gulf of Mexico, Bahamas, Caribbean Sea to Brazil. Also Bermuda. Inshore, in bays and lagoons and around mangroves. Also offshore. BIOLOGY: Spawn year-round. Prey include crustaceans, fishes, and its own species. Other name: Tenpounder. Sought as gamefish.

Megalopidae - Tarpons

Tarpon - *Megalops atlanticus* Valenciennes, 1847

FEATURES: Bluish to greenish dorsally. Silvery below. Snout short, upturned. Large mouth juts upward. Posterior dorsal-fin ray thin, trailing. Anal fin long-based, with concave margin. HABITAT: Nova Scotia (rare) to FL, Gulf of Mexico, Bahamas, Caribbean Sea to Brazil. Also Bermuda. Occur in inshore, marine, brackish, and fresh waters. Also offshore. BIOLOGY: Prey on fishes and invertebrates. Spawn during late spring and summer. Larvae are transparent, ribbon-like. Sought as game fish.

Albulidae - Bonefishes

Bonefish - *Albula vulpes* (Linnaeus, 1758)

FEATURES: Silvery bluish to greenish dorsally with faint saddles. Silvery below. Small, dark blotch at tip of long snout. Black blotch at pectoral-fin base. Mouth subterminal. Anal fin located near tail. HABITAT: In warm, coastal waters over sandy or muddy bottoms. Bay of Fundy (rare) to FL, Gulf of Mexico, Bahamas, Caribbean Sea to Brazil. Also Bermuda. BIOLOGY: Forage singly or in groups for invertebrates and fishes. Prized game fish. NOTE: At least two other possibly undescribed species in FL Keys.

Gars

Alligator gar
to 9.8 ft.

Longnose gar
to 6.5 ft.

Tenpounders

Ladyfish
to 3 ft.

Tarpons

Tarpon
to 7.2 ft.

Bonefishes

Bonefish
to 2.6 ft.

Anguillidae - Freshwater eels

American eel - *Anguilla rostrata* (Lesueur, 1817)

FEATURES: Brownish, greenish to black or gray dorsally. Ventral area whitish. Lower jaw juts beyond upper jaw. Mouth has fleshy 'lips.' Body long, round, slender. Pectoral fin well developed. Dorsal fin originates between gill and anus. HABITAT: Greenland to FL, Gulf of Mexico, Bahamas, Antilles to Trinidad. Also Bermuda. On or near bottom in fresh, estuarine, and marine environments. Inshore and coastal. BIOLOGY: American eel are nocturnal foragers. Catadromous.

Muraenidae - Moray eels

Chain moray - *Echidna catenata* (Bloch, 1795)

FEATURES: Brownish black irregular bars and spots with yellowish chain-like pattern cover head and body. Pattern is variable and colors may be reversed in larger specimens. Dorsal fin begins on head. Body stout, tail tip rounded. HABITAT: S FL, Bahamas, and Caribbean Sea to northern South America. Also Bermuda. Prefer shallow water. Found on coral reefs, in rocky areas, and over sandy bottoms. BIOLOGY: Chain moray are demersal and solitary. Observed leaving water for short periods in pursuit of prey. Feed on crustaceans and small fishes.

Chestnut moray - *Enchelycore carychroa* Böhlke & Böhlke, 1976

FEATURES: Head and body uniformly dark brown; darker toward tail. Blackish corners of mouth, brachial grooves, eye margins, and dorsal-fin folds. White spots around head pores. Snout long, narrow, and arched. Jaws close only at tips. Body moderately compressed. HABITAT: FL Keys, Bahamas, Caribbean Sea to northern South America. Also Bermuda, Flower Garden Banks. From shore to shallow rocky bottoms and on coral reefs. BIOLOGY: Chestnut moray are secretive and harmless.

Viper moray - *Enchelycore nigricans* (Bonnaterre, 1788)

FEATURES: Head and body uniformly or slightly mottled in shades of brown. Young are pale with reticulating pattern. Snout is long, narrow, and arched. Jaws close only at tips. Dorsal fin begins over gill opening. Body moderately compressed, tail tapered. HABITAT: NW Gulf of Mexico, S FL, Bahamas, and Caribbean Sea. Also Bermuda. Around coral reefs and rocky shorelines to about 78 ft. BIOLOGY: Viper moray are secretive, solitary, and harmless. Other names: Mullato, Mottled conger.

Green moray - *Gymnothorax funebris* Ranzani, 1839

FEATURES: Head and body uniform shades of green to brown. Head pores, gill openings, and anus may be dark. Eyes may be reddish. Dorsal fin begins between eye and gill opening. Body stout. HABITAT: FL Keys, Bahamas, and Caribbean Sea to Brazil. Also western Gulf of Mexico and Bermuda. Found in shallow tide pools, in rocky crevices, and on coral reefs. Occasionally in brackish tidal creeks and around mangroves. BIOLOGY: The coloring of the Green moray is a result of yellow mucus overlying gray blue skin. May be defensive or aggressive.

American eel
to 5 ft.

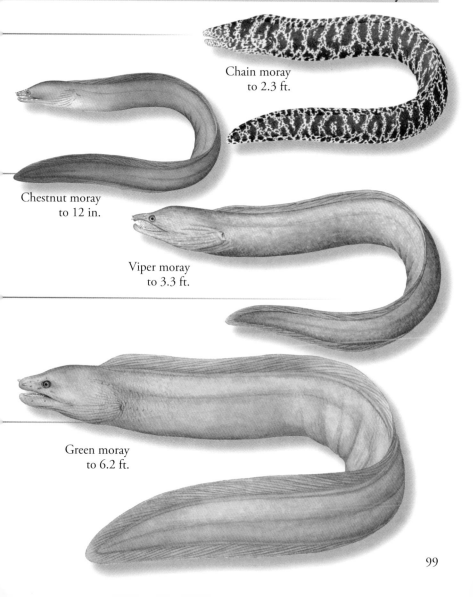

Chain moray
to 2.3 ft.

Chestnut moray
to 12 in.

Viper moray
to 3.3 ft.

Green moray
to 6.2 ft.

Muraenidae - Moray eels, *cont.*

Blacktail moray - *Gymnothorax kolpos* Böhlke & Böhlke, 1980

FEATURES: Head and body greenish brown, grading to almost black at tail. Small, pale, ocellated spots cover head, body, and dorsal fin. Spots become larger and more separated toward tail. Only 3–4 spots on tail tip. Anal fin dark, unmarked. Body stout, tail tapered. HABITAT: NC and GA, Gulf of Mexico to Campeche Bay. Occur on or near muddy or sandy bottoms and banks from about 150 to 750 ft. BIOLOGY: Blacktail moray are harmless. Caught by trawl, trap, and hook-and-line.

Goldentail moray - *Gymnothorax miliaris* (Kaup, 1856)

FEATURES: Color and pattern highly variable. May have brownish background with closely scattered, pale yellowish spots or dots. Spots may be minute, small, or large and blending. Some individuals are uniformly pale yellowish with dark markings (inset). Pattern continuous over entire body in all. Tail always pale. HABITAT: S FL, Bahamas, Caribbean Sea to northern South America. Also Bermuda. On coral reefs and rocky shorelines to about 164 ft. BIOLOGY: Goldentail moray are solitary. May hunt with other predators. Prey include fishes and invertebrates.

Spotted moray - *Gymnothorax moringa* (Cuvier, 1829)

FEATURES: Head, body, and fins covered in small, overlapping dark spots over pale background color. Pattern and intensity of color varies. Pores on lower jaw surrounded by white spots. Snout moderately long, jaws close completely. Dorsal fin often with anterior dark margin and pale posterior margin. HABITAT: NC to FL, Bahamas, and Caribbean Sea to Brazil. Also Bermuda. Found in shallow coral reef, rock, and seagrass bed habitats. BIOLOGY: Spotted moray may be defensive or aggressive when threatened. Feed on fishes and crustaceans.

Blackedge moray - *Gymnothorax nigromarginatus* (Girard, 1858)

FEATURES: Head and body brown with small, well-separated, whitish to yellow spots. Abdomen pale. Eyes with a dark ring. Dorsal fin often with undulating, dark margin. Anal fin with uniformly dark margin. Snout and jaws short. HABITAT: Along coast of Gulf of Mexico from Mobile Bay, AL, to Honduras. Occur over seagrass beds and banks from about 33 to 62 ft. Also found around inshore jetties. BIOLOGY: Blackedge moray may inhabit large, abandoned snail shells. Caught as trawl bycatch of shrimp fisheries.

Honeycomb moray - *Gymnothorax saxicola* Jordan & Davis, 1891

FEATURES: Head and body brown with irregular, pale yellow spots. Spots become larger and less numerous toward tail. Midbody often has row of larger, regularly spaced spots. Dorsal fin has undulating black-and-white marginal saddles. Anal fin begins with black margin, ends in saddles. Snout short, blunt. Jaws short. Body compressed posteriorly. HABITAT: NC to FL. In eastern Gulf of Mexico from FL to Mobile Bay, AL. Also Bermuda. Inhabit seagrass beds and banks. BIOLOGY: Honeycomb moray are reported to forage at night. Taken as bycatch. Other name: Ocellated moray.

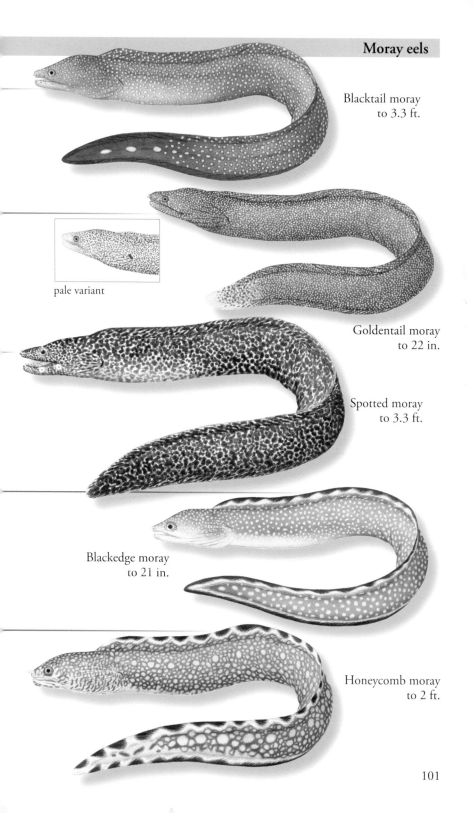

Blacktail moray
to 3.3 ft.

pale variant

Goldentail moray
to 22 in.

Spotted moray
to 3.3 ft.

Blackedge moray
to 21 in.

Honeycomb moray
to 2 ft.

Muraenidae - Moray eels, *cont.*

Purplemouth moray - *Gymnothorax vicinus* (Castelnau, 1855)

FEATURES: Color varies. May be mottled with irregular, dark spots or almost uniformly purplish brown with darker freckles. Dark mark at each corner of mouth. Inside of mouth grayish purple. Posterior one-third of dorsal fin and entire anal fin with pale edge. HABITAT: NC, northern Gulf of Mexico, S FL, Bahamas, and Caribbean Sea to Brazil. Also Bermuda. Demersal over shallow rocky, reef, and seagrass habitats. BIOLOGY: Purplemouth moray are solitary. More active at night.

Reticulate moray - *Muraena retifera* Goode & Bean, 1882

FEATURES: Body brown to dark brown with pale spots variably arranged in rosette pattern. Rosettes peppered with white dots. Pattern may be inconspicuous. Head brown with pale spots. Dark blotch over gill opening. Body stout, tapers toward tail. HABITAT: NC to west coast of FL. Also Bay of Campeche. Occur over continental coastal waters over muddy or sandy bottoms to about 250 ft. BIOLOGY: Uncommon in the area. Caught in trawls and by hook-and-line. Flesh may be poisonous.

NOTE: There are eight other Moray eels in the area, p. 431.

Ophichthidae - Snake eels and Worm eels

Spotted spoon-nose eel - *Echiophis intertinctus* (Richardson, 1848)

FEATURES: Creamy to pale yellow dorsally. Ventral area pale. Head with small dark spots; upper body with irregular, large to small, dark brown spots. Fins pale with dark margins. Head and mouth large. Tail finless. HABITAT: NC to FL, Bahamas, and Caribbean Sea to Brazil. In the Gulf of Mexico from FL to LA and Yucatán. Over sandy bottoms. Near shore to about 210 ft. BIOLOGY: Spotted spoon-nose eel forage at night. They burrow and hide in bottom sediment, then ambush prey.

Goldspotted eel - *Myrichthys ocellatus* (Lesueur, 1825)

FEATURES: Yellowish to greenish tan with two rows of regularly spaced spots. Spots consist of a black ring with a bright yellow center. Snout conical. Nostrils tube-like. Dorsal fin begins at head. Body serpentine. HABITAT: S FL, Bahamas, south to Brazil. Also Bermuda. Demersal over reefs and seagrass beds. Also in lagoons and areas of sandy or rocky rubble. BIOLOGY: Goldspotted eel usually forage at night and bury themselves in the bottom during the day.

Speckled worm eel - *Myrophis punctatus* Lütken, 1852

FEATURES: Dark brown to tan dorsally. Numerous tiny, dark spots dorsally and on sides. Abdomen pale. Color darkens with size. Snout is conical. Nostrils tube-like. Dorsal fin begins above middle of abdomen. Body slender, worm-like, compressed toward tail. HABITAT: NC to FL, Gulf of Mexico, Bahamas, Antilles to Brazil. Also Bermuda. Occur on seagrass beds, around mangroves, in estuaries, and over reefs to about 23 ft. BIOLOGY: Adults spawn offshore.

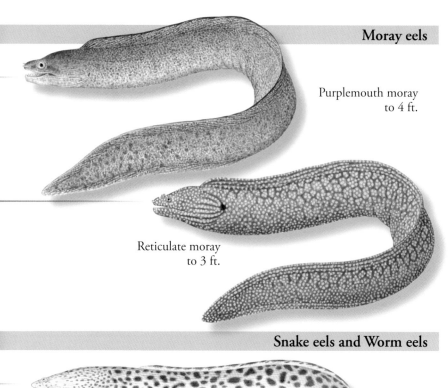

Purplemouth moray
to 4 ft.

Reticulate moray
to 3 ft.

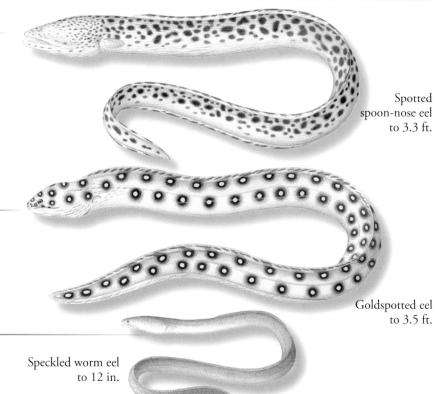

Spotted
spoon-nose eel
to 3.3 ft.

Goldspotted eel
to 3.5 ft.

Speckled worm eel
to 12 in.

103

Ophichthidae - Snake eels and Worm eels, *cont.*

Shrimp eel - *Ophichthus gomesii* (Castelnau, 1855)

FEATURES: Slate gray to grayish brown dorsally. Pale ventrally. Head pores inconspicuous, pale or dark-margined. Posterior margins of dorsal and anal fins are dark. Snout protrudes over lower jaw. Dorsal fin begins behind pectoral fin. Tail is finless. HABITAT: SC to Brazil. Also N Gulf of Mexico, West Indies, and Bermuda. From near shore to about 295 ft. BIOLOGY: Shrimp eel are a common inshore fish where they occur. Commonly caught as bycatch of shrimp fisheries.

Palespotted eel - *Ophichthus puncticeps* (Kaup, 1860)

FEATURES: Brown, dark gray, or gray dorsally. Thin, dark stripe and series of pale spots along lateral line. Small, pale spots on head. Pale ventrally. Snout slighty overhangs lower jaw. Tail finless. HABITAT: NC to Surinam, including West Indies. In Gulf of Mexico from FL to Texas. Occur near bottom from near shore to about 700 ft. BIOLOGY: Palespotted eel have been reported to have toxic flesh.

NOTE: There are 28 other Snake and Worm eels in the area, p. 431.

Congridae - Conger eels

Bandtooth conger - *Ariosoma balearicum* (Delaroche, 1809)

FEATURES: Shades of brown dorsally. Sides with golden or silvery reflections. Paler below. Dorsal and anal fins with dark margins. Snout long, extends beyond lower jaw. Eyes large. Lateral line has series of 46 to 52 pores. HABITAT: NC to FL, Gulf of Mexico, Caribbean Sea to Brazil. Also Bermuda, Mediterranean Sea and Indian Ocean. Found on sandy bottoms of continental shelves from bays and shallow coastal waters to about 2,400 ft. May also be found at the surface. BIOLOGY: Bandtooth conger may burrow in galleries on sandy and muddy bottoms.

Conger eel - *Conger oceanicus* (Mitchill, 1818)

FEATURES: Uniformly gray to brown dorsally. Abdomen slightly paler. Dorsal and anal fins with dark margins. Snout long, fleshy, and longer than lower jaw. Mouth with fleshy 'lips.' Gill openings low. Lateral lines have series of 37 to 42 pores. HABITAT: MA to eastern Gulf of Mexico. Found on or near bottom from shoreline to edge of continental shelves. BIOLOGY: Conger eel are seasonally migratory and spawn offshore. Females produce millions of eggs. Conger eel feed primarily on fishes. Commonly caught around piers, jetties, and docks.

Yellow conger - *Rynchoconger flavus* (Goode & Bean, 1896)

FEATURES: Yellowish, yellowish brown, or brownish. Paler abdomen. Black posterior margins on caudal and anal fins. Eyes large. Snout long, fleshy, and longer than lower jaw. Lateral lines have series of 26 to 34 pores. HABITAT: Gulf of Mexico to mouth of Amazon River. Occur on soft bottoms of coastal waters from about 85 to 600 ft.

NOTE: There are 16 other Conger eels in area, p. 431.

Shrimp eel
to 2 ft.

Palespotted eel
to 3 ft.

Bandtooth conger
to 13 in.

Conger eel
to 6.5 ft.

Yellow conger
to 6.5 in.

Engraulidae - Anchovies

Striped anchovy - *Anchoa hepsetus* (Linnaeus, 1758)

FEATURES: Dusky, translucent dorsally. Abdomen and ventral area pale, translucent. Silver midbody-stripe depth about equal to eye diameter. Caudal fin may have darker margin. Jaw extends past first gill cover. HABITAT: Nova Scotia to FL, northern Gulf of Mexico, Cuba, Venezuela to S Brazil. In bays, in estuaries, and along coast from shore to about 230 ft. BIOLOGY: Schooling and tolerant of varying salinities. Feed on zooplankton. SIMILAR SPECIES: Bigeye anchovy, *Anchoa lamprotaenia*, p. 431.

Dusky anchovy - *Anchoa lyolepis* (Evermann & Marsh, 1900)

FEATURES: Translucent gray dorsally. Silver midbody-stripe depth about equal to eye diameter. Upper margin of band often dark. Dorsal, caudal, and anal fins with dark spots at base and along rays. Head long, somewhat pointed. Jaw extends beyond first gill cover. HABITAT: NY to FL, northern Gulf of Mexico to Venezuela, and Antilles. Occur coastally in salt water from shore to about 75 ft. BIOLOGY: Dusky anchovy often form dense schools. Feed on plankton. Used as bait.

Bay anchovy - *Anchoa mitchilli* (Valenciennes, 1848)

FEATURES: Dusky, translucent with greenish tinge dorsally. Abdomen translucent. Sides whitish or pale. Abdomen translucent. Silvery midbody-stripe depth about as wide as pupil diameter. Fins speckled with tiny spots. Snout blunt. Mouth extends to first gill opening. HABITAT: ME to FL and Gulf of Mexico to Yucatán. Near shore, coastal, in bays and estuaries to about 118 ft. BIOLOGY: Bay anchovy are schooling and feed on zooplankton. An important prey of larger fishes and for birds.

Silver anchovy - *Engraulis eurystole* (Swain & Meek, 1884)

FEATURES: Translucent blue green dorsally. Sides silvery or with broad, silver stripe. Stripe has dark upper edge. Snout pointed. Mouth extends to anterior margin of first gill opening. Dorsal fin near midbody. HABITAT: MA to FL. In Gulf of Mexico from S FL to Mississippi Sound. Also along coast from Venezuela to Brazil. Pelagic in shallow coastal and protected waters to about 200 ft. BIOLOGY: Silver anchovy are plankton-feeders that form large, compact schools. Of minor commercial value.

NOTE: There are four other Anchovies in the area, p. 431.

Clupeidae - Herrings

Blueback herring - *Alosa aestivalis* (Mitchill, 1814)

FEATURES: Dark blue fading to pale blue dorsally; sometimes with dusky lines along scales. Silvery on sides and below. Dark spot behind gill. Spot absent on smaller specimens. Dorsal and caudal fins yellowish to green. Mouth upturned. Upper jaw notched. Body moderately compressed. Lining of the abdominal cavity is sooty to black. HABITAT: Nova Scotia to FL. Pelagic. In rivers, in estuaries, and along coast. BIOLOGY: Blueback herring are anadromous and schooling. In springtime, they ascend rivers to spawn.

Striped anchovy
to 6 in.

Dusky anchovy
to 4.7 in.

Bay anchovy
to 4 in.

Silver anchovy
to 4.7 in.

Blueback herring
to 15 in.

107

Alabama shad - *Alosa alabamae* Jordan & Evermann, 1896

FEATURES: Bluish gray, metallic greenish, or bluish dorsally. Silvery on sides and below. Vague dark spot behind gill. Dorsal and caudal fins dusky. Eyes moderately large with adipose lids. Mouth slightly upturned. Upper jaw with median notch. Body moderately deep and compressed. HABITAT: In the Gulf of Mexico from Suwanee River, FL, to Mississippi River. In marine inshore waters and in rivers along coast. BIOLOGY: Alabama shad are highly tolerant of varying salinities. Anadromous. Spawn in rivers. Do not eat during spawning. Populations declining. Currently listed as Endangered.

Skipjack herring - *Alosa chrysochloris* (Rafinesque, 1820)

FEATURES: Dark blue green dorsally, fading to pale green. Silvery below. Dark spot behind gill absent. Dorsal and caudal fins dusky or yellowish. Tips of caudal fin dark. Mouth upturned. Upper jaw notched. Teeth present at front of mouth. Body moderately deep and compressed. HABITAT: Pensacola, FL, to Corpus Christi, TX. Along coast, in brackish and fresh water. BIOLOGY: Skipjack herring are migratory within rivers. Leap from surface of water. Prey on small fishes.

Hickory shad - *Alosa mediocris* (Mitchill, 1814)

FEATURES: Grayish green dorsally and on upper sides. Dark spots on scales form stripes on upper sides. Lower sides and ventral area silvery. Sides of head brassy. Large, dark spot behind gill. Spot often followed by several faint spots. Mouth upturned, lower jaw juts forward. HABITAT: ME to FL. Coastal. In marine, brackish, and fresh water. BIOLOGY: Hickory shad are anadromous. Spend most of adult life at sea. Ascend coastal streams and rivers in spring to spawn.

Alewife - *Alosa pseudoharengus* (Wilson, 1811)

FEATURES: Grayish to bluish green dorsally. Sides and ventral area silvery. Adults have a dusky to dark spot just behind gill. Faint stripes may be present on upper sides along scales. Fins dusky. Body moderately deep and compressed. Eyes large, with adipose eyelids. Mouth large, upturned. Upper jaw notched. HABITAT: Labrador to northern SC. Occur in coastal marine and tributary waters at varying depths. BIOLOGY: Alewife are migratory, schooling, and anadromous. Spend most of their lives at sea. Found near surface during night. Feed on plankton.

American shad - *Alosa sapidissima* (Wilson, 1811)

FEATURES: Dark blue or blue green dorsally with metallic hues. Sides and ventral area silvery. Large, dark spot behind gill. Spot is followed by a single or double row of fainter spots. Dorsal and caudal fins greenish. Body deep and compressed, becoming deeper with age. Eyes relatively small with adipose eyelids. Upper jaw has median notch. HABITAT: Nova Scotia to FL. Found in marine, brackish, and fresh water along coast. BIOLOGY: American shad are highly tolerant of varying salinities. Adults migrate upriver to spawn. Females scatter up to 600,000 eggs. After spawning, some return to sea, others perish.

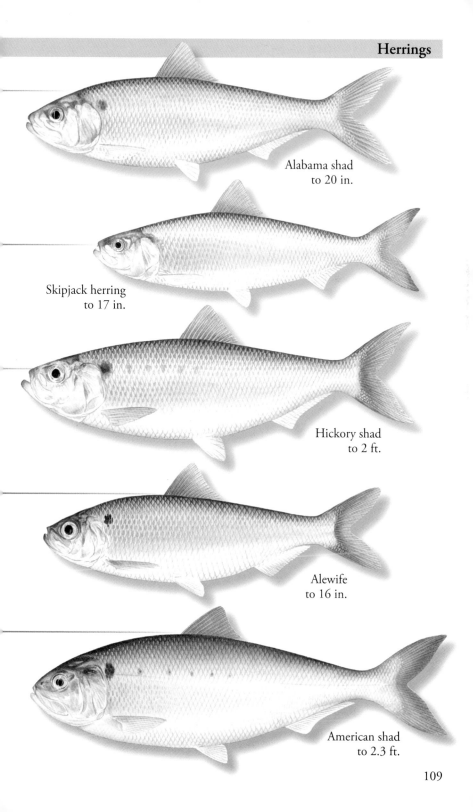

Alabama shad
to 20 in.

Skipjack herring
to 17 in.

Hickory shad
to 2 ft.

Alewife
to 16 in.

American shad
to 2.3 ft.

Clupeidae - Herrings, *cont.*

Finescale menhaden - *Brevoortia gunteri* Hildebrand, 1948

FEATURES: Grayish or dusky dorsally from eye-level to tail. Lower sides and ventral area silvery. Single dark spot behind gill. Dorsal and caudal fins dusky or yellowish. Head large. Eyes relatively large with adipose lids. Mouth upturned. Upper jaw notched. Body deep, compressed. HABITAT: In the Gulf of Mexico from Chandeleur Sound, LA, to Gulf of Campeche, Mexico. Occur inshore along coast in marine and brackish water. BIOLOGY: Finescale menhaden spawn near shore and in bays during winter and early spring. Feed on plankton.

Gulf menhaden - *Brevoortia patronus* Goode, 1878

FEATURES: Green blue to bluish gray dorsally. Sides brassy or silvery. Ventral area silvery. Large, dark spot behind gill. Spot followed by single or double row of smaller, paler spots. Dorsal and caudal fins yellowish to brassy. Caudal fin with dusky margin. Head large. Eyes relatively large with adipose lids. Upper jaw with median notch. Body deep, compressed. HABITAT: In the Gulf of Mexico from Florida Bay to Gulf of Campeche, Mexico. Occur primarily in shallow marine waters along coast. BIOLOGY: Gulf menhaden spawn near or off shore. Eggs hatch at sea and are carried inshore by currents. Caught commercially.

Yellowfin menhaden - *Brevoortia smithi* Hildebrand, 1941

FEATURES: Greenish blue dorsally. Sides and ventral area silvery. Single dark spot behind gill. Dorsal and caudal fins golden. Head large. Eyes large with adipose lids. Mouth upturned. Upper jaw notched. Body deep, compressed. HABITAT: NC to FL. In the Gulf of Mexico from FL to LA. Pelagic. Found in coastal waters. Prefer bays and estuaries. Also enter brackish and fresh water. BIOLOGY: Yellowfin menhaden filter-feed on plankton. Spawn during winter along Atlantic coast of US.

Atlantic menhaden - *Brevoortia tyrannus* (Latrobe, 1802)

FEATURES: Dark blue green dorsally, sides silvery to brassy green. Ventral area silvery. Large, dark spot behind gill followed by up to six irregular rows of smaller, paler spots. Fins brassy to yellowish. Upper jaw with median notch. Body deep, compressed. HABITAT: Nova Scotia to FL. Found in marine and brackish inshore waters. Most abundant near estuaries. BIOLOGY: Atlantic menhaden form large, tight schools near surface. Seasonally migratory. Young form nurseries in estuaries. Feed on plankton. Important food for larger predators and as fishery catch.

Atlantic herring - *Clupea harengus* Linnaeus, 1758

FEATURES: Iridescent greenish blue dorsally. Sides reflective, silvery. Ventral area silvery. Dorsal and caudal fins blue gray to greenish gray. Head relatively small. Upper jaw upturned, lacks notch. Lower jaw juts forward. Body elongate, compressed. HABITAT: Southern Greenland to NC. Also eastern Atlantic. Occur in cold and cold-temperate waters from inshore to about 650 ft. BIOLOGY: Atlantic herring form very large, dense schools that migrate seasonally for food and reproduction. They filter-feed on plankton. Important food for larger predators and as fishery catch.

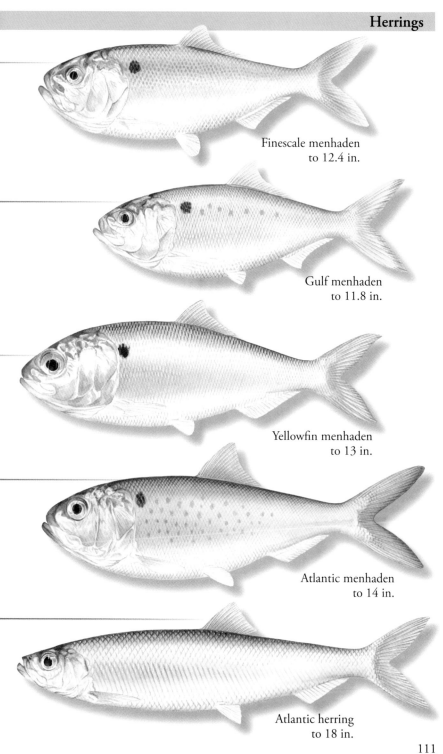

Finescale menhaden
to 12.4 in.

Gulf menhaden
to 11.8 in.

Yellowfin menhaden
to 13 in.

Atlantic menhaden
to 14 in.

Atlantic herring
to 18 in.

Gizzard shad - *Dorosoma cepedianum* (Lesueur, 1818)

FEATURES: Greenish blue dorsally. Lower sides milky white, often with brassy highlights. Upper sides with up to 8 vague stripes along scales. Large, dark spot behind gill. Bulbous snout projects beyond lower jaw. Head small. Last ray of dorsal fin long, filamentous. Body very deep, compressed. HABITAT: St. Lawrence River to FL. In Gulf of Mexico from FL to Mexico. Occur in rivers, bays, and estuaries connected to the coast. BIOLOGY: Gizzard shad are anadromous. Females scatter up to 53,000 adhesive eggs. Many die after spawning. Feed on plankton.

Threadfin shad - *Dorosoma petenense* (Günther, 1867)

FEATURES: Olivaceous to bluish black with golden highlights dorsally. Sides and ventral area silvery. Large, dark spot behind gill. Dorsal fin olivaceous with single long, trailing ray. Head relatively small, eyes large. Mouth small, slightly upturned. Upper jaw notched. HABITAT: Natively in the Gulf of Mexico from FL to N Guatemala. Introduced into various other US river systems. Pelagic, primarily in quiet fresh water. Also found in brackish to saltwater estuaries and bays. BIOLOGY: Threadfin shad are anadromous and schooling. Females cast up to 21,000 adhesive eggs. Life span does not exceed four years. Primarily a filter-feeder. Used as bait.

Round herring - *Etrumeus teres* (DeKay, 1842)

FEATURES: Olive green dorsally. Sides and ventral area silvery. Head small. Eyes large. Dorsal fin tall, slightly forward of mid-body. Body elongate, cylindrical. HABITAT: Bay of Fundy to FL and Gulf of Mexico. Also Cuba and northern coast of South America. Usually occur over deep water off coast and along continental shelves and slopes. Pelagic. BIOLOGY: Round herring migrate vertically in the water column. Spawning occurs at night. Females cast up to 19,000 eggs.

False pilchard - *Harengula clupeola* (Cuvier, 1829)

FEATURES: Blue green with faint streaks along scales dorsally and on upper sides. Silvery below. Faint orange to yellow spot followed by faint dark spot behind gill. Fins transparent. Head and eyes large. Upper jaw lacks notch. HABITAT: FL Keys, Gulf of Mexico, and Caribbean Sea to northern Brazil. Pelagic in shallow coastal waters, estuaries, bays, and lagoons. Occur in turbid and clear water. BIOLOGY: False pilchard tolerate varying salinities. They prey at night on zooplankton.

Scaled sardine - *Harengula jaguana* Poey, 1865

FEATURES: Bluish gray, bluish black, or brownish dorsally with faint streaks along scales. Sides and ventral area silvery. Small, faint to conspicuous spot behind gill. Margin of caudal fin often dusky. Eyes are large. Mouth is upturned and lacks a notch. Body is compressed and moderately deep. HABITAT: NJ to FL, Gulf of Mexico, and Caribbean Sea to Brazil. Pelagic or living near sandy to muddy bottoms along coast. Also in bays and estuaries. BIOLOGY: Scaled sardine are schooling. Often targeted or caught as bycatch.

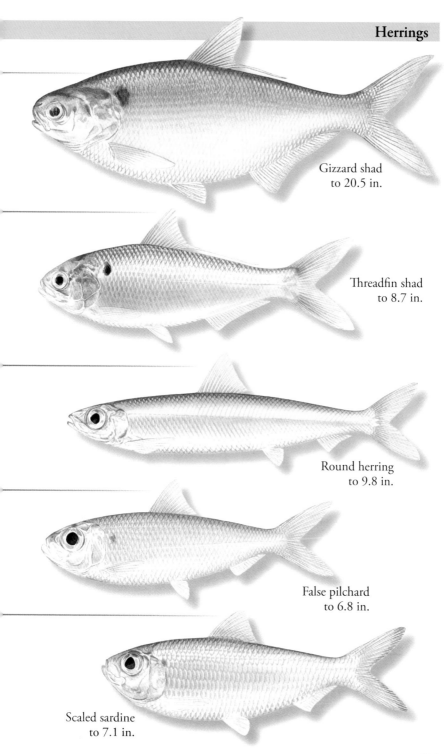

Gizzard shad
to 20.5 in.

Threadfin shad
to 8.7 in.

Round herring
to 9.8 in.

False pilchard
to 6.8 in.

Scaled sardine
to 7.1 in.

Clupeidae - Herrings, *cont.*

Dwarf herring - *Jenkinsia lamprotaenia* (Gosse, 1851)

FEATURES: Pale yellowish green dorsally. Silvery white ventrally. Broad silver stripe with bright blue margin on sides. Small, dark spots on snout, back, and bases of dorsal and anal fins. HABITAT: FL, Gulf of Mexico, and Caribbean Sea to Venezuela. Also Bermuda. Near surface, coastal, inshore, and in bays and estuaries. BIOLOGY: Dwarf herring form large, migratory schools. Feed on zooplankton.

Atlantic thread herring - *Opisthonema oglinum* (Lesueur, 1818)

FEATURES: Blue dorsally and on upper sides; may have dark spots forming lines along scales. Silvery below. Dark spot behind gill. Spot may be followed by row of smaller spots. Dorsal and caudal fins tipped black. Body moderately deep and compressed. Last ray of dorsal fin long and trailing. HABITAT: Gulf of Maine to FL, Gulf of Mexico, and Caribbean Sea to Brazil. Also Bermuda. Pelagic and coastal. BIOLOGY: Atlantic thread herring are migratory and schooling.

Spanish sardine - *Sardinella aurita* Valenciennes, 1847

FEATURES: Iridescent dark blue dorsally. Silvery below. Dark spot on gill margin. Faint, golden spot behind gill followed by faint gold stripe on sides. Dorsal and caudal fins olivaceous. Last two anal-fin rays elongate. HABITAT: MA to FL and Gulf of Mexico to Argentina. Also Bermuda. Coastal and pelagic in warm temperate to tropical marine water from inshore to edge of continental shelves. BIOLOGY: Spanish sardine are schooling and migratory. Follow planktonic concentrations.

ALSO IN THE AREA: Redear sardine, *Harengula humeralis;* Little-eye herring, *Jenkinsia majua*; Shortband herring, *Jenkinsia stolifera*, p. 431.

Ariidae - Sea catfishes

Hardhead catfish - *Ariopsis felis* (Linnaeus, 1766)

FEATURES: Dark blue to brown dorsally. Whitish ventrally. Adipose fin blackish. Head rounded, slightly flattened. Narrow, fleshy groove along top of head. Three pairs of barbels around mouth. Longest pair extends to pectoral fins. Dorsal and pectoral fins with a serrated spine. HABITAT: NC to FL and Gulf of Mexico to Yucatán. Along coast over muddy bottoms in turbid marine and brackish waters. Occasionally in fresh water. BIOLOGY: Prey on bottom-dwelling invertebrates. Spines venomous.

Gafftopsail catfish - *Bagre marinus* (Mitchill, 1815)

FEATURES: Bluish gray to dark brown dorsally. White to pale ventrally. Head slightly arched and flattened. Two pairs of barbels around mouth. Longer pair ribbon-like. Dorsal and pectoral fins long, trailing. Dorsal and pectoral fins with a serrated spine. HABITAT: MA to FL and Gulf of Mexico to Brazil. Occur along coast in marine and brackish waters to about 160 ft. BIOLOGY: Gafftopsail catfish feed on small fishes and invertebrates. Males mouth-brood eggs. Spines are venomous.

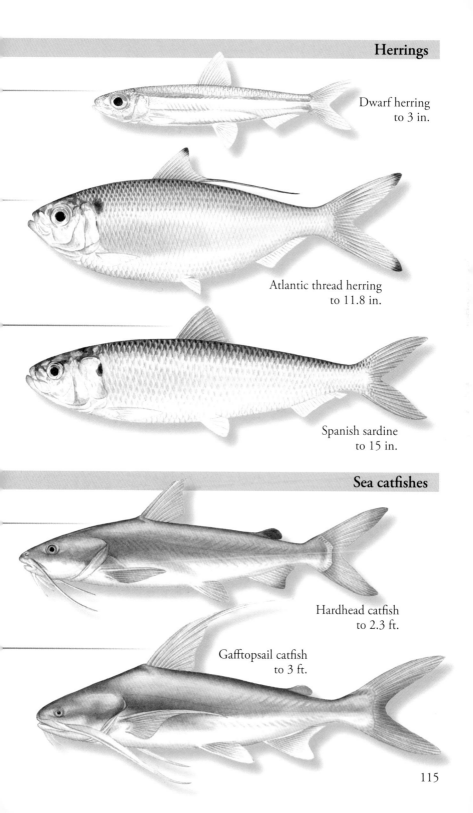

Dwarf herring
to 3 in.

Atlantic thread herring
to 11.8 in.

Spanish sardine
to 15 in.

Hardhead catfish
to 2.3 ft.

Gafftopsail catfish
to 3 ft.

115

Argentinidae - Argentines or Herring smelts

Atlantic argentine - *Argentina silus* (Ascanius, 1775)

FEATURES: Brownish, grayish, or greenish dorsally. Sides steely with brassy luster. Silvery below. Adipose fin yellowish. Mouth small. Eyes large. Body slender and compressed. Scales easlily shed. HABITAT: S Labrador, Nova Scotia to Grand Banks. Over continental slopes to about 547 ft. Also northeastern Atlantic.

ALSO IN THE AREA: Striated argentine, *Argentina striata*; Pygmy argentine, *Glossanodon pygmaeus*, p. 431.

Osmeridae - Smelts

Capelin - *Mallotus villosus* (Müller, 1776)

FEATURES: Greenish to transparent dorsally. Silvery below. All fins transparent. Snout pointed. Lower jaw protrudes. Adipose fin long-based. Males have an expanded anal-fin base. HABITAT: Circumpolar. In NW Atlantic from Hudson Bay to Nova Scotia and Grand Banks. In cold saltwater over deep offshore banks and in coastal areas. BIOLOGY: Capelin are schooling and migratory. Spawning takes place in surf wash on exposed beaches. Economically and environmentally important.

Rainbow smelt - *Osmerus mordax* (Mitchill, 1814)

FEATURES: Transparent greenish to olivaceous dorsally. Sides with broad, irridesent, silvery stripe. Silvery below. Small spots pepper body and fins. Snout pointed. Lower jaw protrudes. Adipose fin small. HABITAT: Labrador to NJ. In bays and estuaries. Also in northern land-locked lakes and ponds. BIOLOGY: Rainbow smelt are schooling and anadromous. They have the ability to tolerate subzero water temperatures.

Salmonidae - Salmonids

Atlantic salmon - *Salmo salar* Linnaeus, 1758

FEATURES: Color varies with age. Adult sea-run fish are brownish, greenish, to bluish dorsally with bronze to silver reflections. Silvery on sides and below. Numerous small black flecks on back and sides. Larger spots on head. Dorsal, adipose, and caudal fins dark. Head small. Body large, somewhat compressed. HABITAT: NE Atlantic and Greenland to NY. Coastal. BIOLOGY: Atlantic salmon are anadromous and migratory. Adults return to natal rivers to spawn. Populations severely depleted.

Aulopidae - Flagfins

Yellowfin aulopus - *Aulopus filamentosus* (Bloch, 1792)

FEATURES: Opalescent with yellowish blotches and dark saddles dorsally. Iridescent below. Males with yellow to reddish markings on fins and elongate filament on dorsal fin. Fins of females not brightly banded. HABITAT: S FL, eastern Gulf of Mexico, and Caribbean Sea. On continental shelves and slopes to about 600 ft.

Argentines or Herring smelts

Atlantic argentine
to 18 in.

Smelts

♂ Capelin
to 10 in.

Rainbow smelt
to 14 in.

Salmonids

Atlantic salmon
to 3 ft.

Flagfins

♂ Yellowfin aulopus
to 13.7 in.

Synodontidae - Lizardfishes

Largescale lizardfish - *Saurida brasiliensis* Norman, 1935

FEATURES: Brownish above with three vague saddles. Pale below. About six blotches along lateral line. Single band on dorsal fin. Lower jaw protrudes. HABITAT: NC to FL Keys, Gulf of Mexico, western Caribbean Sea to Brazil. On or near bottom, offshore on continental shelves and slopes, from about 60 to 1,300 ft. Possibly off eastern Atlantic. BIOLOGY: Prey mostly on fishes. Other name: Brazilian lizardfish.

Shortjaw lizardfish - *Saurida normani* Longley, 1935

FEATURES: Grayish with small dark blotches dorsally. Ventral area silvery. Five to six dark blotches along lateral line. Pelvic fin with dark blotch. HABITAT: NC to FL Keys, Gulf of Mexico, western Caribbean Sea to Guianas. Also Bahamas and off Cuba. On sandy or muddy bottoms from about 160 to 1,800 ft. BIOLOGY: Shortjaw lizardfish reported to lay eggs scattered on substrate.

Inshore lizardfish - *Synodus foetens* (Linnaeus, 1766)

FEATURES: Brownish to greenish dorsally. White ventrally. Throat yellowish. Adults with about eight diamond-shaped blotches on sides. Dorsal and caudal fins dusky. Dark spot on opercle absent. HABITAT: MA to FL, Gulf of Mexico, Caribbean Sea to northern Brazil. Also Bermuda. Near bottom. Along beaches, in bays, inlets, estuaries, and lagoons to about 600 ft. BIOLOGY: Bury themselves, ambush prey.

Sand diver - *Synodus intermedius* (Spix & Agassiz, 1829)

FEATURES: Color and pattern variable. Yellowish stripes through dark saddles dorsally and on sides. Head and jaws blotched and streaked. Dark spot at upper edge of gill opening. Young specimens more brightly colored. HABITAT: NC to FL, Gulf of Mexico, Caribbean Sea to Brazil. Also Bermuda. Near bottom over sandy, rocky, or coralline bottoms to about 1,000 ft. BIOLOGY: Sand diver feed on fishes and shrimps. May ambush fishes that visit shrimp cleaning stations.

Red lizardfish - *Synodus synodus* (Linnaeus, 1758)

FEATURES: Color variable. Broad, reddish to brownish saddles dorsally. Head and jaws blotched. Small, dark spot on tip of snout. HABITAT: W FL, Gulf of Mexico, Caribbean Sea to Uruguay. Also Bermuda. Bottom-dwelling on rocky or reef areas from near shore to about 295 ft. BIOLOGY: They perch on hard surfaces and feed mostly on small fishes. Other name: Redbarred lizardfish or Diamond lizardfish.

Snakefish - *Trachinocephalus myops* (Forster, 1801)

FEATURES: Alternating yellowish and pearly stripes dorsally and on sides. Five to seven indistinct, brown saddles. Head mottled. Dark spot at upper edge of gill. Snout very short. Mouth gaping. HABITAT: Worldwide in warm seas. MA to FL, Gulf of Mexico, Caribbean Sea to Brazil. Demersal over sandy, shelly, muddy, or rocky bottoms to about 1,300 ft. BIOLOGY: Bury themselves, swiftly ambush prey.

ALSO IN THE AREA: Smallscale lizardfish, *Saurida caribbaea*, p. 431.

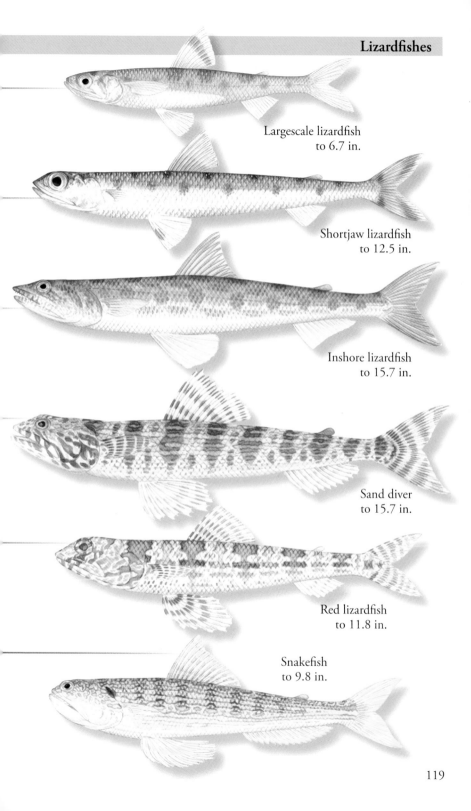

Largescale lizardfish
to 6.7 in.

Shortjaw lizardfish
to 12.5 in.

Inshore lizardfish
to 15.7 in.

Sand diver
to 15.7 in.

Red lizardfish
to 11.8 in.

Snakefish
to 9.8 in.

Chlorophthalmidae - Greeneyes

Shortnose greeneye - *Chlorophthalmus agassizi* Bonaparte, 1840

FEATURES: Grayish with silvery luster. Several dark bars and blotches on upper sides. Eyes reflective green, very large, and wrap around top of head. Snout moderately long, upturned. Adipose fin present. HABITAT: Nova Scotia to FL, Gulf of Mexico, Caribbean Sea to Surinam. On mud and clay bottoms from about 165 to 3,300 ft.

Alepisauridae - Lancetfishes

Longnose lancetfish - *Alepisaurus ferox* Lowe, 1833

FEATURES: Iridescent silvery to coppery. Jaws large, toothy. Dorsal fin long-based, generally tall. Body elongate, laterally compressed. HABITAT: Circumglobal in offshore cold to tropical seas, occasionally in inshore waters. Occur from surface to about 6,000 ft. BIOLOGY: Voracious predator of cephalopods, crustaceans, and fishes. Preyed upon by sharks, opahs, tunas, and seals.

Lampridae - Opahs

Opah - *Lampris guttatus* (Brünnich, 1788)

FEATURES: Iridescent bluish to greenish dorsally, pinkish on sides and below. Body covered in silvery spots. All fins red. Body very deep and compressed. HABITAT: Worldwide in tropical and temperate seas. Nova Scotia to Argentina and Gulf of Mexico. Pelagic in open water from surface to about 650 ft. BIOLOGY: Opah are muscular and predatory. Swim by flapping pectoral fins. Mostly solitary. Feed on invertebrates and fishes. Sought as gamefish and foodfish.

Lophotidae - Crestfishes

Crestfish - *Lophotus lacepede* Giorna, 1809

FEATURES: Silvery. May have pale spots on sides. Fins reddish. First ray of dorsal fin elongate. Prominent crest on head arches forward and upward. Body relatively deep, laterally compressed; tapers toward tail. Caudal and anal fins small. HABITAT: Occur nearly worldwide in warm to tropical seas from surface to about 300 ft. In western Atlantic from FL to Brazil. BIOLOGY: Crestfish feed on squids and fishes.

Trachipteridae - Ribbonfishes

Polka-dot ribbonfish - *Desmodema polystictum* (Ogilby, 1898)

FEATURES: Silvery with faint grayish spots on body. Dorsal fin reddish, becoming blackish posteriorly. Pelvic fins rudimentary to absent. Tail long and narrow. Caudal fin small, not sharply upturned. Juveniles with long, trailing anterior dorsal-fin rays, fan-like pelvic fins, and a short tail. HABITAT: Circumglobal in temperate to tropical seas. Occur from surface to about 1,600 ft.

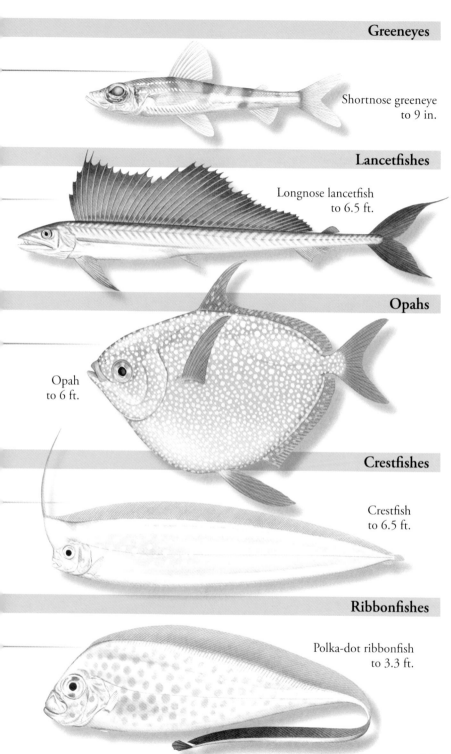

Greeneyes

Shortnose greeneye
to 9 in.

Lancetfishes

Longnose lancetfish
to 6.5 ft.

Opahs

Opah
to 6 ft.

Crestfishes

Crestfish
to 6.5 ft.

Ribbonfishes

Polka-dot ribbonfish
to 3.3 ft.

Regalecidae - Oarfishes

Oarfish - *Regalecus glesne* Ascanius, 1772

FEATURES: Silvery with variably spaced spots and wavy lines on sides. Fins reddish. Head angular. Jaws highly protrusible. First 10–12 dorsal-fin rays elongate. Pelvic fins long, trailing. Body very long, laterally compressed. Tail blunt, with few or no caudal-fin rays. HABITAT: Worldwide in tropical to warm temperate seas. From surface to about 3,300 ft. BIOLOGY: Use undulating fins and body to swim vertically in the water. Feed on variety of invertebrates and fishes. May be longest known bony fish.

Polymixiidae - Beardfishes

Beardfish - *Polymixia lowei* Günther, 1859

FEATURES: Shiny grayish blue dorsally. Sides silvery gray. Silvery below. Dorsal, caudal, and anal fins tipped black in males, dusky in females. Two pinkish barbels on lower jaw. HABITAT: ME to Brazil, Gulf of Mexico, Caribbean Sea, and Bermuda. Over muddy bottoms of shelves and slopes to about 2,100 ft. BIOLOGY: Taken as trawl bycatch. SIMILAR SPECIES: Stout beardfish, *Polymixia nobilis*, p. 431.

Macrouridae - Grenadiers

Marlin-spike - *Nezumia bairdii* (Goode & Bean, 1877)

FEATURES: Gray dorsally, silvery below. Abdomen bluish. Underside of snout pinkish. Snout short, pointed. Eyes large. Body elongate, tapering to pointed tail. First dorsal fin tall with serrated spine; second dorsal fin low. HABITAT: Newfoundland to FL. Near bottom from about 54 to 2,300 ft. BIOLOGY: Feed mostly on crustaceans. Taken as trawl bycatch. NOTE: About eight other shallow water Grenadiers in the area, p. 431.

Merlucciidae - Merlucciid hakes

Offshore hake - *Merluccius albidus* (Mitchill, 1818)

FEATURES: Grayish dorsally, fading to silvery below. Snout depressed. Cheeks and opercles scaled. Second dorsal fin notched near mid-length. Pectoral fins pointed at tips, reach to anus. Caudal-fin margin slightly concave. HABITAT: ME to FL, Gulf of Mexico to Brazil. Near bottom of continental shelves and slopes to about 3,800 ft. BIOLOGY: Female offshore hake carry up to 340,000 eggs.

Silver hake - *Merluccius bilinearis* (Mitchill, 1814)

FEATURES: Grayish dorsally, fading to silvery below. Cheeks and opercles scaleless. Lateral line prominent. Second dorsal fin notched at about two-thirds of the distance from origin. Pectoral fins slightly rounded at tips, reach past second dorsal-fin origin. Caudal-fin margin straight. HABITAT: Newfoundland to FL. Over continental shelves and slopes from about 180 to 3,000 ft. BIOLOGY: Silver hake migrate vertically in water column. Hunt at night. Sought commercially.

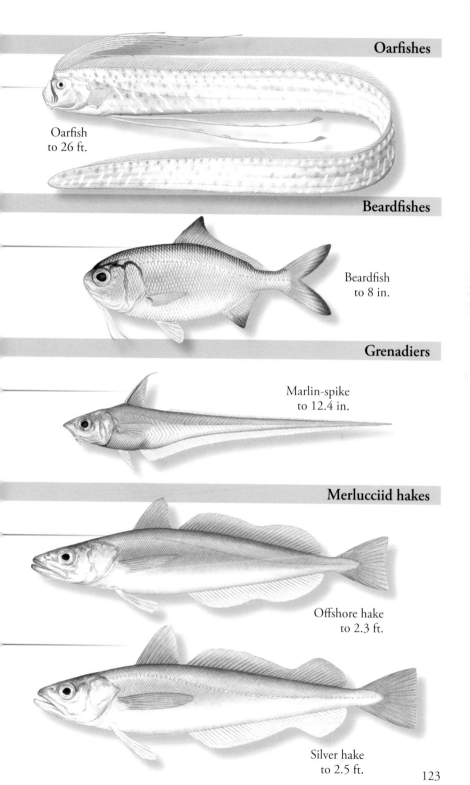

Oarfishes

Oarfish
to 26 ft.

Beardfishes

Beardfish
to 8 in.

Grenadiers

Marlin-spike
to 12.4 in.

Merlucciid hakes

Offshore hake
to 2.3 ft.

Silver hake
to 2.5 ft.

123

Phycidae - Phycid hakes

Fourbeard rockling - *Enchelyopus cimbrius* (Linnaeus, 1766)

FEATURES: Brownish dorsally and on sides. Abdomen usually pale, but may be spotted. Second dorsal fin with one or more dark blotches. Anal fin with elongate, dark blotch. Lower portion of caudal fin dark. Four barbels on snout, one on lower jaw. First dorsal-fin spine tall, followed by about 50 tiny hair-like rays. HABITAT: Greenland to FL and northern Gulf of Mexico. Near shore to about 2,100 ft.

Red hake - *Urophycis chuss* (Walbaum, 1792)

FEATURES: Usually reddish with pale mottling dorsally. May be brownish, olive brown, or uniformly colored. Dark blotch on opercle. First dorsal fin has long, filamentous ray. Pelvic fins reach anal fin. Lateral line with 95–117 scales. HABITAT: Nova Scotia to NC. On or near bottom along coast to about 3,200 ft. BIOLOGY: Adults are host to the commensal Pearlfish. Juveniles may live commensally in Atlantic deepsea scallops. SIMILAR SPECIES: White hake, *Urophycis tenuis*, p. 431.

Gulf hake - *Urophycis cirrata* (Goode & Bean, 1896)

FEATURES: Pale brownish dorsally. Abdomen silvery. Ventral area may be spotted. Diffuse blotch on opercle. Dorsal and anal fins with dark margins. Chin barbel very small or absent. Pectoral-fin rays reach beyond anal-fin origin. HABITAT: E FL, Gulf of Mexico, and northern coast of South America. Along coast, over muddy bottoms from about 88 to 2,200 ft.

Carolina hake - *Urophycis earllii* (Bean, 1880)

FEATURES: Dark brown with pale mottling dorsally and on sides. Paler ventrally. Lateral line may be pale. Dorsal and anal fins with dark margins. First dorsal fin bluntly pointed. Pelvic fins do not reach anal fin. HABITAT: NC to NE FL. Possibly NE Gulf of Mexico, but may be separate species. Occur near bottom over hard bottoms. From near shore to about 265 ft. BIOLOGY: Taken as trawl bycatch.

Southern hake - *Urophycis floridana* (Bean & Dresel, 1884)

FEATURES: Brownish dorsally and on sides. Abdomen silvery. Small dark spots, a vague blotch, and obscure stripes on operculum. Lateral line dark, with a series of pale spots. Pelvic fins reach about to the anus. HABITAT: NC to FL and northern Gulf of Mexico. On or near bottom from near shore to about 1,300 ft. BIOLOGY: Southern hake feed on worms, crustaceans, and fishes. Taken commercially.

Spotted hake - *Urophycis regia* (Walbaum, 1792)

FEATURES: Plain or blotchy brownish dorsally. Pale below. Small dark spots, a vague blotch, and obscure stripes on operculum. Lateral line dark, with a series of pale spots. First dorsal fin with black spot and white margins. Pelvic fins reach to or beyond anal fin origin. HABITAT: MA to FL and NE Gulf of Mexico. On or near bottom from near shore to about 1,300 ft. BIOLOGY: Spotted hake spawn offshore. Young found in estuaries. Feed on crustaceans, fishes, and squids.

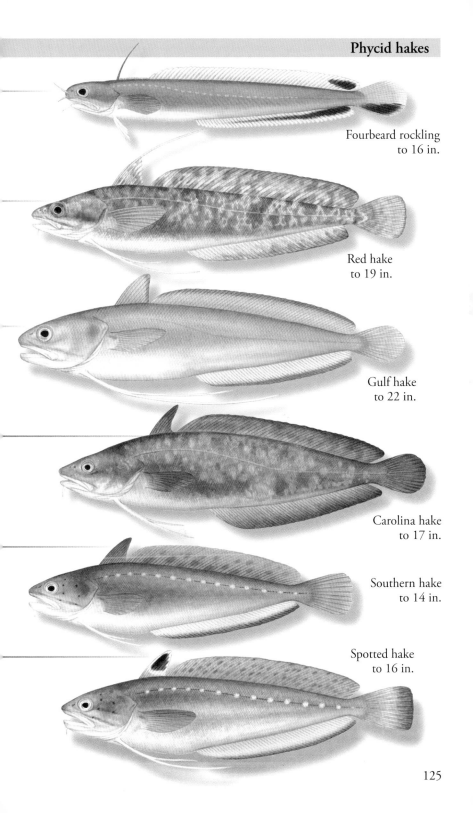

Fourbeard rockling
to 16 in.

Red hake
to 19 in.

Gulf hake
to 22 in.

Carolina hake
to 17 in.

Southern hake
to 14 in.

Spotted hake
to 16 in.

Cusk - *Brosme brosme* (Ascanius, 1772)

FEATURES: Dull reddish brown, gray, to yellowish dorsally and on sides. Pale ventrally. Young may be cross-barred. Dorsal, anal, and caudal fins with white outer margin, blackish inner margin. Pectoral and pelvic fins with dark margins. Single chin barbel. Body elongate, slightly compressed. Dorsal, caudal, and anal fins continuous. HABITAT: Newfoundland to NJ. Near bottom in moderately deep, cool water. Prefer hard bottoms. BIOLOGY: Cusk are solitary, sluggish, and inactive. They feed on crustaceans, echinoderms, and fishes. Sought as foodfish.

Atlantic cod - *Gadus morhua* Linnaeus, 1758

FEATURES: Color widely variable. Brownish, grayish, olivaceous, greenish, or pale. Usually with darker spots dorsally and on sides. May be uniformly colored. Pale ventrally. Lateral line pale. Fins similarly colored to body. Single chin barbel present. Three dorsal fins. Caudal fin straight-edged. HABITAT: Greenland to NC. Also NE Atlantic. From surface to near bottom, over rocky, gravelly, and sandy grounds to about 1,500 ft. BIOLOGY: Spawn at night in the water column. Females may produce up to 8 million eggs. Diet is fairly indiscriminate and varied. Once the most important US commercial fish, now depleted due to overfishing. Listed as Vulnerable.

Haddock - *Melanogrammus aeglefinus* (Linnaeus, 1758)

FEATURES: Usually purplish gray dorsally. Sometimes with several broad bars or blotches. Silvery on sides, white below. Single black blotch on each side above pectoral fin. Lateral line dark. Single small barbel on chin present. Dorsal fin tall, pointed. Caudal-fin margin concave. HABITAT: Newfoundland to NJ. Occur near gravelly and sandy bottoms in cold water from about 147 to 442 ft. BIOLOGY: Prey on a wide variety of bottom organisms, including young haddock. Listed as Vulnerable.

Atlantic tomcod - *Microgadus tomcod* (Walbaum, 1792)

FEATURES: Olivaceous with yellowish cast dorsally. Yellowish with dark mottling and splotches on sides. Abdomen pale. Dorsal and caudal fins similarly marked. Single small chin barbel present. Caudal-fin margin rounded. HABITAT: Labrador to VA. Found in shallow, inshore waters near river mouths, in estuaries, and in harbors. BIOLOGY: Atlantic tomcod produce antifreeze proteins that allow them to tolerate subfreezing water. They migrate up rivers and streams in winter to spawn. Employ barbel and pelvic fins to find food. Taken as accidental catch and as sportfish.

Pollock - *Pollachius virens* (Linnaeus, 1758)

FEATURES: Shades of dark green dorsally, fading to yellowish or grayish on sides. Pale ventrally. Lateral line pale. All fins greenish except pelvic fins. Single tiny chin barbel present; may be absent in large fish. Caudal fin forked. HABITAT: Greenland to NC. Occur around banks and shoals in cold water from surface to about 900 ft. BIOLOGY: Pollock are active, schooling, and migratory. Prey mostly on crustaceans, fishes, and squids. May live to 19 years. Taken commercially and as sportfish.

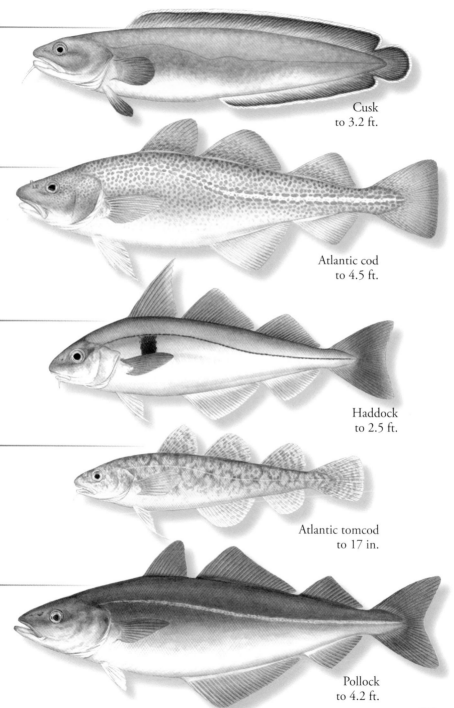

Cusk
to 3.2 ft.

Atlantic cod
to 4.5 ft.

Haddock
to 2.5 ft.

Atlantic tomcod
to 17 in.

Pollock
to 4.2 ft.

Carapidae - Pearlfishes

Pearlfish - *Carapus bermudensis* (Jones, 1874)

FEATURES: Translucent. Silvery cheek, abdomen. Vertebrae dark. Dark spots on sides. Anal fin longer, taller than dorsal fin. Body tapers to pointed tail. HABITAT: NC to Brazil, Gulf of Mexico, and Caribbean Sea. Also Bermuda. Near bottom from shore to about 700 ft. BIOLOGY: Adults live commensally in sea cucumbers. Emerge at night to forage. SIMILAR SPECIES: Chain pearlfish, *Echiodon dawsoni*, p. 431.

Ophidiidae - Cusk-eels

Atlantic bearded brotula - *Brotula barbata* (Bloch & Schneider, 1801)

FEATURES: Brownish to reddish on most of body and fins. Juveniles speckled. Dorsal and anal fins with white margins. Six barbels on snout and lower jaw. Stout-bodied. HABITAT: FL, Gulf of Mexico, Caribbean Sea to South America. Also Bermuda. On sandy and muddy bottoms and on reefs from near shore to upper slopes. BIOLOGY: Female Atlantic bearded brotula lay masses of gelatinous eggs.

Blackedge cusk-eel - *Lepophidium brevibarbe* (Cuvier, 1829)

FEATURES: Tannish to grayish dorsally. Ventral area silvery. Body unmarked. Dorsal fin, and sometimes anal fin, with dark margin. Snout with small spine. Small pelvic fins below lower jaw. Body tapers to pointed tail. HABITAT: SE FL and Gulf of Mexico to Brazil. Demersal from shore to about 260 ft. BIOLOGY: Female Blackedge cusk-eel lay floating, gelatinous masses of eggs. Taken as trawl bycatch.

Mottled cusk-eel - *Lepophidium jeannae* Fowler, 1941

FEATURES: Tannish dorsally and on sides with irregular, darker splotches. Abdomen pale. Small dark spot on gill cover. Dorsal fin with spots at base and black saddles at margin. Anal fin with dark margin. Snout with small spine. Small pelvic fins below lower jaw. HABITAT: NC to S FL and Gulf of Mexico. Near bottom from about 59 to 295 ft. BIOLOGY: Mottled cusk-eel lay floating masses of eggs.

Fawn cusk-eel - *Lepophidium profundorum* (Gill, 1863)

FEATURES: Brownish yellow dorsally and on sides. Two rows of pearly, oval spots run along upper sides. Dorsal and anal fins with dark margins posteriorly. Tip of snout with small spine. HABITAT: ME to FL, eastern and southern Gulf of Mexico to northern South America. Over continental shelves from about 180 to 1,200 ft. BIOLOGY: Fawn cusk-eel are more active at night. Preyed upon by Conger eels.

Stripefin brotula - *Neobythites marginatus* Goode & Bean, 1886

FEATURES: Pale yellowish brown. Two broken, narrow, brown stripes run from snout to posterior portion of body. Dorsal fin with dark inner stripe. Pelvic fins very slender, originate below preopercular margin. HABITAT: Gulf of Mexico and Caribbean Sea to northern South America. Also reported from NC to FL. Found near bottom from about 250 to 3,000 ft. BIOLOGY: Feed on crustaceans.

Pearlfishes

Pearlfish
to 7.8 in.

Cusk-eels

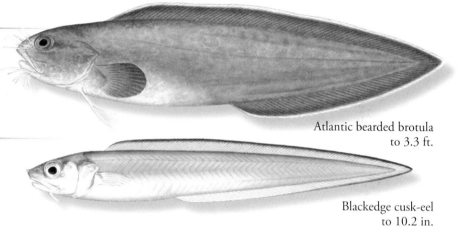

Atlantic bearded brotula
to 3.3 ft.

Blackedge cusk-eel
to 10.2 in.

Mottled cusk-eel
to 11 in.

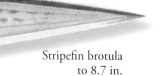

Fawn cusk-eel
to 9 in.

Stripefin brotula
to 8.7 in.

129

Blotched cusk-eel - *Ophidion grayi* (Fowler, 1948)

FEATURES: Pale tan with irregular, dark brown spots and blotches dorsally and on sides. Abdomen pale. Dorsal fin with spots, blotches, and dark margin. Anal fin with uniform, broad, dark margin. Pectoral fins dark. Pelvic fins on lower jaw below eyes. Body comparatively robust. HABITAT: SC to FL and Gulf of Mexico. More common in northeastern than northwestern Gulf of Mexico. Occur near bottom between about 33 and 190 ft. BIOLOGY: Blotched cusk-eel spawn during the winter in the northern Gulf of Mexico.

Bank cusk-eel - *Ophidion holbrookii* Putnam, 1874

FEATURES: Uniformly brownish dorsally and on sides. Lower head and abdomen pale. Dorsal fin, and sometimes anal fin, with dark margins. Lateral line pale. Body otherwise unmarked. Pelvic fins on lower jaw below eyes. Head profile almost straight. HABITAT: NC to FL, Gulf of Mexico to Brazil. More common in northeastern than northwestern Gulf of Mexico. Absent from Bahamas. On soft bottoms from shore to about 250 ft. BIOLOGY: Bank cusk-eel are caught as bycatch of shrimp-trawling fisheries.

Crested cusk-eel - *Ophidion josephi* Girard, 1858

FEATURES: Pale tannish dorsally and on sides with brownish spots that form lines and rows. Abdomen pale. Dorsal fin with black blotch near origin followed by black margin. Anal fin with black margin. Adult males with crest on nape. Females similarly colored, but lack crest. Pelvic fins on lower jaw below eyes. HABITAT: GA to S FL and Gulf of Mexico. Near bottom from shore to about 180 ft. BIOLOGY: In nothern Gulf of Mexico, Crested cusk-eel spawn in the fall. Life span is about two years. Previously known as *Ophidion welshi*.

Striped cusk-eel - *Ophidion marginatum* (DeKay, 1842)

FEATURES: Grayish green dorsally. Sides golden. Abdomen pale. Two or three complete stripes on sides. Dorsal and anal fins with black margins. Adult males with crest on nape. Females similarly colored but lack crest. Pelvic fins on lower jaw below eyes. HABITAT: NY to N FL. In Gulf of Mexico along western coast of FL and the coast of AL. Occur on or near bottom in shallow coastal waters. BIOLOGY: Striped cusk-eel are more active at night. Burrow in bottom sediment. Capable of producing drumming sounds.

Sleeper cusk-eel - *Otophidium dormitator* Böhlke & Robins, 1959

FEATURES: Nearly colorless. May have some golden to orange tint anteriorly. Body otherwise unmarked. Stout triangular spine under skin at snout tip. Pelvic fins on lower jaw below eyes. First pelvic-fin ray much shorter than second. HABITAT: S FL and Bahamas and Lesser Antilles to Yucatán. Usually found over coral sand near reefs. Also reported over turtle-grass beds. Occur from shore to about 50 ft. BIOLOGY: Sleeper cusk-eel burrow in bottom sediment. Reported to be nocturnal.

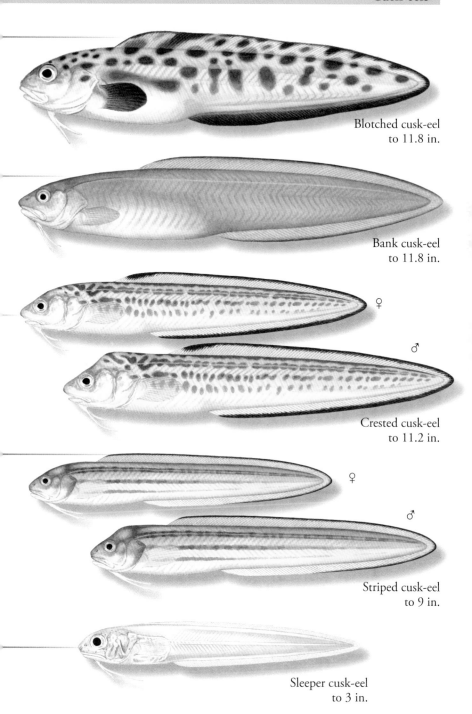

Blotched cusk-eel
to 11.8 in.

Bank cusk-eel
to 11.8 in.

♀

♂

Crested cusk-eel
to 11.2 in.

♀

♂

Striped cusk-eel
to 9 in.

Sleeper cusk-eel
to 3 in.

Ophidiidae - Cusk-eels, *cont.*

Polka-dot cusk-eel - *Otophidium omostigma* (Jordan & Gilbert, 1882)

FEATURES: Tannish with two to three irregular rows of brown spots and blotches on back and sides. Upper row forms saddles. Large spot just behind upper gill cover always darker than other spots. Dorsal fin with several black blotches along margin. Anal-fin base black. Eyes large. A spine is present on the snout and operculum. HABITAT: NC to S FL, northern Gulf of Mexico, and Lesser Antilles. Occur on or near bottom from about 32 to 164 ft.

ALSO IN AREA: Twospot brotula, *Neobythites gilli*; Mooneye cusk-eel, *Ophidion selenops*; Dusky cusk-eel, *Parophidion schmidti*, p. 431.

Bythitidae - Viviparous brotulas

Reef-cave brotula - *Grammonus claudei* (Torre y Huerta, 1930)

FEATURES: Dark blackish brown to dark gray. Fins blackish. Snout bluntly pointed. Rear margin of upper jaw broad, flattened. Preopercular margin with several small, indistinct spines. Lower preopercular corner elongate. Lateral line double; runs below dorsal profile and above ventral profile. HABITAT: Reported from S FL. Also Bahamas to Curaçao and Bermuda. Found in caverns of coral reefs and rocky shores. Uncommon. BIOLOGY: Previously known as *Oligopus claudei*.

Gold brotula - *Gunterichthys longipenis* Dawson, 1966

FEATURES: Pale yellowish to golden brown. Dorsal fin with freckles anteriorly and dark brown stripe posteriorly. Anal fin dark posteriorly. Head unscaled. Upper jaw with overhanging fold. Dorsal and anal fins separate from caudal fin. Males possess pair of hardened claspers. HABITAT: In Gulf of Mexico from FL Keys to TX. On or near bottom in quiet lagoons and estuaries. BIOLOGY: Gold brotula burrow in soft sediment. Previously referred to as *Ogilbia sp.* and *Dinematichthys sp.*

Key brotula - *Ogilbia cayorum* Evermann & Kendall, 1898

FEATURES: Reddish brown to olive brown. Fins similarly colored with pale margins. Head partially scaled. Anterior nostril close to upper lip. Dorsal and anal fins separate from caudal fin. Males possess pair of hardened claspers. HABITAT: Occur in shallow waters around the FL Keys and Bermuda. Possibly more widely distributed in tropical western Atlantic. Reef associated. BIOLOGY: Key brotula are secretive. Females give birth to live young.

Black brotula - *Stygnobrotula latebricola* Böhlke, 1957

FEATURES: Brownish black to black. Fins brownish black to blackish. Snout bluntly rounded. Rear margin of jaw narrow. Preopercle lacks spines. Lateral line interrupted; upper portion runs above pectoral fins, lower portion runs along midline. HABITAT: Reported from S FL. Also Bahamas to Curaçao. Found over shallow rocky ledges and reefs. BIOLOGY: Hide in caves during the day. Females give birth to live young.

Polka-dot cusk-eel
to 4 in.

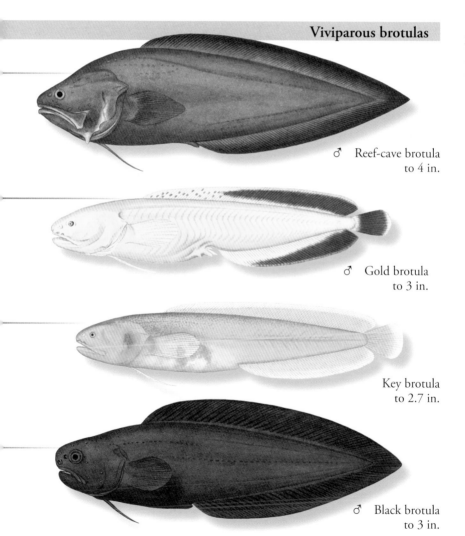

♂ Reef-cave brotula
to 4 in.

♂ Gold brotula
to 3 in.

Key brotula
to 2.7 in.

♂ Black brotula
to 3 in.

133

Batrachoididae - Toadfishes

Gulf toadfish - *Opsanus beta* (Goode & Bean, 1880)

FEATURES: Color variable. Brownish to grayish with pale marbling and mottling. Posterior pale blotches often form rosettes. Dorsal, anal, and caudal fins banded. Pectoral fins with spotty bands. Head broad, flattened. Mouth large. Fleshy tabs around mouth, on cheeks, and over eyes. Pectoral fins with 18–19 rays. HABITAT: SE FL, Little Bahama Bank, and Gulf of Mexico. Demersal in shallow lagoons and bays and over seagrass beds and rocky areas. BIOLOGY: Seek shelter in variety of cavities.

Leopard toadfish - *Opsanus pardus* (Goode & Bean, 1880)

FEATURES: Color variable. Pale brownish to yellowish with small dark spots over head and body. Spots may blend together. Dorsal and anal fins with spotted to blotched bands. Pectoral fins spotty. Head broad, flattened. Mouth large. Fleshy tabs around jaws, on cheeks, and over eyes. Pectoral fins with 20–22 rays. HABITAT: Gulf of Mexico. Demersal. Offshore over rocky bottoms and reefs on continental shelf. BIOLOGY: Leopard toadfish lie in wait among bottom organisms. Ambush prey.

Oyster toadfish - *Opsanus tau* (Linnaeus, 1766)

FEATURES: Color variable. Brownish, greenish, to yellowish with irregular, dark spots, mottling, and bars. Head speckled. Fins banded. Head broad, flattened. Mouth large. Fleshy tabs around jaws and cheeks and over eyes. Pectoral fins with 19–20 rays. HABITAT: ME to FL. Demersal on sandy and muddy bottoms along coast and in estuaries. BIOLOGY: Oyster toadfish can change color with surroundings. Use grunts and whistles during courtship. Used for scientific research.

Atlantic midshipman - *Porichthys plectrodon* Jordan & Gilbert, 1882

FEATURES: Metallic straw to dark brown or gray with dark spots dorsally. Larger spots along midbody lateral line. Lower sides silvery. Ventral area golden. Anal fin with dark margin. Series of photophores on head and along lateral lines. Two small spines precede second dorsal fin. HABITAT: VA to Brazil and Gulf of Mexico. On soft bottoms to about 840 ft. BIOLOGY: Atlantic midshipman are one of few shallow water fishes to possess photophores. Bioluminescence may be used during courtship.

Lophiidae - Goosefishes

Goosefish - *Lophius americanus* Valenciennes, 1837

FEATURES: Shades of brown dorsally with variable darker and paler spots and blotches. Whitish ventrally. Posterior edges of pectoral fins whitish. Body flattened, plate-like. Mouth very large. Tall, filamentous spines on head. Dorsal surface spiny. Fleshy tabs edge body and tail. HABITAT: Newfoundland to FL. Demersal on hard and soft bottoms from shore to about 2,750 ft. BIOLOGY: Voracious. May eat prey as large as themselves. Feed on fishes, invertebrates, and birds. Use first spine as lure.

ALSO IN THE AREA: Reticulate goosefish, *Lophiodes reticulatus*; Blackfin goosefish, *Lophius gastrophysus*, p. 431.

Gulf toadfish
to 12.7 in.

Leopard toadfish
to 12.7 in.

Oyster toadfish
to 15 in.

Atlantic midshipman
to 8.6 in.

Goosefishes

Goosefish
to 4 ft.

Antennariidae - Frogfishes

Longlure frogfish - *Antennarius multiocellatus* (Valenciennes, 1837)

FEATURES: Color blends with habitat. Shades of yellow, orange, red, pink, white, tan, gray, or black; may be somewhat to highly mottled. May have whitish saddles on shoulder and peduncle. Variably-sized dark ocelli at base of dorsal and anal fins. Always with three ocelli on caudal fin. May have smaller ocelli on fins and body. Lure very long, fringed at tip. HABITAT: FL Keys, Gulf of Mexico, Caribbean Sea to Brazil. Also Bermuda. From shore to about 370 ft. Prefer areas with sponges.

Ocellated frogfish - *Antennarius ocellatus* (Bloch & Schneider, 1801)

FEATURES: Brownish, yellowish, grayish, to reddish with variable darker and paler blotching. Small spots and smudges pepper head, body. Always with large, dark ocelli at dorsal-fin base, on caudal fin, and on midbody. Lure short, densely clustered at tip. HABITAT: NC to FL, E Gulf of Mexico, Bahamas, Bermuda, and W Caribbean Sea to Venezuela. From shore to about 450 ft. BIOLOGY: Use lure to bait prey.

Dwarf frogfish - *Antennarius pauciradiatus* Schultz, 1957

FEATURES: Yellowish, whitish, to brown. Red specimens recently reported. Small, vague spot at base of second dorsal fin. Body may have paler and darker spots and blotches. Lure very short, branched at tip. Second dorsal-fin spine branched at tip. HABITAT: E FL, Dry Tortugas, E Gulf of Mexico, Bahamas to Belize and Colombia. Also Bermuda. Occur over patch reefs from about 20 to 240 ft.

Singlespot frogfish - *Antennarius radiosus* Garman, 1896

FEATURES: Color variable. Beige, pale brown to grayish brown. Single, dark ocellus under base of second dorsal fin. Body may have reticulating pattern. Fins may be barred. Lure about as long as second dorsal-fin spine; tip small, simple. HABITAT: NY to FL Keys, Gulf of Mexico, and Greater Antilles. On offshore banks and shelf waters from about 65 to 900 ft. BIOLOGY: Other name: Big-eye frogfish.

Striated frogfish - *Antennarius striatus* (Shaw, 1794)

FEATURES: Shades of brown, yellow, beige. May be uniformly colored or densely banded and spotted. Some are black with pale mottling. Body and fins may have few to numerous fleshy filaments. Lure short; tip fleshy, branched, worm-like. HABITAT: NJ to FL, Gulf of Mexico, Caribbean Sea to Brazil. Also Bermuda. Over reefs and soft bottoms. BIOLOGY: Lie in wait, ambush prey. Other name: Split-lure frogfish.

Sargassumfish - *Histrio histrio* (Linnaeus, 1758)

FEATURES: Yellowish, grayish, olivaceous, to brownish with irregular, darker and paler lines and spots. Fleshy tabs on head, body, and fins. Lure short, fleshy at tip. Second and third dorsal spines fleshy. Pelvic fins long. HABITAT: Worldwide in warm seas. ME to FL, Gulf of Mexico, Bahamas, Caribbean Sea to Uruguay. Also Bermuda. BIOLOGY: Usually found in floating *Sargassum* seaweed. Well camouflaged. Use fins to cling and crawl. Lure used to bait prey. Larvae are pelagic.

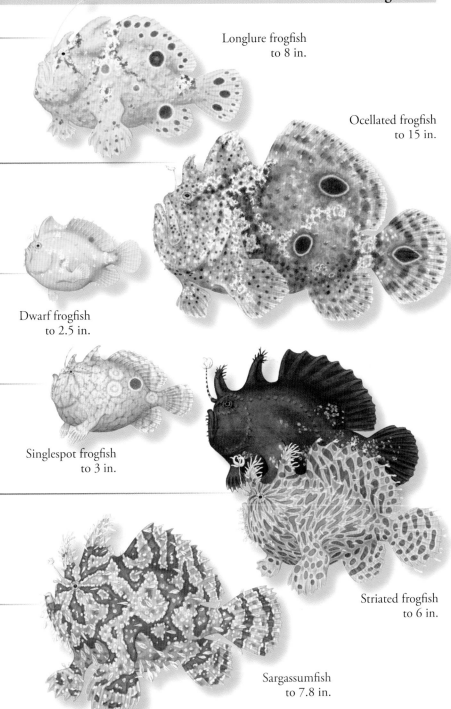

Longlure frogfish
to 8 in.

Ocellated frogfish
to 15 in.

Dwarf frogfish
to 2.5 in.

Singlespot frogfish
to 3 in.

Striated frogfish
to 6 in.

Sargassumfish
to 7.8 in.

Ogcocephalidae - Batfishes

Atlantic batfish - *Dibranchus atlanticus* Peters, 1876

FEATURES: Uniformly brownish to reddish gray dorsally, slightly paler ventrally. Dorsal surface covered with tubercles. Outer margin of body with a row of multi-spined tubercles. Rear margins of disk with an elongate spine with multiple spinelets. Second dorsal fin with six or seven rays. Body depressed, rounded in outline. Tail long, tapered. HABITAT: RI to Brazil, Gulf of Mexico, and Lesser Antilles. Also eastern Atlantic. Demersal. Occur on soft bottoms from about 130 to 4,200 ft. BIOLOGY: Atlantic batfish possess photophores as adults. Use lure to attract invertebrates.

Pancake batfish - *Halieutichthys aculeatus* (Mitchill, 1818)

FEATURES: Pale tan, olive, or yellowish gray dorsally with variable darker reticulating pattern. Pectoral and caudal fins banded with yellow margins. Dorsal surface covered with irregularly-shaped tubercles. Fleshy tabs around margin of body. Second dorsal fin with four or five rays. Body flattened, almost circular in outline. Tail comparatively short. HABITAT: NC to FL, Gulf of Mexico, and Caribbean Sea to northern South America. Demersal. Occur on sandy bottoms from about 100 to 1,400 ft. BIOLOGY: Pancake batfish may lie on the bottom, partially covered with sand.

Longnose batfish - *Ogcocephalus corniger* Bradbury, 1980

FEATURES: Brown to gray with uniformly scattered, small pale spots dorsally. Spots may form reticulating pattern. Pale ventrally. Pectoral fins with broad, dark reddish brown to blackish margins and white tips. Caudal fin with broad dark reddish brown to blackish margin. Dorsal surface covered in tubercles and bucklers. Rostrum always long, protrudes straight from head. Second dorsal fin with three to five rays. HABITAT: NC to FL, Bahamas, and Greater Antilles. Also northeastern and southern Gulf of Mexico. Demersal. Found from about 95 to 750 ft.

Polka-dot batfish - *Ogcocephalus cubifrons* (Richardson, 1836)

FEATURES: Tannish, brown, or reddish to grayish dorsally. Pale-ringed dark spots on shoulders and along sides. May have patches of yellow to orange dorsally. Pectoral fins always with wavy spots. Tubercles and bucklers scattered on dorsal surface. Fleshy tabs along margin of body and tail. Rostrum shortens with age. Pectoral fins with 11 to 12 rays, lack fleshy pads on ventral surface. HABITAT: NC to FL, northeastern Gulf of Mexico, Yucatán, and Bahamas. Demersal. From shore to about 223 ft.

Slantbrow batfish - *Ogcocephalus declivirostris* Bradbury, 1980

FEATURES: Uniformly dark brown to brownish dorsally, paler ventrally. Small specimens may have indistinct round spots on head, shoulders, and sides of tail. Dorsal surface covered with tubercles that are paler than dorsal surface. Area above eyes arched, slopes downward to end of rostrum. Fleshy tabs along margin of body and tail. Pectoral fins unmarked; lack fleshy pads on ventral surface. Tail long, broad, rounded at end. HABITAT: Northern and western Gulf of Mexico. Rare east of the Mississippi River. Occur on bottoms from about 11 to 590 ft.

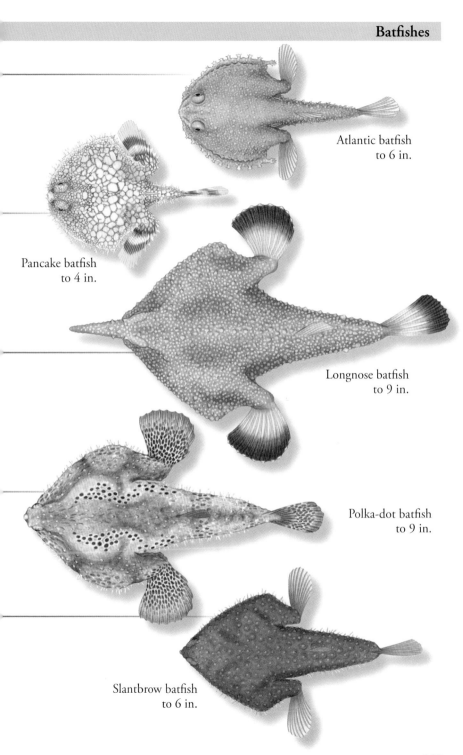

Atlantic batfish
to 6 in.

Pancake batfish
to 4 in.

Longnose batfish
to 9 in.

Polka-dot batfish
to 9 in.

Slantbrow batfish
to 6 in.

Ogcocephalidae - Batfishes, *cont.*

Shortnose batfish - *Ogcocephalus nasutus* (Cuvier, 1829)

FEATURES: Tan to brownish dorsally with small clusters of spots on shoulders. May be somewhat uniformly colored or blotched. May also have spots under eyes and on lateral sides of tail. Pectoral fins dusky to almost black, darker toward margins. Caudal fin banded. Rostrum variable in length but relatively short and upturned. Body covered in close-set tubercles. Pectoral fins with 12 to 15 rays and fleshy pads on ventral surface. HABITAT: S FL, northern Gulf of Mexico, Bahamas, Caribbean Sea to mouth of Amazon River. Occur on bottom from shore to about 900 ft.

Spotted batfish - *Ogcocephalus pantostictus* Bradbury, 1980

FEATURES: Tannish to grayish dorsally with small, close-set darker spots covering entire dorsal surface, including pectoral fins. Darker patches around shoulders and along tail. Caudal fin dark at margin. Dorsal surface covered in close-set tubercles and bucklers. Rostrum variable in length, relatively long in juveniles, shortens to a knob with age. Pectoral fins with fleshy pads on ventral surface. Tail long, broad, tapers at very end. HABITAT: In western Gulf of Mexico from Mobile Bay, AL, to Tampico, Mexico. Found on bottom from about 30 to 100 ft.

Roughback batfish - *Ogcocephalus parvus* Longley & Hildebrand, 1940

FEATURES: Color and pattern highly variable. Tannish, pinkish, or yellowish to brownish dorsally. May be mottled or uniformly colored. Usually with patches of dark spots at shoulders. Pectoral fins pale to white at base with broad dark margins and white tips. Caudal fin colorless or with dark margin. Dorsal surface covered in irregular tubercles and bucklers. May also be covered dorsally with numerous irregular fleshy tabs. Rostrum finger-shaped to cone-shaped, variable in length. Pectoral fins with fleshy pads on ventral surface. HABITAT: NC to FL, E Gulf of Mexico, Caribbean Sea to Brazil. Reef associated. On bottom from about 95 to 410 ft.

Palefin batfish - *Ogcocephalus rostellum* Bradbury, 1980

FEATURES: Tannish with paler areas. Clusters of pale-ringed dark spots and orange blotches on shoulders. Pectoral fins unmarked or indistinctly mottled or spotted at base, becoming yellowish to pale on middle portion and pale to dark at tips. Rostrum very short in adults. Area above eyes shelf-like. Body covered in close-set tubercles. Cirri sometimes present, but inconspicuous. Pectoral fins usually with 13 rays and with fleshy pads on ventral surface. Body comparatively flattened. HABITAT: Cape Hatteras, NC, to Dry Tortugas. Occur on bottom from about 90 to 750 ft.

Tricorn batfish - *Zalieutes mcgintyi* (Fowler, 1952)

FEATURES: Uniformly pale to dark brown, or olive brown dorsally. Pectoral and caudal fins pale and unmarked. Tubercles and bucklers cover dorsal surface of body and are paler than dorsal color. Rostrum short with prominent, recurved horns on either side. Pectoral fins lack fleshy pads on ventral surface. Tail long, tapered at end. HABITAT: FL, Gulf of Mexico, Caribbean Sea to northern South America. Found on bottom from about 295 to 590 ft.

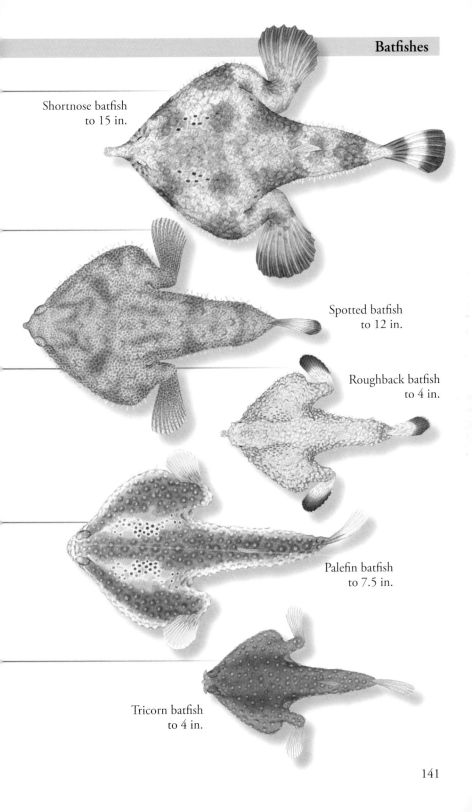

Shortnose batfish
to 15 in.

Spotted batfish
to 12 in.

Roughback batfish
to 4 in.

Palefin batfish
to 7.5 in.

Tricorn batfish
to 4 in.

Mugilidae - Mullets

Mountain mullet - *Agonostomus monticola* (Bancroft, 1834)

FEATURES: Brownish dorsally, whitish below. Sometimes with silvery band on sides. First dorsal fin yellowish with dark areas. Second dorsal and caudal fins yellowish with dusky margins. Pectoral and caudal fins with dark spot at base. Nape profile convex. HABITAT: NC, FL, Gulf of Mexico to Venezuela. Also Bahamas and Greater Antilles. Inshore and in brackish and fresh waters with access to the sea. BIOLOGY: Adult mountain mullet are anadromous. Adults live in freshwater streams. Spawn during heavy rainfall in lower rivers and at sea. Omnivorous.

Striped mullet - *Mugil cephalus* Linnaeus, 1758

FEATURES: Grayish olive to grayish brown dorsally. Sides silvery. Abdomen whitish to pale yellow. Dark spots on scales form stripes along sides. Dorsal and caudal fins dusky. Pectoral fins with dark spot at base. Second dorsal and anal fins with small scales on lower anterior portions only. HABITAT: Worldwide in tropical to warm temperate coastal, brackish, and fresh waters. Nova Scotia to FL, Gulf of Mexico to Brazil. Found near bottom to about 390 ft. BIOLOGY: Striped mullet form large schools. Catadromous. Feed on plankton and detritus. Caught and raised commercially.

White mullet - *Mugil curema* Valenciennes, 1836

FEATURES: Bluish green to olive dorsally. Sides and abdomen silvery to whitish. Yellowish to orangish blotch at upper edge of opercle. Pectoral fins with dark spot at base. Caudal fin with yellowish base and blackish margin. Second dorsal and anal fins almost entirely scaled. HABITAT: MA to FL, Gulf of Mexico and Caribbean Sea to Brazil. Also Bermuda. In tropical to warm temperate, coastal, and brackish waters. Occasionally in fresh water. BIOLOGY: White mullet are schooling. Catadromous, sometimes enter rivers. Feed on plankton and detritus. Caught commercially.

Liza - *Mugil liza* Valenciennes, 1836

FEATURES: Dusky bluish dorsally. Sides and abdomen silvery to white. Indistinct, dark stripes follow scale rows on upper sides. Pelvic fins pale or yellowish; other fins dusky. Head slender, triangular in profile. Second dorsal and anal fins with small scales on lower anterior portions only. Caudal fin deeply forked. HABITAT: Southern FL, Bahamas, and Antilles to northern coast of South America. Also Bermuda. Occur in inshore marine and brackish waters. BIOLOGY: Adult Liza may occasionally enter fresh water. Schooling. Catadromous. Feed on algae and detritus.

Fantail mullet - *Mugil trichodon* Poey, 1875

FEATURES: Dusky olive dorsally with bluish reflections. Silvery to whitish below. Pectoral fins with dark bluish spot at base. Dorsal and caudal fins dusky. Caudal fin with dusky margin. Second dorsal and anal fins scaled. HABITAT: Southern FL, Gulf of Mexico, and Caribbean Sea to NE Brazil. Also Bermuda. Occur in inshore marine and brackish waters and in river mouths. BIOLOGY: Fantail mullet prefer clear waters. Caught commercially. Previously known as *Mugil gyrans*.

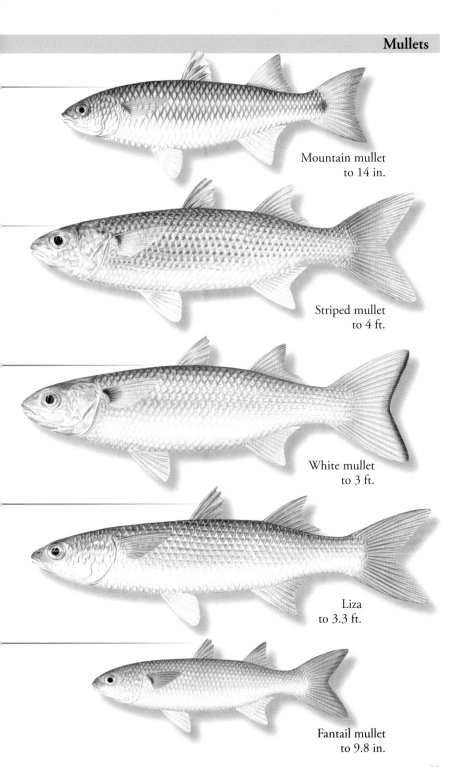

Mountain mullet
to 14 in.

Striped mullet
to 4 ft.

White mullet
to 3 ft.

Liza
to 3.3 ft.

Fantail mullet
to 9.8 in.

Atherinopsidae - New World silversides

Rough silverside - *Membras martinica* (Valenciennes, 1835)

FEATURES: Translucent greenish, bluish to yellowish dorsally, with dusky cross-hatching or peppering. Pale below. Silver stripe with narrow, dark margin on sides. Anal fin weakly concave. Scale edges scalloped, rough to the touch. HABITAT: NY to Mexico. Along coast and in saline bays, estuaries, and inlets. BIOLOGY: Rough silverside typically school near the surface. Preyed on by other fishes and birds.

Inland silverside - *Menidia beryllina* (Cope, 1867)

FEATURES: Pale yellowish to olivaceous dorsally; may have stippling or cross-hatching. Pale below. Silver stripe on sides. May have reddish tinge during breeding. Swim bladder extends over anal fin. Anal-fin margin concave; 15–18 rays. HABITAT: MA to FL, N Gulf of Mexico. In fresh water, inland waterways, and estuaries and along coast in low salinities. BIOLOGY: Form large schools. Tolerate varying salinities. Omnivorous. SIMILAR SPECIES: Texas silverside, *Menidia clarkhubbsi*, p. 431.

Atlantic silverside - *Menidia menidia* (Linnaeus, 1766)

FEATURES: Translucent greenish to grayish green dorsally. Pale below. May have cross-hatching. Silver stripe with narrow, dark margin on sides. Anal fin long-based, weakly concave with 19–29 rays. HABITAT: Gulf of St. Lawrence to FL. Near shore along sandy beaches, inlet mouths, and estuaries. BIOLOGY: Atlantic silverside are schooling omnivores. Follow tides. Preyed on by Striped bass and Bluefish.

Tidewater silverside - *Menidia peninsulae* (Goode & Bean, 1879)

FEATURES: Translucent yellow green to greenish dorsally. Pale to translucent below. Silver stripe with narrow, dark margin on sides. Anal fin with nearly straight margin and 13–19 rays. Body comparatively deep. HABITAT: NE FL to MS. Galveston Bay, TX, to Mexico. Along coast and in high-salinity tidal creeks and marshes.

Atherinidae - Old World silversides

Hardhead silverside - *Atherinomorus stipes* (Müller & Troschel, 1848)

FEATURES: Translucent bluish green dorsally. Pale below. Silver stripe with black margin on sides. Scales flecked with black. Head and eyes large. Outer caudal-fin lobes blackish in large specimens. First dorsal fin with 4–6 spines. HABITAT: S FL, E and S Gulf of Mexico, Bahamas, and Caribbean Sea to Brazil. Found in coastal waters. BIOLOGY: Hardhead silverside are schooling planktivores.

Reef silverside - *Hypoatherina harringtonensis* (Goode, 1877)

FEATURES: Translucent greenish dorsally. Silvery below. Broad silver stripe with black margin bordered by iridescent green. Black flecks on scales do not reach stripe. Caudal-fin tips dusky to black. First dorsal fin with 5–7 spines. HABITAT: S FL, E and S Gulf of Mexico, Bahamas, Antilles, Bermuda, to northern South America. Pelagic in coastal and offshore waters. BIOLOGY: Reef silverside form large schools.

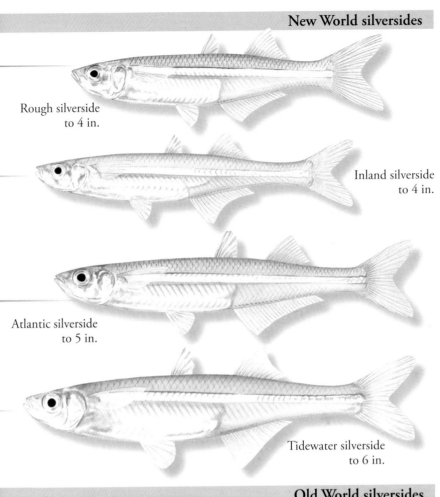

Rough silverside
to 4 in.

Inland silverside
to 4 in.

Atlantic silverside
to 5 in.

Tidewater silverside
to 6 in.

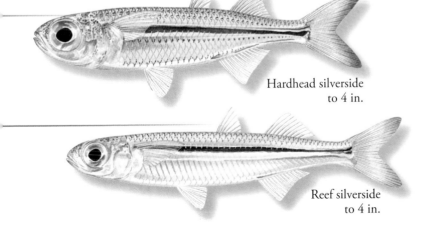

Hardhead silverside
to 4 in.

Reef silverside
to 4 in.

Margined flyingfish - *Cheilopogon cyanopterus* (Valenciennes, 1847)

FEATURES: Dark iridescent blue dorsally. Silvery white below. Pectoral fins bluish black with a thin, pale margin. Dorsal fin grayish with a large black blotch along central margin; blotch may cover entire fin in smaller specimens. Caudal fin dark gray to black. Pelvic and anal fins transparent white. Pectoral fins extend past middle of dorsal-fin base. HABITAT: NJ to FL, Gulf of Mexico, and Caribbean Sea to Brazil. Also Bermuda. Found near surface, along coast, to about 400 miles from shore. Absent inshore. BIOLOGY: Margined flyingfish feed on plankton and small fishes. Leap and glide for long distances. Attracted to lights at night.

Bandwing flyingfish - *Cheilopogon exsiliens* (Linnaeus, 1771)

FEATURES: Dark dorsally. Silvery white below. Pectoral fin dark bluish black with a pale inner band that is broader ventrally. Dorsal fin grayish with a large, black blotch. Upper lobe of caudal fin pale, lower lobe blackish. Pelvic fins grayish; may be blackish at margin. HABITAT: In the Gulf Stream from NJ to Brazil. Also in Gulf of Mexico and off Bermuda. Oceanic. May enter clear, coastal waters. BIOLOGY: Bandwing flyingfish feed on zooplankton. Leap and glide across surface. Eggs are pelagic.

Atlantic flyingfish - *Cheilopogon melanurus* (Valenciennes, 1847)

FEATURES: Dark iridescent bluish dorsally. Silvery white below. Pectoral fins grayish with pale inner band and a thin, pale margin. Dorsal and caudal fins grayish. Anal and pelvic fins transparent white. Pectoral fins reach beyond middle of dorsal-fin base. Anal fin originates under fifth to seventh dorsal-fin ray. HABITAT: MA to FL, Gulf of Mexico, and Caribbean Sea to Brazil. Also Bermuda. Occur at surface over continental shelves to about 400 miles from shore. BIOLOGY: Atlantic flyingfish feed on zooplankton. Leap and glide across surface.

Clearwing flyingfish - *Cypselurus comatus* (Mitchill, 1815)

FEATURES: Iridescent bluish dorsally. Silvery white below. Pectoral fins uniform pale gray to gray. Dorsal and caudal fins grayish. Pelvic fins transparent white, grayish base. Snout relatively pointed. Pectoral fins reach to end of dorsal-fin base. Anal fin originates under fourth to sixth dorsal-fin ray. HABITAT: Eastern FL to Brazil. Rare in Gulf of Mexico. Occur near surface over continental shelves. BIOLOGY: Clearwing flyingfish feed on zooplankton. Leap and glide over surface.

Oceanic two-wing flyingfish - *Exocoetus obtusirostris* Günther, 1866

FEATURES: Iridescent bluish dorsally. Silvery white below. Pectoral fins brownish gray with a broad, pale margin. Dorsal and caudal fins grayish. Pelvic and anal fins transparent white. Pectoral fins reach past dorsal-fin base. Pelvic fins small, originate forward of mid-body line. Anal fin originates slightly anterior to dorsal-fin origin. HABITAT: NJ to FL, Gulf of Mexico, and Caribbean Sea to Brazil. Also Bermuda. Oceanic in warm waters. Found near surface. BIOLOGY: Oceanic two-wing flyingfish feed on zooplankton. Preyed upon by predatory fishes, squids, and sea birds.

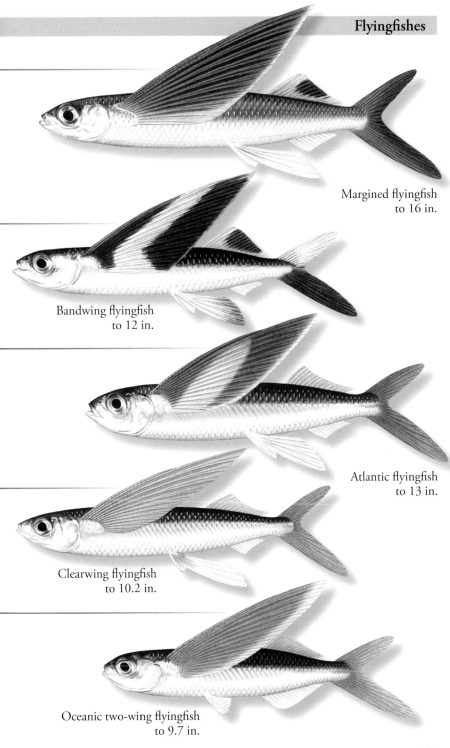

Margined flyingfish
to 16 in.

Bandwing flyingfish
to 12 in.

Atlantic flyingfish
to 13 in.

Clearwing flyingfish
to 10.2 in.

Oceanic two-wing flyingfish
to 9.7 in.

Tropical two-wing flyingfish - *Exocoetus volitans* Linnaeus, 1758

FEATURES: Dark iridescent blue dorsally. Silvery white below. Pectoral fins grayish. Pelvic and anal fins transparent white. Caudal fin gray. Pectoral fins reach past dorsal fin. Pelvic fin small, forward of mid-body line. Anal fin originates under first to third dorsal-fin ray. HABITAT: Worldwide in tropical seas. NJ to FL, eastern Gulf of Mexico, Caribbean Sea to S Brazil. Rare in Gulf of Mexico. Occur near surface of offshore waters. Absent inshore. BIOLOGY: Tropical two-wing flyingfish feed mainly on copepods. Leap and glide across surface. Spawn intermittently. Eggs are pelagic.

Fourwing flyingfish - *Hirundichthys affinis* (Günther, 1866)

FEATURES: Dark iridescent bluish dorsally. Silvery white below. Pectoral fins gray with pale inner band and thin, pale margin. Caudal fin grayish. Pelvic and anal fins transparent white. Pectoral fins extend beyond end of dorsal-fin base. Anal-fin origin slightly anterior or posterior to dorsal-fin origin. HABITAT: VA to FL, Bermuda, Gulf of Mexico, Caribbean Sea to Brazil. Found near surface over continental shelves. BIOLOGY: Preyed upon by dolphinfishes. Sought commercially in the Caribbean.

Blackwing flyingfish - *Hirundichthys rondeletii* (Valenciennes, 1847)

FEATURES: Dark iridescent bluish dorsally. Silvery white below. Pectoral fin bluish black with thin, pale margin. Dorsal and caudal fins grayish. Anal fin transparent whitish. Pelvic fins usually with large, black blotch. HABITAT: Nearly worldwide in temperate seas. MA to FL, Gulf of Mexico, Bahamas, Caribbean Sea to Brazil. Also Bermuda. Found at surface of open oceanic waters; also inshore. Juveniles in bays. BIOLOGY: Blackwing flyingfish leap and glide for long distances across the surface. Feed on zooplankton.

Smallwing flyingfish - *Oxyporhamphus micropterus similis*, Bruun, 1935

FEATURES: Greenish blue dorsally. Silvery sides and abdomen. Pectoral fin does not reach pelvic-fin origin. Pelvic fins small, triangular. HABITAT: Circumtropical. NC to FL, Gulf of Mexico, and Caribbean Sea to the equator. Also Bermuda. Marine, pelagic, and near surface. Inshore and offshore. BIOLOGY: Smallwing flyingfish are considered a halfbeak by some authors. Juveniles have slightly elongate lower jaw that is lost with age.

Sailfin flyingfish - *Parexocoetus hillianus* (Gosse, 1851)

FEATURES: Iridescent greenish blue dorsally. Silvery below. Dorsal fin expanded with large, black blotch. Pectoral fins transparent. Other fins pale to transparent. Pectoral fins reach to about middle of dorsal fin. HABITAT: MA to N Brazil. Gulf of Mexico, Bermuda, and Caribbean Sea. Occur in inshore and offshore waters. BIOLOGY: Prey on crustacean plankton. They are preyed upon by other fishes and sea birds. Previously known as *Parexocoetus brachypterus*.

ALSO IN THE AREA: Spotfin flyingfish, *Cheliopogon furcatus*; Western bluntnose flyingfish, *Prognichthys occidentalis*, p. 431.

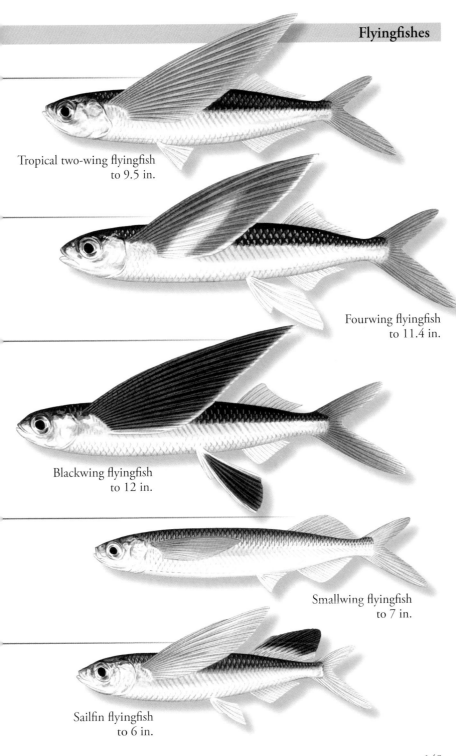

Tropical two-wing flyingfish
to 9.5 in.

Fourwing flyingfish
to 11.4 in.

Blackwing flyingfish
to 12 in.

Smallwing flyingfish
to 7 in.

Sailfin flyingfish
to 6 in.

Hemiramphidae - Halfbeaks

Hardhead halfbeak - *Chriodorus atherinoides* Goode & Bean, 1882

FEATURES: Translucent olive or tan dorsally, silvery below. Darker spots on scales form streaks. Silver stripe on sides. Lower jaw short, rounded. Caudal fin symmetrical. HABITAT: S FL, Bahamas, Cuba, Yucatán, and Belize. In clear nearshore waters and in brackish lakes or fresh water. BIOLOGY: Hardhead halfbeak feed on algae.

Flying halfbeak - *Euleptorhamphus velox* Poey, 1868

FEATURES: Olivaceous dorsally. Ventral area silvery. Body elongate, ribbon-like. Lower jaw very long and pointed. Pectoral fins long. HABITAT: RI to FL, Gulf of Mexico, Caribbean Sea to Brazil. Also Bermuda and in eastern Atlantic. Pelagic. Found along coast and offshore. BIOLOGY: Flying halfbeak may leap and glide for short distances. Preyed upon by sea birds.

Balao - *Hemiramphus balao* Lesueur, 1821

FEATURES: Bluish dorsally. Silvery white below. Both upper and lower caudal-fin lobes bluish violet. Lower jaw long and dark with red tip. HABITAT: NY to FL, Gulf of Mexico, Bahamas, and Caribbean Sea to Brazil. Found at the surface of inshore waters. BIOLOGY: Balao form large schools and feed primarily on planktonic invertebrates. Mature in one year. Sought as bait for offshore game fishing.

Ballyhoo - *Hemiramphus brasiliensis* (Linnaeus, 1758)

FEATURES: Dark bluish green dorsally. Silvery white below. Upper lobe of caudal fin yellowish orange. Lower jaw long and dark with red tip. HABITAT: MA to FL, Gulf of Mexico, Bahamas, and Caribbean Sea to Brazil. Occur at the surface of inshore waters. BIOLOGY: Ballyhoo form large schools. Feed on seagrasses and invertebrates. Few exceed 2 years of age. Sought as bait. Often caught with Balao.

False silverstripe halfbeak - *Hyporhamphus meeki* Banford & Collette, 1993

FEATURES: Transparent green dorsally. Silvery below. Silvery stripe on sides. Three thin black lines on back from head to dorsal fin. Fleshy tip of lower jaw red. Caudal-fin margin blackish. 31 to 40 gill rakers on first arch. HABITAT: ME to FL, northern Gulf of Mexico, and Yucatán. Found inshore at surface over sandy bottoms. In bays, estuaries, and harbors. BIOLOGY: Omnivorous and schooling. Previously mistaken as *Hyporhamphus unifasciatus*. Differ from *Hyporhamphus unifasciatus* by slightly shorter snout, smaller eyes, smaller size, and higher number of gill rakers.

Atlantic silverstripe halfbeak - *Hyporhamphus unifasciatus* (Ranzani, 1841)

FEATURES: Transparent greenish dorsally. Silvery below. Silvery stripe on sides. Three thin black lines on back from head to dorsal fin. Fleshy tip of lower jaw red. Caudal-fin margin blackish. 26 to 35 gill rakers on first arch. HABITAT: SW FL, Bahamas, Caribbean Sea, Veracruz, and Bermuda. Occur inshore at surface and enter estuaries. BIOLOGY: Atlantic silverstripe halfbeak are omnivorous and schooling. Feed at surface on sea grasses, small invertebrates.

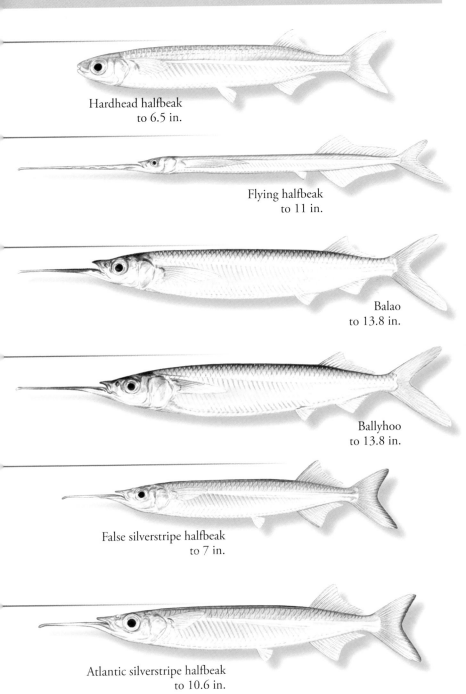

Hardhead halfbeak
to 6.5 in.

Flying halfbeak
to 11 in.

Balao
to 13.8 in.

Ballyhoo
to 13.8 in.

False silverstripe halfbeak
to 7 in.

Atlantic silverstripe halfbeak
to 10.6 in.

Belonidae - Needlefishes

Flat needlefish - *Ablennes hians* (Valenciennes, 1846)

FEATURES: Bluish dorsally, silvery below. Dark bars on sides. Dorsal fin with expanded, black, posterior lobe. Body laterally compressed. HABITAT: Worldwide in tropical and subtropical seas. MA to FL, Gulf of Mexico, Bahamas, Caribbean Sea to Brazil. Also Bermuda. Common offshore. BIOLOGY: Flat needlefish feed on fishes.

Keeltail needlefish - *Platybelone argalus* (Lesueur, 1821)

FEATURES: Greenish dorsally, silvery below. Narrow stripe on sides. Fins transparent. Lower jaw up to 20 percent longer than upper jaw. Well-developed, bluish black caudal keels present. Body round in cross-section. HABITAT: Worldwide in tropical and warm temperate seas. VA to FL, Gulf of Mexico, Bahamas, and Caribbean Sea to Trinidad. BIOLOGY: Keeltail needlefish feed on small fishes. Attracted to lights.

Atlantic needlefish - *Strongylura marina* (Walbaum, 1792)

FEATURES: Bluish green dorsally, silvery below. Black area behind eyes. Blue stripe on sides. Caudal keels absent. Body oval in cross-section. HABITAT: MA to FL, Gulf of Mexico to Brazil. Occur coastally and in brackish and fresh water. BIOLOGY: Atlantic needlefish prey on small fishes and crustaceans. Sought for bait.

Redfin needlefish - *Strongylura notata* (Poey, 1860)

FEATURES: Bluish green dorsally, silvery below. Black bar on preopercular margin. Dorsal, caudal, and anal fins with areas of red to orange. HABITAT: FL, Bahamas, Antilles, and coast of Central America. Found inshore and in bays and estuaries. BIOLOGY: Redfin needlefish feed on fishes and shrimps.

Timucu - *Strongylura timucu* (Walbaum, 1792)

FEATURES: Bluish green to dusky green above, silvery below. Dusky stripe on sides. Dark pigment in front of and behind eyes. Caudal keel absent. Body oval in cross-section. HABITAT: FL, Bahamas, Caribbean Sea, Yucatán to Brazil. In coastal waters, lagoons, and estuaries. BIOLOGY: Feed mainly on fishes. Young hide in seagrasses.

Atlantic agujon - *Tylosurus acus* (Lacepède, 1803)

FEATURES: Dark bluish dorsally, silvery below. Dark blue stripe on sides. Posterior dorsal-fin lobe blackish; elevated in juveniles. Small, blackish caudal keels present. Beak about twice the length of the head. HABITAT: Worldwide in tropical to warm temperate seas. MA to FL, Bermuda, Gulf of Mexico, Bahamas, Antilles to Brazil. Occur coastally and offshore. BIOLOGY: Feed on fishes. Other name: Agujon.

Houndfish - *Tylosurus crocodilus* (Péron & Lesueur, 1812)

FEATURES: Dark bluish green to greenish dorsally, silvery below. Blue stripe on sides. Juveniles with black, elevated posterior dorsal-fin lobe. Small, black caudal keel present. Beak about 1½ times length of head. HABITAT: Worldwide in warm seas. NC to FL, south to Brazil. Also Gulf of Mexico. BIOLOGY: Prey on fishes.

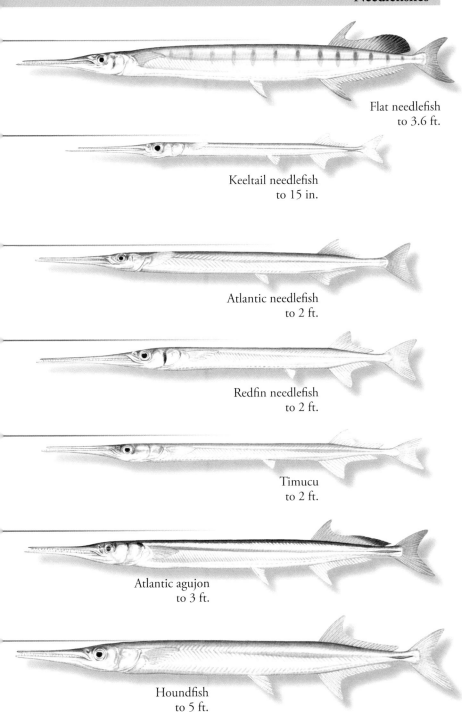

Flat needlefish
to 3.6 ft.

Keeltail needlefish
to 15 in.

Atlantic needlefish
to 2 ft.

Redfin needlefish
to 2 ft.

Timucu
to 2 ft.

Atlantic agujon
to 3 ft.

Houndfish
to 5 ft.

Scomberesocidae - Sauries

Atlantic saury - *Scomberesox saurus* (Walbaum, 1792)

FEATURES: Dark bluish green dorsally. Sides and abdomen silvery. Silver stripe on sides. Small dark spot above pectoral-fin base. Both jaws slender, elongate. Five or six finlets follow both dorsal and anal fins. Body laterally compressed. HABITAT: Newfoundland to NC. Scattered from NC to FL, Bahamas and Bermuda. Pelagic in warm surface waters of open ocean. BIOLOGY: Atlantic saury are schooling omnivores. Feed on a variety of zooplankton, fish eggs, and algae. They are a vital prey item for squids, billfishes, sharks, marine mammals, and birds.

Rivulidae - New World rivulines

Mangrove rivulus - *Kryptolebias marmoratus* (Poey, 1880)

FEATURES: Shades of brown, tan, to olive with darker blotches and mottling. Spotted on ventral area. Dark blotch above pectoral fin. Females with ocellus on peduncle. Males have an orange cast and lack ocellus. HABITAT: E FL to Tampa Bay, FL, Bahamas, and West Indies to Brazil. Found in poorly oxygenated brackish marshes and swamps. BIOLOGY: Mangrove rivulus are typically hermaphroditic. They are self-fertilizing and lay fertilized eggs. Males are rare. Feed on invertebrates.

Fundulidae - Topminnows

Diamond killifish - *Adinia xenica* (Jordan & Gilbert, 1882)

FEATURES: Females: greenish above with about five to seven pale stripes on sides; abdomen pale; fins pale. Males: dark greenish above with eight or more pale stripes on sides; dark stripes may have indistinct stripes in between; abdomen pale; fins spotted. Both with pointed snout. Body and peduncle deep. HABITAT: Florida Bay to southern TX. In freshwater to very saline estuaries, swamps, and lagoons.

Gulf killifish - *Fundulus grandis* Baird and Girard, 1853

FEATURES: Females: olivaceous above, silvery below with pearly flecks and 12–15 narrow bars on sides. Males: olivaceous to brownish above, with numerous pearly flecks on sides; bars faint to absent; abdomen may be golden. Both with convex snout profile, ten pores on lower jaw. Caudal-fin margin almost straight. HABITAT: NE FL and Gulf of Mexico to Veracruz. In coastal bays, marshes, pools, ditches, and nearby fresh water. BIOLOGY: Feed on invertebrates and small fishes. Used as bait.

Mummichog - *Fundulus heteroclitus* (Linnaeus, 1766)

FEATURES: Females: olivaceous above with pale abdomen; may have dusky bars; fins unmarked. Males: olivaceous above, pale below with yellow tinge; sides spotted, with up to 15 pale bars; fins spotted, with yellowish margins; dark spot usually present on dorsal fin. Both with convex snout profile and eight pores on lower jaw. Caudal-fin margin broadly rounded. HABITAT: Labrador to NE FL. In coastal bays, marshes, and channels and over seagrass flats. BIOLOGY: Feed on invertebrates, plants, and fishes.

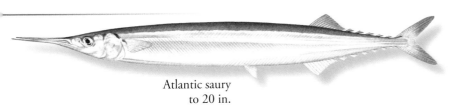

Atlantic saury
to 20 in.

New World rivulines

Mangrove rivulus   ♀
to 2 in.

Topminnows

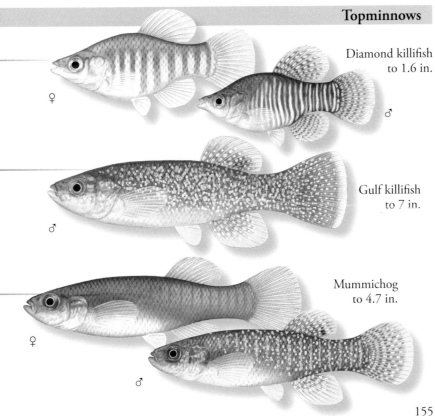

Diamond killifish
to 1.6 in.

♀

♂

Gulf killifish
to 7 in.

♂

Mummichog
to 4.7 in.

♀

♂

Saltmarsh topminnow - *Fundulus jenkinsi* (Evermann, 1892)

FEATURES: Males and females similar in shape and color. Olivaceous dorsally. Pale below. Others described as yellowish. Dark spots on sides arranged in two irregular rows. Scales with dark specks along edges. Dark blotch on upper opercle. Fins unmarked. Anal-fin origin slightly anterior to dorsal-fin origin. Body comparatively shallow. HABITAT: In Gulf of Mexico from west coast of FL to TX. Occur near bottom in low-salinity brackish streams and bays.

Spotfin killifish - *Fundulus luciae* (Baird, 1855)

FEATURES: Females (not shown) with a few trace bars; dorsal-fin ocellus absent. Males: brownish to greenish with broad, pale bars on sides; abdomen pale; dorsal fin with dark ocellus. Both lack dark spots on body. Anal-fin base longer and more anterior than dorsal-fin base. HABITAT: MA to GA. Found near bottom in brackish, intertidal marshes. BIOLOGY: Spotfin killifish tolerate a wide range of salinities, temperatures, and oxygen levels. Omnivorous. Preyed upon by wading birds.

Striped killifish - *Fundulus majalis* (Walbaum, 1792)

FEATURES: Females: olivaceous above, silvery below; dark stripes on sides, bars on caudal peduncle; fins may have yellow cast. Males: olivaceous dorsally, silvery to bronze below; 10–15 dark bars on sides; may have dark spot on dorsal fin; anal fin longer than female anal fin. Both with long snout, straight to slightly concave head profile. HABITAT: NH to SE FL. In Gulf of Mexico from west coast of FL to Tampico, Mexico. Occur in inlets, bays, estuaries, and marshes. Also along beaches. BIOLOGY: Striped killifish prefer high salinities. Gulf populations of females tend to have more bars than stripes but fewer bars than males do.

Bayou killifish - *Fundulus pulvereus* (Evermann, 1892)

FEATURES: Females: olivaceous dorsally and on sides with irregular dark spots; abdomen silvery; dark spot on dorsal fin usually present. Males: olivaceous dorsally, pale ventrally; 12–17 alternating dark and silvery bars on sides; dorsal, caudal, and anal fins spotted, may have yellowish margins; dark spot on dorsal fin usually present. Both with convex head profile. Snout length about equal to eye diameter. HABITAT: In Gulf of Mexico from AL to Corpus Christi, TX. In brackish bays and estuaries. BIOLOGY: Bayou killifish prefer medium to low salinities. May survive in fresh water. SIMILAR SPECIES: Marsh killifish, *Fundulus confluentus*, p. 431.

Longnose killifish - *Fundulus similis* (Baird & Girard, 1853)

FEATURES: Males and females similarly patterned. Olivaceous dorsally. Sides silvery, with 9–13 blackish bars. Blackish spot usually present on peduncle. Dorsal fin may have dark spot posteriorly. Snout long. Pectoral fins broad. HABITAT: NE FL and FL Keys. Possibly to Tampico. Found near bottom in shallow water surrounding mangroves and tidal flats. BIOLOGY: Longnose killifish are classified by some as identical to, or as a subspecies of, the Striped killifish, *Fundulus majalis*. At present, *Fundulus similis* is recognized as a valid species by the American Fisheries Society.

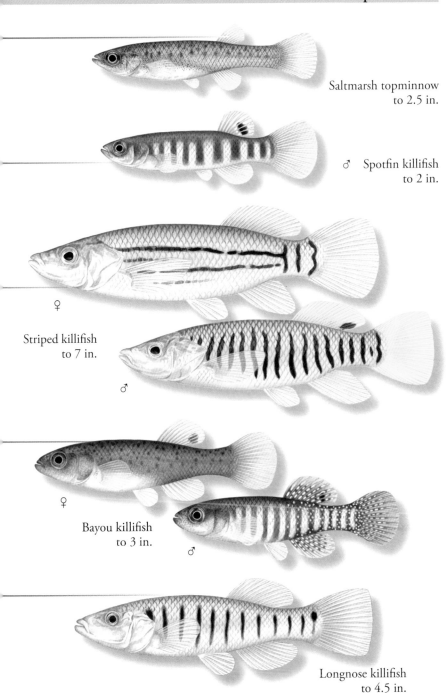

Saltmarsh topminnow
to 2.5 in.

♂ Spotfin killifish
to 2 in.

♀

Striped killifish
to 7 in.

♂

♀

Bayou killifish
to 3 in.

♂

Longnose killifish
to 4.5 in.

Fundulidae - Topminnows, *cont.*

Rainwater killifish - *Lucania parva* (Baird & Girard, 1855)

FEATURES: Tannish, greenish to silvery above. Abdomen silvery. Scales cross-hatched. Females: less colorful than males; lack dark pigment in fins. Males: more colorful than females; dorsal, anal, and pelvic fins may have orange and black pigment; black anterior spot on dorsal fin may be present. HABITAT: Cape Cod to FL Keys and in eastern and northern Gulf of Mexico. Along shorelines and in brackish bays and estuaries. BIOLOGY: Rainwater killifish may leap out of the water to avoid predators. Introduced into western US freshwater habitats.

Cyprinodontidae - Pupfishes

Sheepshead minnow - *Cyprinodon variegatus* Lacepède, 1803

FEATURES: Olive, gray, to brown dorsally and on sides with alternating, irregular bars. May be very pale. Females: one or two ocelli on dorsal fin. Males: blue highlights dorsally; margin of caudal fin blackish. Both with convex snout profile and deep body. Caudal fin of both broad with nearly straight margin. HABITAT: Cape Cod to FL, Gulf of Mexico, Bahamas, Cuba, Jamaica to Venezuela. In shallow, coastal, vegetated, fresh, brackish, and marine waters. BIOLOGY: Male Sheepshead minnow defend territories. Females may have multiple spawns with up to 140 eggs.

Goldspotted killifish - *Floridichtys carpio* (Günther, 1866)

FEATURES: Greenish dorsally. Sides silvery with numerous gold to orange spots and dashes. Females (not shown): dorsal, caudal fins unmarked; anal fin small. Males: dorsal, caudal fins with spots that form bands. Both with deep body, convex snout profile. HABITAT: East central FL. E Gulf of Mexico and Yucatán Peninsula. In shallow salt and brackish water. BIOLOGY: May form large schools.

Poeciliidae - Livebearers

Pike killifish - *Belonesox belizanus* Kner, 1860

FEATURES: Olivaceous, brownish, or grayish dorsally. Pale to slivery below. Sides with rows of small, black spots. Dark spot present at caudal-fin base. Fins translucent. Males with elongate anal fin. Both with pointed snout and enlarged jaws. HABITAT: Tampa Bay and Florida Everglades. Also Veracruz to Yucatán. In fresh and coastal brackish water. BIOLOGY: Pike killifish were introduced into Florida waters.

Western mosquitofish - *Gambusia affinis* (Baird & Girard, 1853)

FEATURES: Shades of olive, tan, brown, silver, or yellow dorsally. Pale below. Scale margins flecked, forming diamond-shaped pattern. Dorsal fin with two to three rows of dark spots. Females: gravid spot on abdomen; body deep; larger than males. Males: anal fin elongate at tip; body slender. HABITAT: In Gulf of Mexico from Mobile Bay, AL, to Tampico, Mexico. Occur in slow-moving fresh water and protected brackish and marine waters. BIOLOGY: Feed on crustaceans and insects and their larvae.

Rainwater killifish
to 2.4 in.

♀

♂

Pupfishes

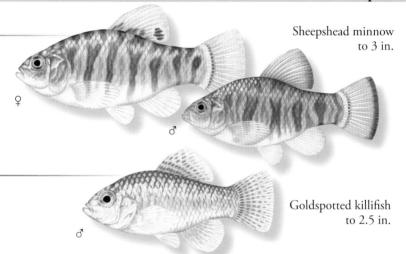

Sheepshead minnow
to 3 in.

♀

♂

Goldspotted killifish
to 2.5 in.

♂

Livebearers

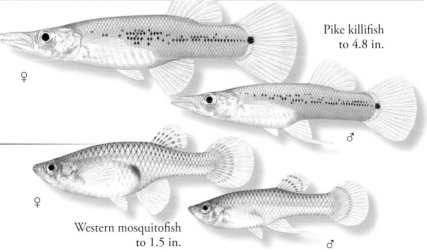

Pike killifish
to 4.8 in.

♀

♂

♀

Western mosquitofish
to 1.5 in.

♂

159

Poeciliidae - Livebearers, *cont.*

Eastern mosquitofish - *Gambusia holbrooki* Girard, 1859

FEATURES: Shades of olive, tan, brown, or silver dorsally. Pale below. May have a bar below eyes. Scale margins dusky, forming a diamond-shaped pattern. Dorsal fin with two to three rows of dark spots. Females: distinct gravid spot on abdomen; body deep; larger than males. Males: anal fin elongate at tip; body slender. HABITAT: NJ to S FL and in Gulf of Mexico to Mobile Bay, AL. Also Bermuda. Occur in quiet brackish and fresh water. BIOLOGY: Eastern mosquitofish hybridize with Western mosquitofish where ranges overlap. Some authors do not recognize *G. holbrooki* as a valid species.

Mangrove gambusia - *Gambusia rhizophorae* Rivas, 1969

FEATURES: Tannish, grayish to olivaceous. Distinct blackish spots follow scale rows on sides. Dorsal fin yellowish with dark spots. Females: may have a gravid spot on abdomen; body deep; larger than males. Males: anal fin elongate at tip; body slender. HABITAT: S FL, northern Cuba. Primarily occur around Red mangroves in marine and hypersaline waters. Also found somewhat inland in brackish water with White and Black mangroves. BIOLOGY: Mangrove gambusia are tolerant of high water temperatures. Feed on spiders and ants.

Least killifish - *Heterandria formosa* Agassiz, 1855

FEATURES: Pale to dark tannish or olivaceous dorsally. Pale below. Six to nine dark bars on upper sides. Dark stripe on lower side. Dark spot at caudal-fin base. Scales cross-hatched. Females: blackish spot on dorsal and anal fins; body deep; larger than males. Males: black spot on dorsal fin below yellow area; anal fin elongate; body slender. HABITAT: SC to FL, Gulf of Mexico to E TX. In shallow, densely vegetated brackish and freshwater pools, ditches, and streams. BIOLOGY: Omnivorous.

Sailfin molly - *Poecilia latipinna* (Lesueur, 1821)

FEATURES: Olivaceous to brownish dorsally. Pale below. Scales form diamond-shaped pattern. Spots on scales form stripes along sides. Females: dorsal fin low, banded; peduncle moderately deep. Males: dorsal fin tall, sail-like, and banded with a yellow to orange margin; peduncle deep. HABITAT: NC to FL Keys, in Gulf of Mexico from SW FL to Yucatán. In weedy, quiet, fresh, brackish, and marine waters. Prefer clear waters. BIOLOGY: Feed on algae, detritus, insects, and crustaceans. Tolerant of polluted waters. Southern populations are larger and more colorful than northern populations. Color variants are popular in the aquarium trade.

Shortfin molly - *Poecilia mexicana* Steindachner, 1863

FEATURES: Color variable. Females (not shown): body color similar to males; fins transparent; caudal peduncle moderately deep. Males: olivaceous, brownish, to grayish dorsally with orangish spots on scales; dorsal and caudal fins may be blackish and often with yellowish to orange margins; membranes may be spotted or blotched; caudal peduncle deep. Both with short-based dorsal fin. HABITAT: S TX to Colombia. Also southern islands of the West Indies. Possibly also FL. Occur in fresh to salt water, mostly in standing or slow-moving currents.

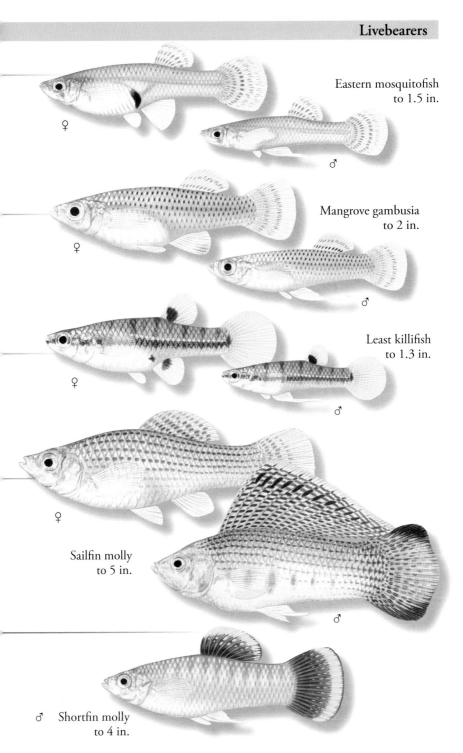

♀

Eastern mosquitofish
to 1.5 in.

♂

♀

Mangrove gambusia
to 2 in.

♂

♀

Least killifish
to 1.3 in.

♂

♀

Sailfin molly
to 5 in.

♂

♂ Shortfin molly
to 4 in.

161

Trachichthyidae - Roughies

Big roughy - *Gephyroberyx darwinii* (Johnson, 1866)

FEATURES: Reddish dorsally. Silvery red to pink below. Low ridges around eyes. Opercles with large spines at lower and upper corners. Abdomen with row of heavy scutes. Body deep, compressed. HABITAT: DE to Panama, northern Gulf of Mexico, and western Caribbean Sea. On or near bottom of slopes from about 240 to 2,000 ft. Young near coast. BIOLOGY: Young Big roughy feed on shrimps and fishes.

Silver roughy - *Hoplostethus mediterraneus* Cuvier, 1829

FEATURES: Pinkish dorsally. Sides and ventral area silvery grayish to blackish. All fins reddish. Snout short. Inside of mouth black. Low ridges and spines on head. Lateral-line scales enlarged. Row of 8-12 scutes along abdomen. Body deep, compressed. HABITAT: NJ to Greater Antilles, and Gulf of Mexico. Also Bermuda and in Indian and western Pacific oceans. Occur near bottom in tropical to warm temperate waters from about 1,000 to 4,800 ft. BIOLOGY: Feed on crustaceans. Caught commercially. SIMILAR SPECIES: Western roughy, *Hoplostethus occidentalis*, p. 431.

Berycidae - Alfonsinos

Splendid alfonsino - *Beryx splendens* Lowe, 1834

FEATURES: Pinkish red dorsally. Silvery below. Fins pinkish. Eyes large. Opercle with two spines. Pectoral fin erect. Single, mid-body dorsal fin. Caudal fin deeply forked. HABITAT: Scattered worldwide in warm seas. ME to Brazil. Near bottom from about 650 to 3,300 ft. BIOLOGY: Form dense schools. Feed on fishes, crustaceans, and cephalopods. SIMILAR SPECIES: Red bream, *Beryx decadactylus*, p. 431.

Holocentridae - Squirrelfishes

Spinycheek soldierfish - *Corniger spinosus* Agassiz, 1831

FEATURES: Head and body uniformly colored in shades of deep red. All fins red. Mouth and eyes large. Three long spines below each eye. Snout, cheeks, and opercles very spiny. Scales serrated. Body deep. Soft dorsal, caudal, and anal fins relatively small. HABITAT: SC, FL, NE Gulf of Mexico, Cuba, and Brazil. Also in the eastern Atlantic at St. Helena. Over deep rocky slopes and reefs. BIOLOGY: Spinycheek soldierfish scatter their eggs over the bottom or in open water.

Squirrelfish - *Holocentrus adscensionis* (Osbeck, 1765)

FEATURES: Body with alternating reddish and silvery stripes. May have pale blotches or bars. Snout red. White, diagonal streak from jaw to corner of preopercle. Spiny dorsal fin yellowish, other fins pale pinkish. Jaws extend to rear margin of pupil. Soft dorsal fin and upper lobe of caudal fin elongate. HABITAT: VA to FL, Bahamas, Caribbean Sea to Brazil. Also Bermuda and portions of Gulf of Mexico. Over shallow inshore coral reefs to deeper offshore waters to about 295 ft. BIOLOGY: Nocturnal. They seek shelter in crevices and under ledges during the day, forage at night.

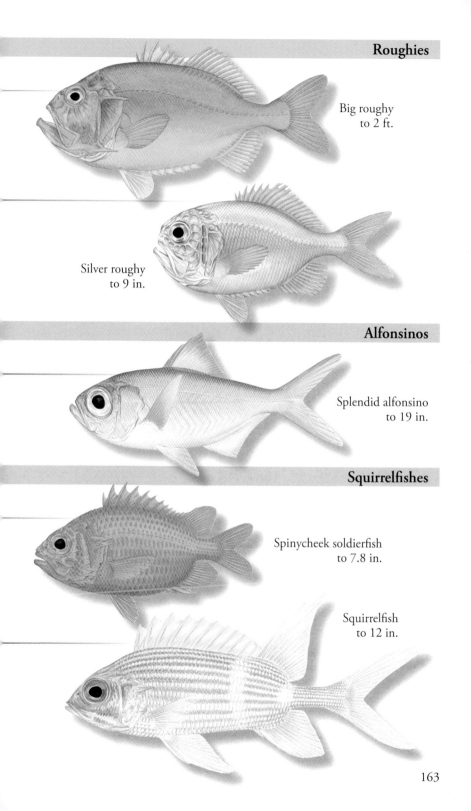

Roughies

Big roughy
to 2 ft.

Silver roughy
to 9 in.

Alfonsinos

Splendid alfonsino
to 19 in.

Squirrelfishes

Spinycheek soldierfish
to 7.8 in.

Squirrelfish
to 12 in.

163

Longspine squirrelfish - *Holocentrus rufus* (Walbaum, 1792)

FEATURES: Reddish dorsally with alternating dark and pale stripes. Usually with irregular, pale bars on sides. Top of head deep red. Eyes dark red. White stripe from jaw to preopercular corner and along preopercular margin. Spiny dorsal fin with a white blotch at upper margin of each membrane. Other fins pinkish. Jaws short. HABITAT: NC to FL, Bahamas, Caribbean Sea to Central and South America. Also Bermuda and portions of the Gulf of Mexico. Over shallow reefs to about 105 ft. BIOLOGY: Longspine squirrelfish hide in crevices by day, feed over seagrass beds at night.

Blackbar soldierfish - *Myripristis jacobus* Cuvier, 1829

FEATURES: Reddish dorsally. Ventral area pale to silvery. Broad, dark bar from top of opercle to rear pectoral-fin base. Spiny dorsal fin red with white at tips and between spines. Anterior margins of soft dorsal, anal, pelvic, and caudal fins white. Other fin portions reddish. Body and fins may appear pale to grayish. Eyes and mouth large. Small spine on opercle. HABITAT: NC to FL, Bahamas, Caribbean Sea to Brazil. Also Bermuda and portions of Gulf of Mexico. Over shallow coral reefs to deeper waters to about 295 ft. BIOLOGY: Nocturnal. Form aggregations around reefs.

Longjaw squirrelfish - *Neoniphon marianus* (Cuvier, 1829)

FEATURES: Alternating reddish and yellowish stripes dorsally and on sides. Ventral area silvery. Upper head and snout reddish. White bar on preopercle and yellowish bar on opercle. Spiny dorsal fin yellow with white at margins and between spines. Other fins reddish to pinkish. Lower jaw protrudes. Upper jaw extends to below center of pupil. HABITAT: FL Keys, W Gulf of Mexico, Bahamas, Antilles to Trinidad and coast of Central America. Over patch reefs from surface to about 200 ft., more commonly below 50 ft. BIOLOGY: Longjaw squirrelfish are nocturnal. Feed on shrimps and crabs. Previously known as *Holocentrus marianus*.

Bigeye soldierfish - *Ostichthys trachypoma* (Günther, 1859)

FEATURES: Reddish with indistinct pale stripes dorsally and on sides. Paler below. Spiny dorsal fin red. Mouth large. Upper jaw extends beyond posterior margin of eyes. Preopercle serrated. Opercle with single spine at upper margin. Lobes of soft dorsal, caudal, and anal fins angular. HABITAT: NY to FL, Gulf of Mexico, and Caribbean Sea to Brazil. Usually near bottom between about 124 and 1,600 ft. BIOLOGY: Bigeye soldierfish are nocturnal, feed at night, seek shelter during the day.

Cardinal soldierfish - *Plectrypops retrospinis* (Guichenot, 1853)

FEATURES: Uniformly red. May be paler below. Spiny dorsal fin red; membranes pale at tips. Other fins reddish. Irises may be red, yellow, or silvery. Spines under and behind eyes overlap upper jaw. Opercles serrated and spiny. Body relatively deep. HABITAT: S FL, Bahamas, Caribbean Sea to Brazil. In Gulf of Mexico from the Flower Garden Banks. Also Bermuda. Occur around shallow reefs. Also in deeper water. BIOLOGY: Cardinal soldierfish are reclusive. Hide in crevices during the day. May swim upside down.

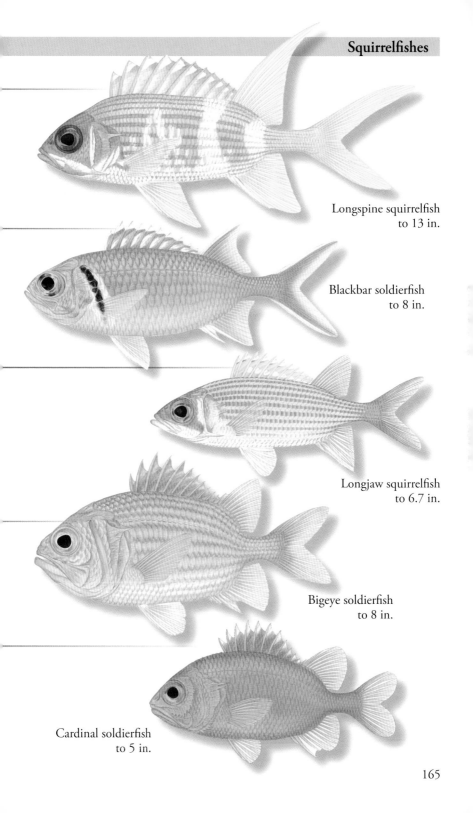

Longspine squirrelfish
to 13 in.

Blackbar soldierfish
to 8 in.

Longjaw squirrelfish
to 6.7 in.

Bigeye soldierfish
to 8 in.

Cardinal soldierfish
to 5 in.

165

Deepwater squirrelfish - *Sargocentron bullisi* (Woods, 1955)

FEATURES: Orange yellow dorsally and on sides with white stripes bordered above and below by narrow, brownish stripes. Head reddish orange above, white below, with reddish band from lower eye to preopercular corner. Dorsal fin yellow with white areas at base and a black spot between first and second spines. Other fins reddish yellow to clear. HABITAT: NC to FL, Gulf of Mexico, Bahamas, Lesser Antilles, Caribbean Sea, and Bermuda. In offshore waters from about 120 to 390 ft. BIOLOGY: Deepwater squirrelfish are nocturnal. Seek deep crevices during the day, forage at night.

Reef squirrelfish - *Sargocentron coruscum* (Poey, 1860)

FEATURES: Alternating red and white stripes dorsally and on sides. Head red above, white below, with red band from lower eye to preopercular corner. Dorsal fin red with white at tips and at base. Black splotch between first and third or first and fourth dorsal-fin spines. Other fins reddish to clear. HABITAT: S FL, Bahamas, and Caribbean Sea. Also Bermuda. Occur over reefs and adjacent areas from surface to about 100 ft. BIOLOGY: Reef squirrelfish are nocturnal. Hide in reef crevices during the day, forage over sandy areas at night.

Dusky squirrelfish - *Sargocentron vexillarium* (Poey, 1860)

FEATURES: Alternating broad dusky red and whitish stripes dorsally and on sides. Breast and abdomen silvery white. Head dusky red above, white below with dusky red band from lower eye to corner of preopercle. Body may have small, dark specks. Dorsal fin red with white areas around spines. Other fins reddish to clear. HABITAT: FL, Gulf of Mexico, Bahamas, Caribbean Sea to northern South America. Also Bermuda. Occur over inshore reefs to about 65 ft. BIOLOGY: Dusky squirrelfish are nocturnal. Feed at night on a variety of invertebrates and small fishes.

Grammicolepididae - Tinselfishes

Thorny tinselfish - *Grammicolepis brachiusculus* Poey, 1873

FEATURES: Silvery overall. Head, snout short. Eyes large, mouth very small. Caudal fin long. Scales narrow. Body deep, compressed. Juveniles with irregular bars/blotches; pronounced thorns on sides and elongate dorsal and anal-fin spines; very deep body. Thorns are lost, fin spines shorten, and body becomes shallower with age. HABITAT: Atlantic, Pacific, and Mediterranean. Georges Bank to FL, Cuba, and Bahamas to Venezuela. In midwater and over bottoms from about 980 to 3,000 ft.

Spotted tinselfish - *Xenolepidichthys dalgleishi* Gilchrist, 1922

FEATURES: Body and fins silvery. Diffuse dark spots on sides. Caudal-fin margin dark. Eyes large, mouth small. Scales narrow. Body very deep, compressed. First dorsal- and anal-fin rays very elongate in juveniles; shorten with age. HABITAT: Scattered worldwide. In western Atlantic from Nova Scotia to FL. Also Bermuda and portions of Central and South America. Found near bottom from about 300 to 3,000 ft.

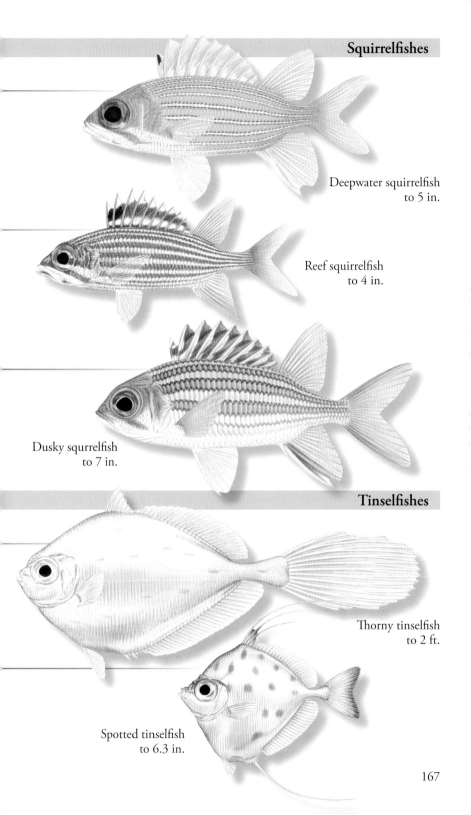

Squirrelfishes

Deepwater squirrelfish
to 5 in.

Reef squirrelfish
to 4 in.

Dusky squrrelfish
to 7 in.

Tinselfishes

Thorny tinselfish
to 2 ft.

Spotted tinselfish
to 6.3 in.

167

Zeidae - Dories

Red dory - *Cyttopsis rosea* (Lowe, 1843)

FEATURES: Head and body silvery red. Head, eyes, and mouth large. Head profile convex. Pelvic fins large, with blackish membranes. Spiny scutes absent. Body deep, compressed. HABITAT: In warm waters of Atlantic, Indian, and Indo-West Pacific oceans. Canada, SC, N Gulf of Mexico, portions of Caribbean Sea. Near bottom from about 800 to 2,000 ft. BIOLOGY: Red dory form schools. Feed on fishes.

Buckler dory - *Zenopsis conchifera* (Lowe, 1852)

FEATURES: Overall silvery with vague, dark blotch on sides. Juveniles spotted. Head and mouth large. Head profile concave. Anterior dorsal-fin spines tall, with filaments. Bucklers along base of dorsal and anal fins. Body deep, compressed. HABITAT: Nova Scotia to FL, Gulf of Mexico to Brazil. Over bottoms and in mid-water from about 320 to 1,300 ft. BIOLOGY: Buckler dory form small schools. Feed on fishes.

Gasterosteidae - Sticklebacks

Fourspine stickleback - *Apeltes quadracus* (Mitchill, 1815)

FEATURES: Grayish to olive green dorsally with small black spots that form irregular bands and blotches. Abdomen silvery white. Four or five dorsal spines. Lateral body plates absent. Peduncle lacks keel. HABITAT: Newfoundland to VA. In shallow brackish inlets, tidal creeks, estuaries, and lagoons. BIOLOGY: Fourspine stickleback prefer low salinities. Males build and care for multiple nests.

Threespine stickleback - *Gasterosteus aculeatus* Linnaeus, 1758

FEATURES: Dark bluish dorsally while in saltwater, turning greenish, grayish to brownish dorsally after spawning in fresh water. Silvery to reddish below. Three dorsal spines. Body with conspicuous lateral plates. Peduncle with lateral keel. HABITAT: Found in nearly all coastal waters of the Northern Hemisphere. In western Atlantic from Baffin Island to NC. Open ocean to freshwater habitats. BIOLOGY: Threespine stickleback may be coastal or anadromous. Males build nests and defend young.

Blackspotted stickleback - *Gasterosteus wheatlandi* Putnam, 1867

FEATURES: Pale olive, bluish, to yellowish dorsally. Ventral area silvery. Small black spots may be absent or inconspicuous or form mottling. Cusps surround base of pelvic spine. Peduncle lacks keel. HABITAT: Newfoundland to NY. Offshore to shallow brackish waters. Sometimes in fresh water. BIOLOGY: Blackspotted stickleback spend most of their lives offshore. Males build and protect nests.

Ninespine stickleback - *Pungitius pungitius* (Linnaeus, 1758)

FEATURES: Olive brown, grayish to blackish dorsally. Ventral area silvery to blackish. May be barred or blotched. Dorsal fin with 9–13 spines. Peduncle slim, keeled. HABITAT: Baffin Island to NJ. In shallow, vegetated estuaries and marshes. Also in fresh water. BIOLOGY: Male Ninespine stickleback build nests and court multiple females.

Dories

Red dory
to 8.6 in.

Buckler dory
to 3 ft.

Sticklebacks

Fourspine stickleback
to 2.5 in.

Threespine stickleback
to 4 in.

Blackspotted stickleback
to 3 in.

Ninespine stickleback
to 3 in.

169

Syngnathidae - Pipefishes and Seahorses

Pipehorse - *Amphelikturus dendriticus* (Barbour, 1905)

FEATURES: Shades of greenish to brownish with dark to pale blotches. May be uniformly colored. Slender, elongate, with head bent slightly downward. Small caudal fin at end of tail. Dermal flaps on head and body may be simple or branched. HABITAT: New Brunswick to FL, Bermuda, N Gulf of Mexico, and Bahamas to Brazil. On bottom or near surface in shallow to oceanic waters. BIOLOGY: Male Pipehorse brood eggs in pouch under the tail. Previously known as *Acentronura dendritica*.

Lined seahorse - *Hippocampus erectus* Perry, 1810

FEATURES: Color and pattern highly variable. Pale and dark dots on neck and along back. Pale and dark lines on head and body. Snout shorter than distance from eye to opercular margin. Body may be covered in fleshy filaments. HABITAT: Nova Scotia (rare) to FL, Caribbean Sea to Venezuela, Gulf of Mexico, and Bermuda. In vegetated waters to about 240 ft. BIOLOGY: Males brood up to 400 eggs. Vulnerable.

Longsnout seahorse - *Hippocampus reidi* Ginsburg, 1933

FEATURES: Color and pattern highly variable. Small dark spots on head and body. May have paler or darker blotches. Snout usually longer than distance from eye to opercular margin. HABITAT: NC through Caribbean Sea to Brazil. E Gulf of Mexico and Bermuda. Over seagrass beds, around pilings, and on reefs to about 50 ft. BIOLOGY: Exhibit bright coloring during courtship. Males brood eggs.

Dwarf seahorse - *Hippocampus zosterae* Jordan & Gilbert, 1882

FEATURES: Usually tannish. May have paler and darker mottling. Dorsal fin with submarginal stripe. Snout short. Head large relative to body. May have fleshy tabs on body. HABITAT: E FL to Bahamas, and in Gulf of Mexico. Also Bermuda. In shallow brackish to marine waters over seagrass flats and occasionally in floating weeds. BIOLOGY: Dwarf seahorse feed on crustaceans. Males and females are monogamous during their breeding periods. Males brood eggs in pouch.

Fringed pipefish - *Anarchopterus criniger* (Bean & Dresel, 1884)

FEATURES: Tan to brown with whitish bars on body rings. Always with three black spots on upper trunk. Three pale streaks radiate from eyes. Snout short, head profile straight to slightly concave. Anal fin absent. Trunk with 14–16 rings. HABITAT: NC to FL, Gulf of Mexico to Yucatán, and Bahamas. Near bottom to about 36 ft. and with floating weeds. BIOLOGY: Fringed pipefish feed on tiny crustaceans.

Insular pipefish - *Anarchopterus tectus* (Dawson, 1978)

FEATURES: Tannish to brownish. Body with either darker spots on body rings or fairly evenly spaced, diffuse dark bars on body rings. May have dark band from eye to snout. Snout short, head profile slightly concave. Anal fin absent. Trunk with 17–18 rings. HABITAT: FL Keys, eastern Gulf of Mexico, Bahamas to northern South America. Reported on turtle-grass beds from near shore to about 85 ft.

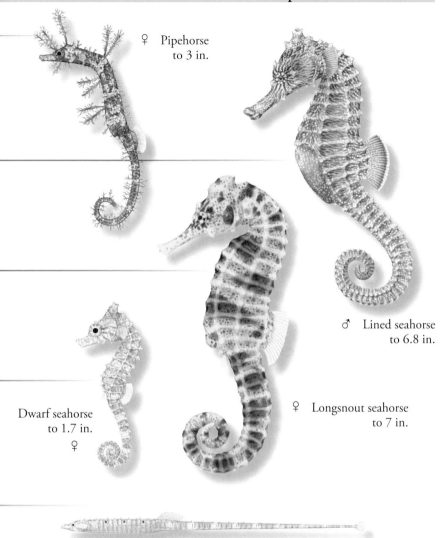

♀ Pipehorse
to 3 in.

♂ Lined seahorse
to 6.8 in.

Dwarf seahorse
to 1.7 in.
♀

♀ Longsnout seahorse
to 7 in.

♀ Fringed pipefish
to 3 in.

♀ Insular pipefish
to 4.5 in.

Pugnose pipefish - *Bryx dunckeri* (Metzelaar, 1919)

FEATURES: Reddish brown to pale. May be uniformly mottled or may have narrow to broad pale bars. Snout very short. Anal fin absent. Trunk with 16–17 rings. HABITAT: NC to FL, Bermuda, Caribbean Sea to northern South America. In shallow estuaries and over seagrass beds and reefs. Also found with *Sargassum* seaweed.

Whitenose pipefish - *Cosmocampus albirostris* (Kaup, 1856)

FEATURES: Grayish to brownish or whitish with broad, diffuse bars on body. Snout long, always whitish. Head may be darker than body. Body short, stout. Body may or may not have dermal flaps. HABITAT: SC to FL, Gulf of Mexico, Bahamas, and Antilles to northern South America. Over shallow weedy or reef bottoms.

Crested pipefish - *Cosmocampus brachycephalus* (Poey, 1868)

FEATURES: Brownish with darker bars and mottling. Underside of head banded. Snout short. Prominent bony crest on head. Body short, stout. HABITAT: FL, Bahamas, and Antilles to Venezuela. Over shallow, coastal seagrass flats.

Shortfin pipefish - *Cosmocampus elucens* (Poey, 1868)

FEATURES: Variable. Blackish, purplish, brownish to pinkish. Two bands radiate from eyes. Evenly spaced pale bars over length of body. May or may not have dermal flaps on head and body. Dorsal fin short-based with 21 to 25 rays. Trunk with 16–18 rings. HABITAT: NJ to FL, eastern Gulf of Mexico, and Caribbean Sea. Also Bermuda.

Banded pipefish - *Halicampus crinitus* (Jenyns, 1842)

FEATURES: Variably barred in distinct shades of yellow and brick red to purplish brown. Bars usually form rings around body. Snout short. HABITAT: S FL, Bahamas to Brazil. Also Bermuda. Reef associated from shore to about 70 ft. BIOLOGY: Some authors believe there are two distinct species and possibly nine color variations. Other names: Harlequin pipefish, *Micrognathus ensenadae, Micrognathus crinitus*.

Opossum pipefish - *Microphis brachyurus lineatus* (Kaup, 1856)

FEATURES: Brownish with a series of red blotches on each trunk ring. Silvery stripe on sides. Lower snout with red and black bars. Caudal fin red with black stripe. Juveniles less colorful. Trunk with16–20 rings. HABITAT: NJ to FL, Gulf of Mexico, and Caribbean Sea to Brazil. Occur in shallow fresh to salt water. BIOLOGY: Opossum pipefish are the only North American pipefish to brood eggs on the trunk.

Dusky pipefish - *Syngnathus floridae* (Jordan & Gilbert, 1882)

FEATURES: Greenish, brownish to whitish. May be mottled or blotched dorsally. Markings do not form bars. Dorsal fin with 26–35 rays (usually 32). Trunk with 16–19 rings. HABITAT: VA to FL, northern Gulf of Mexico, Bahamas, western Caribbean Sea. Also Bermuda. Over shallow seagrass beds and in estuaries and channels.

♀ Pugnose pipefish
to 3 in.

♂ Whitenose pipefish
to 8 in.

♀ Crested pipefish
to 4 in.

variant

♂ Shortfin pipefish
to 6 in.

variant

♀ Banded pipefish
to 9 in.

♀ Opossum pipefish
to 8 in.

♂ Dusky pipefish
to 10 in.

Syngnathidae - Pipefishes and Seahorses, *cont.*

Northern pipefish - *Syngnathus fuscus* Storer, 1839

FEATURES: Color varies. Greenish to brownish (rarely reddish) dorsally with darker bars and mottling. Pale to yellowish ventrally. Dorsal fin with 33–49 rays; may be banded. Trunk with18–21 rings. HABITAT: Gulf of St. Lawrence to NE FL. Found in shallow salt and brackish waters. Occur around seaweeds and seagrasses in bays, harbors, marshes, creeks, and river mouths to nearshore channels.

Chain pipefish - *Syngnathus louisianae* Günther, 1870

FEATURES: Tannish to whitish or brownish. Row of dark, diamond-shaped marks along lower sides. Dark stripe from snout to pectoral fin. Body ridges rather low. Trunk with19–21 rings. HABITAT: NJ to S FL and Gulf of Mexico from surface to about 125 ft. Found near shore over seagrass beds and around marsh grasses and offshore with *Sargassum* seaweed rafts. BIOLOGY: Chain pipefish feed on crustaceans.

Sargassum pipefish - *Syngnathus pelagicus* Linnaeus, 1758

FEATURES: Pale tan to brown with pale and dark bars on body. Snout with dark stripe. Dark bands may radiate from eyes. Dorsal fin banded. Head with low ridge. Body ridges rather low. 15–18 trunk rings. HABITAT: Nova Scotia to FL, Bermuda, Gulf of Mexico, Bahamas, Caribbean Sea to Colombia. Primarily oceanic. Associated with *Sargassum* seaweed. BIOLOGY: Sargassum pipefish hide, well-camouflaged, in rafts of floating *Sargassum* seaweed. Males brood eggs in pouch under tail.

Gulf pipefish - *Syngnathus scovelli* (Evermann & Kendall, 1896)

FEATURES: Dark brownish, olive brown to tannish. Males and females with Y-shaped silvery white bars on sides; bars may be more distinct in females. Dorsal fin banded. Caudal fin rounded. Females with keeled abdomen. Male trunk almost square in cross-section. 16–17 trunk rings. HABITAT: NE FL, Gulf of Mexico to Brazil. Occur over seagrass beds and in marshes, vegetated streams, and river shorelines to about 20 ft. BIOLOGY: Highly tolerant of varying salinities.

ALSO IN THE AREA: Dwarf pipefish, *Cosmocampus hildebrandi*; Texas pipefish, *Syngnathus affinis*; Bull pipefish, *Syngnathus springeri*, p. 431.

Aulostomidae - Trumpetfishes

Atlantic trumpetfish - *Aulostomus maculatus* Valenciennes, 1841

FEATURES: Brownish, olivaceous, yellowish, to reddish with blackish spots and whitish lines on head and body. May have faint bars. Second dorsal and anal fins with dark bars or spots. Caudal fin with one or two black spots or blotches. Lower jaw with short barbel. Body elongate, slightly compressed. First dorsal-fin spines short, separate. HABITAT: S FL, Gulf of Mexico, Bahamas, and Caribbean Sea to Brazil. Also Bermuda. Over reefs and among weeds in clear, shallow water. BIOLOGY: Often hover and drift vertically. Dart toward and suck in prey. Feed on crustaceans and fishes.

♂ Northern pipefish
to 12 in.

♀ Chain pipefish
to 15 in.

♀ Sargassum pipefish
to 8 in.

♀

♂

Gulf pipefish
to 7 in.

Trumpetfishes

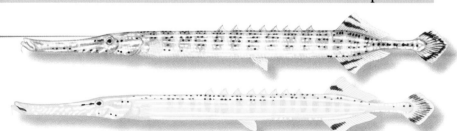

Atlantic trumpetfish
to 3 ft.

variant

Fistulariidae - Cornetfishes

Red cornetfish - *Fistularia petimba* Lacepède, 1803

FEATURES: Reddish to orange brown dorsally. Ventral area silvery. May have faint, dark bars. Snout, head, and body slender, elongate. First dorsal fin absent. Caudal fin forked with long, trailing filament. HABITAT: Circumtropical. MA to FL, Gulf of Mexico, and Caribbean Sea to Brazil. Found over coastal soft bottoms and reefs from about 30 to 650 ft. BIOLOGY: Red cornetfish feed on small fishes and shrimps.

Bluespotted cornetfish - *Fistularia tabacaria* Linnaeus, 1758

FEATURES: Brownish dorsally, pale ventrally. Pale blue spots and lines along snout and body. May have faint bars. Snout, head, and body elongate. First dorsal fin absent. Caudal fin forked with long, trailing filament. HABITAT: Nova Scotia to FL, Gulf of Mexico, and Caribbean Sea to Brazil. Also Bermuda. Over seagrass beds and reefs. BIOLOGY: Bluespotted cornetfish feed on fishes.

Macroramphosidae - Snipefishes

Slender snipefish - *Macroramphosus gracilis* (Lowe, 1839)

FEATURES: Bluish to grayish or pinkish with silvery tint. Snout long, tube-like. Body compressed, comparatively shallow. First dorsal fin with elongate, serrated spine that is shorter than snout length. Scales finely toothed. HABITAT: Worldwide in tropical seas. In western Atlantic from FL, northern Gulf of Mexico, Bahamas, and Cuba. Adults near bottom from about 656 to 985 ft. Juveniles pelagic. BIOLOGY: Slender snipefish feed on plankton. Some authors consider this species a junior synonym of Longspine snipefish, *Macroramphosus scolopax*. Other name: Shortspine snipefish.

Longspine snipefish - *Macroramphosus scolopax* (Linnaeus, 1758)

FEATURES: Reddish to orange dorsally. Sides and abdomen silvery pink. Snout long, tube-like. Body compressed, comparatively deep. First dorsal fin with elongate, serrated spine that is about as long as snout. Scales finely toothed. HABITAT: Gulf of Maine to Argentina. Also Gulf of Mexico and Bermuda. Occur over sandy lower continental shelves. Adults near bottom from about 164 to 1,150 ft. Juveniles pelagic. BIOLOGY: Longspine snipefish feed on plankton and small fishes. Schooling.

Dactylopteridae - Flying gurnards

Flying gurnard - *Dactylopterus volitans* (Linnaeus, 1758)

FEATURES: Typically brownish. Bright blue spots cover body and pectoral fins. Head short. Body elongate. Bony shield and long, keeled spine extend from head to first dorsal fin. Opercles spiny. Pectoral fins greatly expanded. HABITAT: MA to FL, Gulf of Mexico, Bahamas, and Caribbean Sea to Argentina. Also Bermuda and eastern Atlantic. Bottom-dwelling in sandy and muddy coastal waters to about 330 ft. BIOLOGY: Flying gurnard 'walk' across the bottom using pelvic and pectoral rays. Spread pectoral fins when alarmed. Feed on crustaceans, mollusks, and fishes.

Cornetfishes

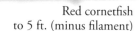

Red cornetfish
to 5 ft. (minus filament)

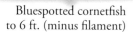

Bluespotted cornetfish
to 6 ft. (minus filament)

Snipefishes

Slender snipefish
to 6 in.

Longspine snipefish
to 7 in.

Flying gurnards

Flying gurnard
to 17 in.

177

Scorpaenidae - Scorpionfishes

Blackbelly rosefish - *Helicolenus dactylopterus* (Delaroche, 1809)

FEATURES: Reddish dorsally. Ventral area pinkish to white. Usually with six diffuse to distinct darker bars on upper body. Bars below soft dorsal fin form a V-shape. Cheeks may have dark grayish cast. May have dark blotch at rear of spiny dorsal fin. Pectoral fins with 19 rays; upper margin squarish; tips of lower rays free. HABITAT: Nova Scotia to FL. Scattered in Gulf of Mexico, around Bahamas, and along northern coast of South America. Over soft bottoms and continental shelves from about 650 to 2,000 ft. BIOLOGY: Blackbelly rosefish feed on crustaceans and fishes.

Spinycheek scorpionfish - *Neomerinthe hemingwayi* Fowler, 1935

FEATURES: Dark reddish with dark brown spots and mottling dorsally and on sides. All fins spotted except pelvic fins. Always with three spots on posterior portion of lateral line. Snout moderately long. Pectoral fins with 17 rays; third to seventh or eighth rays branched. HABITAT: NJ to S FL, and northern and southern Gulf of Mexico. Found near bottom over middle to outer continental shelf from about 100 to 750 ft. BIOLOGY: Spinycheek scorpionfish have venomous spines.

Longspine scorpionfish - *Pontinus longispinis* Goode & Bean, 1896

FEATURES: Reddish to pinkish dorsally, pale below. Dusky red spots or blotches on upper sides and along lateral line. Soft dorsal, pectoral, and caudal fins spotted. Fleshy tentacle over eyes small or absent. Third dorsal-fin spine elongate in larger specimens. Pectoral fins with 17 unbranched rays. HABITAT: SC to FL, Gulf of Mexico, and scattered along coasts of Central and South America. Occur over soft or semihard bottoms from about 260 to 1,400 ft. BIOLOGY: Longspine scorpionfish feed on invertebrates and fishes. May be venomous.

Highfin scorpionfish - *Pontinus rathbuni* Goode & Bean, 1896

FEATURES: Body with four diffuse, reddish bars on pale background. Bars become less distinct with age. Irregular blotches and spots on head and body and along lateral line. Fins with indistinct smudges and bands. Fleshy tentacle over each eye. Spiny dorsal fin tall with third spine taller than others. Pectoral fins with 17 unbranched rays. HABITAT: VA to FL, and NE Gulf of Mexico. Scattered from Yucatán to Brazil. Occur near bottom from about 240 to 1,200 ft. BIOLOGY: May be venomous.

Red lionfish - *Pterois volitans* (Linnaeus, 1758)

FEATURES: Head and body with alternating reddish brown and whitish bars. Spiny dorsal fin and pectoral fins banded. Other fins transparent with reddish spots. Fleshy tabs above eyes and around mouth. Dorsal and pectoral fins expanded, fan-like. HABITAT: Native to the Pacific and Indian oceans. In the western Atlantic from NY to Bahamas, Caribbean Sea, and Bermuda. Range expanding. Typically occur over near and offshore coral and rocky reefs to about 160 ft. Also reported from around dock pilings. BIOLOGY: Believed to have been introduced to Florida waters in the 1990's. Red lionfish stalk, corner, and engulf prey in one swift gulp. Spines are highly venomous. May be aggressive when threatened.

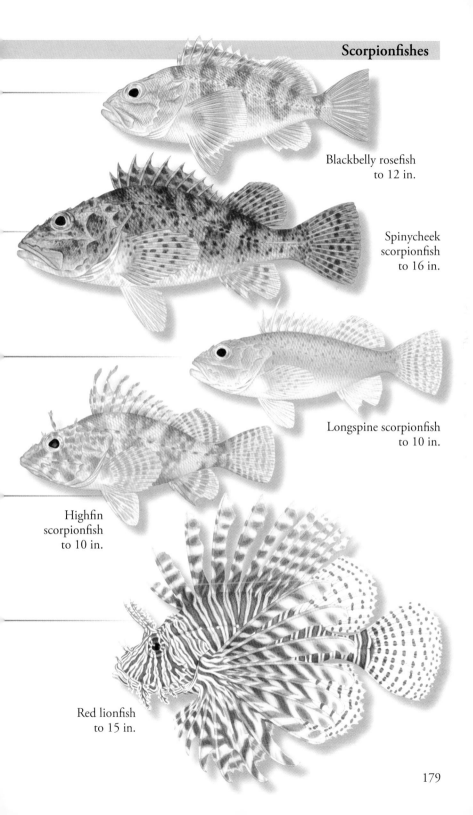

Blackbelly rosefish
to 12 in.

Spinycheek
scorpionfish
to 16 in.

Longspine scorpionfish
to 10 in.

Highfin
scorpionfish
to 10 in.

Red lionfish
to 15 in.

Scorpaenidae - Scorpionfishes, *cont.*

Coral scorpionfish - *Scorpaena albifimbria* Evermann & Marsh, 1900

FEATURES: Body and head mottled in shades of red, pink, and brown. Wide, diffuse dark bar behind head. Anal fin dark red at base. Pelvic fins with dark outer margin. Eyes large with a pair of short to long cirri. Many fleshy tabs on head and body. Pectoral fins with 19–21 rays; uppermost and twelfth to fifteenth rays unbranched. HABITAT: S FL, Bahamas, and scattered through Caribbean Sea. Also Bermuda. Occur inshore over reefs from about 3 to 120 ft. BIOLOGY: Change color with age.

Goosehead scorpionfish - *Scorpaena bergi* Evermann & Marsh, 1900

FEATURES: Speckled and barred in shades of dark reddish brown to brown. Dark, irregular spot on spiny dorsal fin between third, fourth, or fifth spines to seventh or eighth spines. Pelvic fins with dusky margin. Anal fin with three dark bands. Caudal fin with three dark bands. Fleshy tabs over eyes moderate in size. Pectoral fins with 16–17 rays, some branched. HABITAT: Scattered from NY to FL, Bahamas, Caribbean Sea to Brazil. Occur in clear water over turtle grass, coral reefs, and rocky and sandy bottoms. BIOLOGY: Change color and proportions with growth.

Barbfish - *Scorpaena brasiliensis* Cuvier, 1829

FEATURES: Color and pattern highly variable in shades of red, brown and yellow. Always with two dark blotches behind head and small dark spots at pectoral-fin base. Usually with dark spots on abdomen. Caudal fin with two dark bands. Other fins banded. Fleshy tabs on eyes well developed. Pectoral fins with 18–20 rays. HABITAT: VA to FL, Gulf of Mexico, and scattered through Caribbean Sea to Brazil. Possibly Bermuda. Over shallow, soft bottoms to about 165 ft. Sometimes on coral reefs. BIOLOGY: Possess potent venomous spines. Feed on crustaceans and fishes.

Plumed scorpionfish - *Scorpaena grandicornis* Cuvier, 1829

FEATURES: Mottled and barred in shades of brown to reddish. Small white spots on lower opercles, on chest, and at pectoral-fin base. Underside of pectoral-fin base dark with white specks. Fins irregularly banded and mottled. Large, well-developed fleshy tabs over eyes. Pectoral fins with 18–19 rays; upper and lower rays unbranched. HABITAT: S FL, Bahamas, and scattered in Caribbean Sea. Possibly Bermuda. Over sandy bottoms and seagrass beds. Also in channels and bays. BIOLOGY: Plumed scorpionfish lie motionless. Reported to be preyed upon by sharks.

Mushroom scorpionfish - *Scorpaena inermis* Cuvier, 1829

FEATURES: Color variable. Mottled, speckled, and barred in shades of red to brown. Tiny, dark spots at pectoral-fin base. Soft dorsal, anal, and pectoral fins with dark band at margin. Caudal fin with two dark bands. Eyes with small, inverted, mushroom-shaped cirri over pupils. Pectoral fins with 19–21 rays. HABITAT: GA to S FL, Bahamas, and scattered through Caribbean Sea to Venezuela. In clear waters over sandy and grassy bottoms and coral reefs from shore to about 240 ft. BIOLOGY: Mushroom scorpionfish are common on shrimp grounds. Feed on crustaceans and fishes. Caught as bycatch.

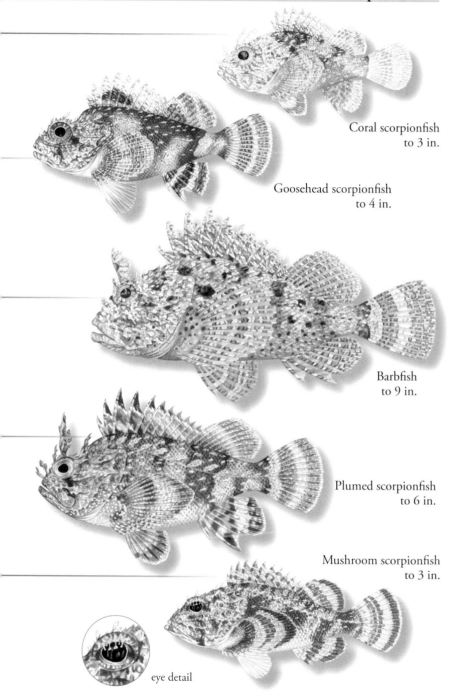

Coral scorpionfish
to 3 in.

Goosehead scorpionfish
to 4 in.

Barbfish
to 9 in.

Plumed scorpionfish
to 6 in.

Mushroom scorpionfish
to 3 in.

eye detail

Spotted scorpionfish - *Scorpaena plumieri* Bloch, 1789

FEATURES: Color, mottling highly variable in shades of red to brown with greens and/ or yellows. Broad, pale bar on caudal peduncle. Fins banded and blotched. Underside of pectoral-fin base black with large, white spots. Head and body may have numerous fleshy tabs. Pectoral fins with 18–21 rays. HABITAT: MA to FL, Gulf of Mexico, Bahamas, and Caribbean Sea. Also Bermuda. More common in southern waters. Over shallow, rocky, and reef areas. Also around pilings, jetties, and oil platforms. BIOLOGY: Lie motionless in wait for prey. Spines are venomous.

Reef scorpionfish - *Scorpaenodes caribbaeus* Meek & Hildebrand, 1928

FEATURES: Densely spotted and blotched in shades of red to brown. May have wide, whitish area on forebody. Spots on fins form bands. Spiny dorsal fin with a dark blotch at posterior end. Caudal peduncle pale in small specimens, spotted in larger specimens. Two rows of spines below eyes. Pectoral fins with 18–20 rays. HABITAT: FL, Bahamas, and Caribbean Sea. Also Bermuda. Occur inshore in clear water over coral or rocky areas. From near shore to about 60 ft. BIOLOGY: Reef scorpionfish hide in protective crevices or hover above the bottom. Spines are venomous.

Deepreef scorpionfish - *Scorpaenodes tredecimspinosus* (Metzelaar, 1919)

FEATURES: Mottled reddish or brownish; predominantly reddish at deeper depths. Fins either spotted or mostly transparent. Black blotch at posterior end of spiny dorsal fin. Single row of spines below eyes. Pectoral fins long, pointed at tip, with 16–18 rays. HABITAT: NC to FL, Bahamas, and Caribbean Sea. Over coral reefs and rocky bottoms from about 26 to 255 ft. BIOLOGY: Deepreef scorpionfish are venomous.

Acadian redfish - *Sebastes fasciatus* Storer, 1854

FEATURES: Shades of red dorsally and on sides. Abdomen paler. May have a dusky blotch on opercle and dark patches along back. All fins red, unmarked. Well-developed knob on lower jaw. Head profile slightly concave. HABITAT: Iceland to NJ. On rocky, hard, and muddy bottoms. In cold marine waters from about 3 to 1,000 ft. BIOLOGY: Acadian redfish are slow growing and long lived. Numbers declining. Commerically sold as Ocean perch. Currently listed as Endangered.

Golden redfish - *Sebastes norvegicus* (Ascanius, 1772)

FEATURES: Shades of red dorsally and on sides. Abdomen pinkish. Dark blotch on opercle. May have dusky patches dorsally. All fins red, unmarked. Knob on lower jaw small or absent. HABITAT: North Sea and North Atlantic to Grand Banks. In fjords and bays and offshore, in cold marine waters from about 320 to 1,300 ft. BIOLOGY: Golden redfish are sought commercially. Sometimes referred to as *Sebastes marinus*.

ALSO IN THE AREA: Longfin scorpionfish, *Scorpaena agassizii*; Smoothhead scorpionfish, *Scorpaena calcarata*; Hunchback scorpionfish, *Scorpaena dispar*; Atlantic thornyhead, *Trachyscorpia cristulata*, p. 431.

underside of fin

Spotted scorpionfish
to 12 in.

Reef scorpionfish
to 10 in.

Deepreef
scorpionfish
to 2.5 in.

Acadian redfish
to 12 in.

Golden redfish
to 20 in.

183

Horned searobin - *Bellator militaris* (Goode & Bean, 1896)

FEATURES: Reddish brown to brownish dorsally. Pale below. Two to three yellow stripes on lower sides. First dorsal fin with yellowish spots. Second dorsal and caudal fins with yellowish bands. Pectoral fins with blackish spots along uppermost rays. Mature males with filamentous first and second dorsal-fin rays. HABITAT: NC to FL, and Gulf of Mexico to Colombia. Occur over outer continental shelves from about 65 to 700 ft. BIOLOGY: Horned searobin feed on crustaceans and fishes.

Spiny searobin - *Prionotus alatus* Goode & Bean, 1883

FEATURES: Shades of brown with darker spots dorsally. First dorsal fin with black, ocellated spot between fourth and fifth spines. Pectoral fins with dark bands; middle posterior margins concave, lower rays elongate. Head very spiny. HABITAT: VA to eastern Gulf of Mexico. Also Bahamas and off Yucatán Peninsula. Occur over shelly, fragmented, and sedimentary bottoms from about 180 to 1,500 ft. BIOLOGY: Spiny searobin feed on worms, crustaceans, and fishes. May hybridize with Mexican searobin, *Prionotus paralatus,* where ranges overlap.

Northern searobin - *Prionotus carolinus* (Linnaeus, 1771)

FEATURES: Brownish dorsally with darker speckles and spots. May also have diffuse, pale bars along back. First dorsal fin with dark, often ocellated spot between fourth and fifth spines. Second dorsal fin spotted. Pectoral fins with broad dark bands; spotted on upper rays. Caudal-fin margin concave. HABITAT: Nova Scotia to E FL. From shore in shallow estuaries to edge of continental shelf. BIOLOGY: Northern searobin migrate offshore during winter. Spawn from late spring through summer. Feed on invertebrates.

Striped searobin - *Prionotus evolans* (Linnaeus, 1766)

FEATURES: Shades of brown to gray dorsally. Two distinct, dark lines on sides: one along lateral line, the second along lower side. May have several vague, dark bars along back. Dark blotch on first dorsal fin between fourth and sixth spines. Outer margins of pectoral fins pale; inner area dark, often with many fine, wavy lines. HABITAT: Nova Scotia to E FL. Possibly Little Bahama Bank. Rare north of Cape Cod. Occur over sandy bottoms from shoreline of estuaries to about 525 ft. BIOLOGY: Migrate offshore in winter. Spawn inshore during summer. Feed on crustaceans.

Bandtail searobin - *Prionotus ophryas* Jordan & Swain, 1885

FEATURES: Variably mottled and spotted in shades of black, brown, red, and gold. First dorsal fin with irregular spots forming bands; distinct spot absent. Second dorsal fin with a dark blotch on anterior portion. Upper pectoral-fin rays with pale and dark bands, lower portion dark with darker spots. Caudal fin dark at base, with two distinct dark bands. A single filament over the nostrils and a fringed tentacle over the eyes are present. Body robust. HABITAT: Cape Hatteras to FL, Bahamas, and Gulf of Mexico to Venezuela. Occur over soft bottoms from about 60 to 210 ft. BIOLOGY: Bandtail searobin feed on crustaceans. Spawning occurs between January and June.

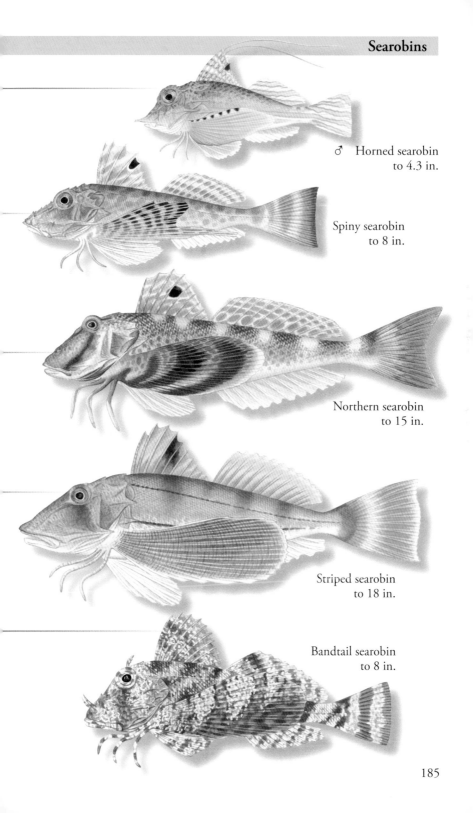

♂ Horned searobin
to 4.3 in.

Spiny searobin
to 8 in.

Northern searobin
to 15 in.

Striped searobin
to 18 in.

Bandtail searobin
to 8 in.

185

Bluespotted searobin - *Prionotus roseus* Jordan & Evermann, 1887

FEATURES: Brownish dorsally with darker blotches. First dorsal fin with a dark spot between fourth and fifth spines. Pectoral fins dusky to dark with blue spots. Caudal fin with two dark bands. HABITAT: NC to FL, Gulf of Mexico, Bahamas, Caribbean Sea to Brazil. On or near soft bottoms from about 30 to 600 ft. BIOLOGY: Bluespotted searobin feed on invertebrates and fishes. Spawn from April to September.

Blackwing searobin - *Prionotus rubio* Jordan, 1886

FEATURES: Brownish to grayish brown dorsally with irregular darker spots and blotches. First dorsal fin with dark spot between third and fourth spines. Spot fades with age. Pectoral fins dark with diffuse spotting and a bright blue lower margin. HABITAT: NC to FL, Cuba, and the Gulf of Mexico from FL to TX. On or near bottom from shore to about 700 ft.; more commonly between about 30 and 180 ft. BIOLOGY: Blackwing searobin feed on invertebrates, fishes, and worms.

Leopard searobin - *Prionotus scitulus* Jordan & Gilbert, 1882

FEATURES: Tannish with dark brownish spots dorsally and on sides. Darker, diffuse bands may be present. Pectoral, dorsal, and caudal fins spotted. First dorsal fin with two blackish spots: one between the first and second spines, a second between the fourth and fifth spines. Body comparatively elongate. HABITAT: VA to FL, and Gulf of Mexico to Venezuela. Usually found near bottom in shallow bays and from shore to about 150 ft. BIOLOGY: Leopard searobin feed on a variety of invertebrates. Spawn in early fall and spring. Prefer high-salinity waters.

Shortwing searobin - *Prionotus stearnsi* Jordan & Swain, 1885

FEATURES: Silvery black to gray dorsally. Ventral area silvery. Pectoral fins and margin of caudal fin blackish. First dorsal fin lacks distinct spot. Small, bony knob at tip of lower jaw. Pectoral fins short. HABITAT: NC to FL, and Gulf of Mexico to French Guiana. Found offshore over soft and semihard bottoms from about 120 to 360 ft. BIOLOGY: Shortwing searobin feed on crustaceans and small fishes. Spawn during winter and spring.

Bighead searobin - *Prionotus tribulus* Cuvier, 1829

FEATURES: Grayish brown dorsally with darker, oblique bars and pale flecks. Pale ventrally. First dorsal fin with dark blotch between fourth and fifth spines. Pectoral fin dark with irregular cross-bars. Caudal fin with single broad, dark bar. HABITAT: NY to FL, Gulf of Mexico to Bay of Campeche. From shore to about 600 ft. Juveniles occur in estuaries. BIOLOGY: Bighead searobin feed on crustaceans, fishes, and worms. Spawn from fall to spring. Taken by hook-and-line.

ALSO THE AREA: Shortfin searobin, *Bellator brachychir*; Streamer searobin, *Bellator egretta*; Bigeye searobin, *Prionotus longispinosus*; Barred searobin, *Prionotus martis*; Mexican searobin, *Prionotus paralatus,* p. 431.

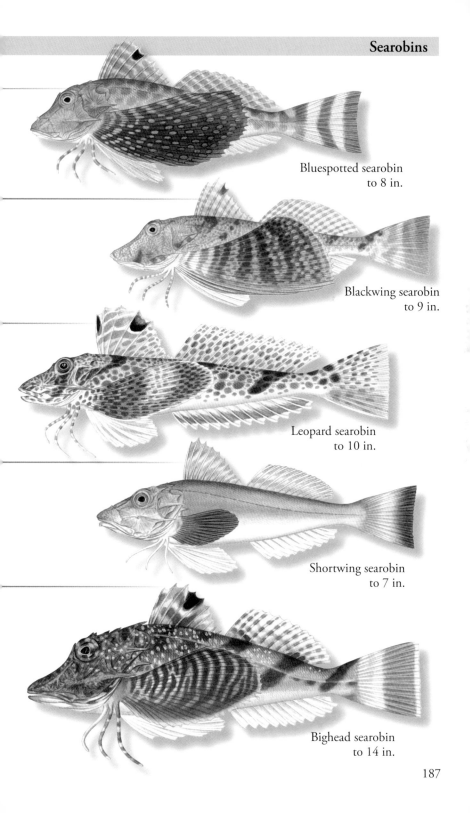

Bluespotted searobin
to 8 in.

Blackwing searobin
to 9 in.

Leopard searobin
to 10 in.

Shortwing searobin
to 7 in.

Bighead searobin
to 14 in.

Peristediidae - Armored searobins

Slender searobin - *Peristedion gracile* Goode & Bean, 1896

FEATURES: Reddish to straw-colored dorsally. Silvery white below. Dark stripe through middle of first and second dorsal fins. Caudal fin forked, with dark margin. Rostral projections very long and slender. Series of short and somewhat elongate barbels on chin. Body slender. HABITAT: VA to FL, and Gulf of Mexico to Surinam. Demersal over continental shelves and slopes from about 240 to 1,500 ft.

Armored searobin - *Peristedion miniatum* Goode, 1880

FEATURES: Reddish dorsally. White below. Margins of dorsal fins blackish red. Rostrum with short projections. Several short and two long filamentous barbels on lower jaw. Sides of head with spiny, wing-like projections. HABITAT: Canada and Georges Bank to FL, Gulf of Mexico, and Honduras to Brazil. Demersal over outer continental shelves and slopes from about 200 to 3,000 ft. BIOLOGY: Armored searobin feed on crustaceans. Preyed upon by spiny dogfish.

Rimspine searobin - *Peristedion thompsoni* Fowler, 1952

FEATURES: Yellowish orange dorsally, pale below. Dorsal fins unmarked. Caudal fin with dark margin. Rostrum with short projections; each bears a small spine on base. Lips and chin with many short, clustered barbels. HABITAT: NC to FL Keys, northern Gulf of Mexico, and Honduras to Brazil. On or near bottom over continental shelves and slopes from about 370 to 1,500 ft. BIOLOGY: Rimspine searobin use free pectoral rays to 'walk' along bottom and barbels to locate prey.

ALSO IN THE AREA: Alligator searobin, *Peristedion greyae*, p. 431.

Cottidae - Sculpins

Atlantic hookear sculpin - *Artediellus atlanticus* Jordan & Evermann, 1898

FEATURES: Dark brownish dorsally with paler mottling. Pale blotches on abdomen. Dark band on caudal peduncle. All fins dark with pale bands. Head often with cirri. Preopercle with distinct, hook-like spine that is partially covered with skin in larger specimens. Males larger and with taller, darker dorsal fins than females. HABITAT: Greenland to Cape Cod. Also in Arctic waters of Europe and Siberia to the Barents Sea. Usually found over soft bottoms from shore to about 2,600 ft. BIOLOGY: Atlantic hookear sculpin feed on invertebrates. Preyed upon by cods.

Arctic staghorn sculpin - *Gymnocanthus tricuspis* (Reinhardt, 1830)

FEATURES: Brownish to gray dorsally with dark bars and/or spots and mottling. White to yellowish below. First dorsal fin with three dark bands, second dorsal fin with four or five bands. Preopercle with three close-set prongs at upper margin. Males are more brightly colored than females. HABITAT: Arctic and North Atlantic oceans. Labrador to Gulf of St. Lawrence; rarely to ME. On or near bottom over rocky or sandy bottoms from about 6 to 570 ft. BIOLOGY: Arctic staghorn sculpin feed on invertebrates and fishes. Preyed upon by sea birds and seals.

Slender searobin
to 8 in.

Armored searobin
to 12 in.

Rimspine searobin
to 10 in.

♂ Atlantic hookear sculpin
to 5 in.

♂ Arctic staghorn sculpin
to 10 in.

Cottidae - Sculpins, *cont.*

Grubby - *Myoxocephalus aenaeus* (Mitchill, 1814)

FEATURES: Color varies. Tannish, grayish to greenish with darker, irregular bars on sides. First dorsal fin with mottled banding. All other fins banded. Head ridged, with two spines over each eye and between nostrils. Preopercle and opercle with a total of six short spines. Anal fin short-based with 10-11 rays. HABITAT: Gulf of St. Lawrence to NJ. Over a variety of bottoms from shore to about 165 ft. BIOLOGY: Grubby are omnivorous. Produce antifreeze proteins during winter.

Longhorn sculpin - *Myoxocephalus octodecimspinosus* (Mitchill, 1814)

FEATURES: Color varies with surroundings. Sides always with four irregular crossbars. First dorsal fin mottled, other fins banded. Head blunt and flattened with spines above nostrils and eyes. Uppermost preopercular spine very long and naked at tip. Bottom preopercular spine turned downward. Anal fin with 12–15 rays. HABITAT: Newfoundland to NJ; sometimes to VA. Near bottom over a variety of bottoms. In harbors, estuaries, and river mouths and over offshore banks to about 630 ft. BIOLOGY: Longhorn sculpin are omnivorous. Females produce up to 8,000 eggs.

Shorthorn sculpin - *Myoxocephalus scorpius* (Linnaeus, 1758)

FEATURES: Shades of brown dorsally. May have darker bars. Females with whitish to yellowish abdomen. Abdomen of males reddish orange with white spots. Uppermost preopercular spine moderately long. Anal fin with 9–16 rays. HABITAT: In western Atlantic from Greenland to NJ. Also in Arctic seas along northern coasts of Europe, Siberia. Demersal over a variety of soft and hard bottoms from shore to about 360 ft. BIOLOGY: Shorthorn sculpin are sluggish and omnivorous.

Moustache sculpin - *Triglops murrayi* Günther, 1888

FEATURES: Brownish to olivaceous with four dark blotches dorsally. White, yellowish, or orange below. Dark 'moustache' at upper edge of mouth. First dorsal fin of males with a dusky blotch behind first dorsal ray and a dark spot on posterior margin. Females lack dorsal-fin spots. Distinct, oblique folds of skin below lateral line. HABITAT: In North Atlantic and Arctic Ocean from Hudson Bay to Cape Cod. Also to the Barents and White seas. Near bottom in cold water from about 60 to 1,000 ft. BIOLOGY: Moustache sculpin feed on invertebrates.

Hemitripteridae - Searavens

Sea raven - *Hemitripterus americanus* (Gmelin, 1789)

FEATURES: Uniformly colored or mottled in shades of brown, red to yellow. Usually with yellowish abdomen. Fins variably barred. Pectoral and anal fins often with yellowish rays. Head large with humps, ridges, and fleshy tabs. First dorsal fin with fleshy tabs and irregular margins, first few spines tall. Skin prickly. HABITAT: Newfoundland to NJ. Over hard bottoms from surface to about 630 ft. BIOLOGY: Sea raven produce antifreeze proteins year-round. Females carry up to 5,000 eggs.

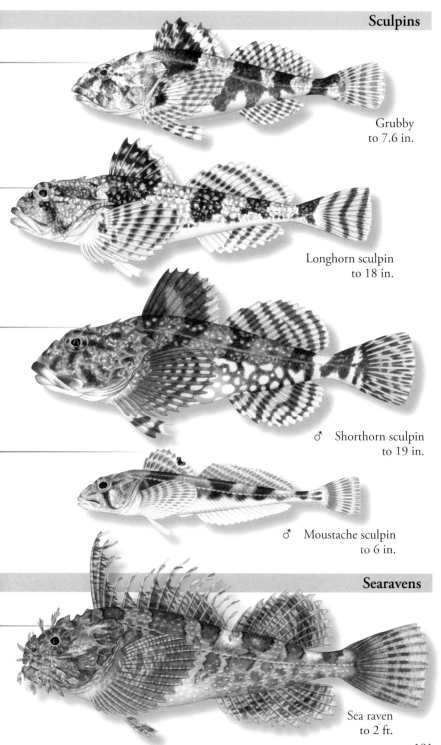

Grubby
to 7.6 in.

Longhorn sculpin
to 18 in.

♂ Shorthorn sculpin
to 19 in.

♂ Moustache sculpin
to 6 in.

Searavens

Sea raven
to 2 ft.

Agonidae - Poachers

Alligatorfish - *Aspidophoroides monopterygius* (Bloch, 1786)

FEATURES: Brownish above, paler below. About eight dark bars on upper body and sides. Dorsal and pectoral fins barred. Body elongate, octagonal in cross-section. Rows of smooth plates cover head and body. Two curved spines over nostrils. HABITAT: Greenland to Cape Cod. Strays to NJ. Bottom-dwelling over a variety of grounds from about 60 to 1,000 ft. BIOLOGY: Alligatorfish feed on crustaceans.

Psychrolutidae - Fathead sculpins

Polar sculpin - *Cottunculus microps* Collett, 1875

FEATURES: Pale tannish with dark, irregular bars and some dark mottling. Body robust anteriorly, laterally compressed posteriorly. Bony knobs on the top and sides of the head. Dorsal fin continuous, taller posteriorly. HABITAT: In western Atlantic from Greenland to NJ. Also NE Atlantic. In cold water over continental shelves and slopes from about 720 to 2,900 ft. BIOLOGY: Polar sculpin feed on invertebrates.

Cyclopteridae - Lumpfishes

Lumpfish - *Cyclopterus lumpus* Linnaeus, 1758

FEATURES: Color highly variable. Bluish, olive, yellowish to brownish. Mature males more vividly colored than females, turn reddish when breeding. Body stout with irregularly spaced, small tubercles. Two ridges on sides and abdomen with rows of larger tubercles. First dorsal fin covered in skin and tubercles in mature specimens. HABITAT: Greenland to NJ. Also in eastern North Atlantic. Semipelagic and bottom-dwelling in cold to temperate waters from about 255 to 1,200 ft. BIOLOGY: Female Lumpfish carry up to 200,000 eggs. Males build nests, care for eggs.

Atlantic spiny lumpsucker - *Eumicrotremus spinosus* (Fabricius, 1776)

FEATURES: Brownish to olivaceous with darker mottling. Body rounded. First dorsal fin free. Body covered in large, closely set tubercles. HABITAT: Greenland and Hudson Bay to Gulf of Maine. Bottom-dwelling in cold northern waters over muddy, gravelly, or rocky bottoms from about 16 to 270 ft. BIOLOGY: Atlantic spiny lumpsucker feed on invertebrates. Preyed upon by Atlantic cod.

Liparidae - Snailfishes

Atlantic seasnail - *Liparis atlanticus* (Jordan & Evermann, 1898)

FEATURES: Dark brown, black, or gray. Fins often barred in white, blue or pink. Dorsal fin notched; reaches to caudal fin but does not overlap. Anterior dorsal-fin rays tall, separate at tips; more so in males. Pectoral fins bilobed. Body tadpole-shaped. HABITAT: Ungava Bay, Quebec, to NY. Bottom-dwelling. Usually along coast over a variety of firm intertidal and subtidal surfaces, but also offshore. BIOLOGY: Male Atlantic seasnail prepare spawning area and aerate and defend eggs.

Poachers

Alligatorfish
to 7 in.

Fathead sculpins

Polar sculpin
to 8 in.

Lumpfishes

Lumpfish
to 2.3 ft.

Atlantic spiny lumpsucker
to 4.5 in.

Snailfishes

Atlantic seasnail
to 5.7 in.

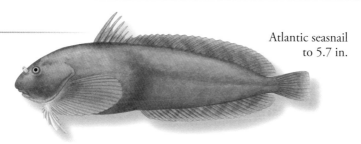

Liparidae - Snailfishes, *cont.*

Gulf snailfish - *Liparis coheni* Able, 1976

FEATURES: Usually uniformly pale brownish. May be striped or patterned like *Liparis inquilinus*. Posterior portions of dorsal and anal fins darker brown. Caudal fin may be banded. Dorsal fin unnotched, overlaps one-quarter to one-third of caudal fin. Body tadpole-shaped. HABITAT: Gulf of St. Lawrence to Gulf of Maine. Bottom-dwelling in estuaries and near shore from about 13 to 690 ft. BIOLOGY: Gulf snailfish spawn in winter and early spring. Previously confused with other species.

Inquiline snailfish - *Liparis inquilinus* Able, 1973

FEATURES: Color and pattern highly variable. May be speckled, striped, blotched, plain, or a combination of all. Dorsal fin notched, overlaps one-tenth to one-quarter of caudal fin. Pectoral fin bilobed. Body tadpole-shaped. HABITAT: Gulf of St. Lawrence to Cape Hatteras. On bottoms from about 16 to 320 ft. BIOLOGY: Inquiline snailfish are commensal with the Sea scallop *Plactopecten magellanicus* during the day. Leave the scallop at night to feed.

Centropomidae - Snooks

Swordspine snook - *Centropomus ensiferus* Poey, 1860

FEATURES: Yellow brown to brownish green above, silvery on sides and below. Lateral line dark. Fins dusky. Pelvic fins may have dark blotch. Usually fewer than nine scales between second dorsal-fin origin and lateral line. Pectoral fins reach to about anus. Pelvic fins reach to or past anus. Second anal-fin spine extends to or beyond caudal-fin base. HABITAT: S FL, southern Gulf of Mexico to Brazil, Greater and Lesser Antilles. Along coast and in estuaries, lagoons, and fresh water. BIOLOGY: Swordspine snook prefer lower salinities. Feed on fishes and crustaceans.

Smallscale fat snook - *Centropomus parallelus* Poey, 1860

FEATURES: Brownish green to yellow brown above, silvery on sides and below. Lateral line dark. Fins dusky. Pectoral fins do not reach anus. Pelvic fins reach to or past anus. Second anal-fin spine may reach caudal-fin base. Body comparatively deep. HABITAT: S FL, southern Gulf of Mexico, Greater and Lesser Antilles to Brazil. Along coast and in estuaries, lagoons, and fresh water. BIOLOGY: Smallscale fat snook prefer low salinities. Feed on fishes and crustaceans.

Tarpon snook - *Centropomus pectinatus* Poey, 1860

FEATURES: Brownish green to yellow brown above, silvery on sides and below. Lateral line dark. Fins dusky. Dark blotch on pelvic-fin tips. Pectoral fins distinctly shorter than pelvic fins. Pelvic fins reach to or past anus. Second anal-fin spine slightly outcurved at tip, not reaching caudal-fin base. HABITAT: S FL, southern Gulf of Mexico, Greater and Lesser Antilles to Brazil. Also in Pacific from Mexico to Colombia. Along coast and in estuaries, lagoons, and fresh water. BIOLOGY: Tarpon snook prefer low salinities. Feed on fishes and crustaceans.

Gulf snailfish
to 4.6 in.

Inquiline snailfish
to 2.8 in.

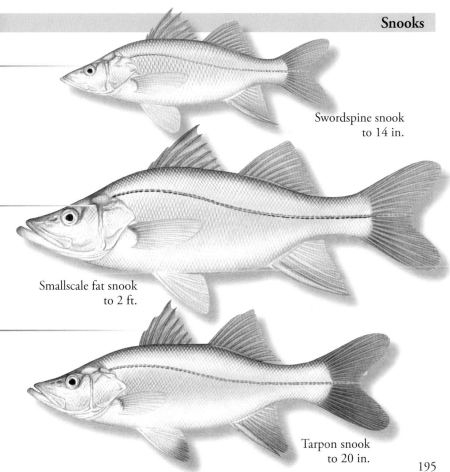

Swordspine snook
to 14 in.

Smallscale fat snook
to 2 ft.

Tarpon snook
to 20 in.

Centropomidae - Snooks, *cont.*

Common snook - *Centropomus undecimalis* (Bloch, 1792)

FEATURES: Yellow brown to brownish green above, silvery on sides and below. Lateral line dark. Fins dusky to yellowish. Pelvic fins do not reach anus. Anal-fin spines do not reach caudal-fin base. Body comparatively elongate. HABITAT: NC to FL, Gulf of Mexico, Greater and Lesser Antilles to Brazil. Along coast and in estuaries, marshes, lagoons, mangroves, river mouths, and freshwater streams to about 65 ft. BIOLOGY: Common snook are intolerant of cool water temperatures. Feed on fishes and crustaceans. Congregate in river mouths during summer spawning. Sought commercially and for sport. Other names: Snook, Robalo, Saltwater pike.

Moronidae - Temperate basses

White perch - *Morone americana* (Gmelin, 1789)

FEATURES: Dark silvery gray, grayish green to olivaceous dorsally; shading to silvery on sides and below. Head may have bluish luster. Fins dusky. Juveniles with pale stripes on sides. Body comparatively deep, compressed. HABITAT: Gulf of St. Lawrence to SC. Occur in freshwater rivers, brackish estuaries, and marine coastal waters. Some populations are landlocked. Found in shallow water to about 125 ft. BIOLOGY: White perch are anadromous. Feed on fishes and invertebrates. Sought commercially and for sport.

Striped bass - *Morone saxatilis* (Walbaum, 1792)

FEATURES: Grayish green to steel blue dorsally with brassy highlights. Sides silvery, with about seven black to dark olive stripes along scales. Pelvic fins white, other fins dusky green. Head profile relatively straight. Body oblong, slightly compressed. HABITAT: St. Lawrence River, Canada, to N FL, and W FL to E TX. In rivers, estuaries, and nearshore waters. Also in shallow bays, along beaches, and in rocky areas. Some populations are landlocked in fresh water. BIOLOGY: Striped bass are anadromous and seasonally migratory. Successfully introduced in many areas. A voracious predator of fishes and invertebrates. Feeding ceases just before spawning. Sought commercially and for sport. Other names: Rockfish, Striper.

Polyprionidae - Wreckfishes

Wreckfish - *Polyprion americanus* (Bloch & Schneider, 1801)

FEATURES: Dark brownish to dark grayish dorsally, paler below. Juveniles blackish with whitish mottling. Lower jaw protrudes. Bony knob over eyes and a distinct ridge at nape. Upper opercle with a horizontal ridge. HABITAT: Newfoundland to FL and Bermuda. Also in eastern Atlantic, Mediterranean, southwest Pacific, and southern Indian Ocean. Adults near bottom over rocky slopes and seamounts and around caves and wrecks in temperate waters from about 160 to 2,600 ft. Juveniles pelagic and under floating objects. BIOLOGY: Wreckfish spawn in summer. Feed on fishes and crustaceans. Sought commercially and for sport. Vulnerable to overfishing.

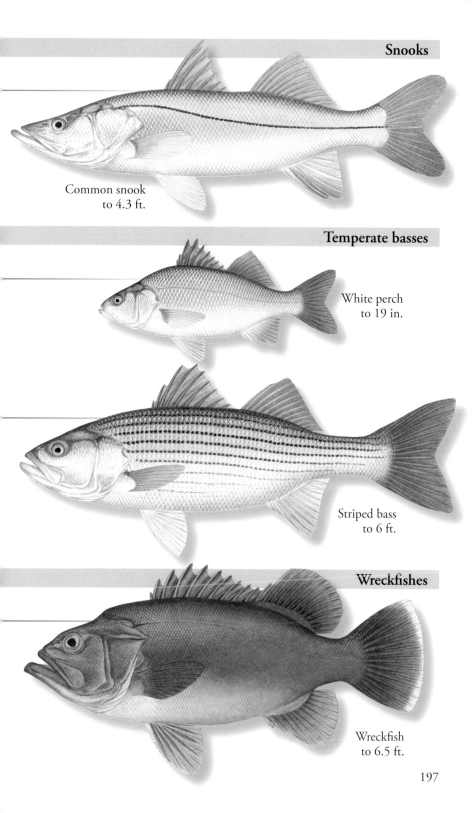

Snooks

Common snook
to 4.3 ft.

Temperate basses

White perch
to 19 in.

Striped bass
to 6 ft.

Wreckfishes

Wreckfish
to 6.5 ft.

197

Serranidae - Sea basses

Mutton hamlet - *Alphestes afer* (Bloch, 1793)

FEATURES: Variably colored. Shades of red, brown, or gray to olive with darker, irregular bars and paler spots and blotches. Small, orange spots on cheeks, sides, and abdomen. Usually with a dark band from eyes to dorsal-fin origin. Fins spotted and mottled. Snout very short. Head profile somewhat steep, nearly straight. HABITAT: S FL, Bahamas, and Antilles to Panama and Brazil. Also Bermuda. Over seagrass beds and around rocks and sponges from shore to about 115 ft. BIOLOGY: Mutton hamlet are sedentary during day, predatory at night. Will hide in crevices or lie partially covered in sand to avoid detection. Feed on crustaceans.

Yellowfin bass - *Anthias nicholsi* Firth, 1933

FEATURES: Reddish lavender with yellow patches dorsally. Irregular yellow stripes on sides. Silvery below. Yellow stripe from snout to pectoral fin and from eye to opercular margin. Fins bright yellow with reddish to lavender markings and margins. Eyes large. Spiny dorsal fin with short, trailing filaments. Caudal fin deeply forked. HABITAT: Nova Scotia to FL, NE Gulf of Mexico, Nicaragua, and Guyana to Brazil. Found near bottom from about 180 to 1,400 ft. BIOLOGY: Yellowfin bass are hermaphroditic. Females mature into males. Spawn in the spring.

Threadnose bass - *Anthias tenuis* Nichols, 1920

FEATURES: Rosy to orange red dorsally. Pale below. Dorsal and anal fins spotted. Nostrils each with a thread-like filament. Eyes with yellow irises. Dorsal fin spines with short, trailing filaments. Caudal fin with pointed lobes and red tips. HABITAT: NC, W FL, Yucatán, Puerto Rico, and Colombia to Venezuela. Also Bermuda. Found near bottom from about 180 to 3,000 ft. Most occur above 490 ft. BIOLOGY: Threadnose bass are planktivorous and form large schools.

Bank sea bass - *Centropristis ocyurus* (Jordan & Evermann, 1887)

FEATURES: Cream to yellowish white dorsally with six or seven dark bars on upper sides. Each bar merges with a dark patch at lateral line. Head yellowish with pale blue lines. Dorsal and caudal fins with blackish spots. Spiny dorsal fin with short, trailing flaps. Caudal fin rounded in juveniles, trilobed or double-concave in adults. HABITAT: Cape Hatteras to FL Keys, Gulf of Mexico to Yucatán. Occur over hard bottoms and sandy-shell bottoms near reefs from about 36 to 330 ft. BIOLOGY: Bank sea bass are hermaphroditic. Feed on invertebrates and fishes.

Rock sea bass - *Centropristis philadelphica* (Linnaeus, 1758)

FEATURES: Brownish to olive gray dorsally. About seven dark oblique bars on upper sides. Midbody bar merges with a dark spot at base of dorsal fin. Head with blue and rusty lines. Dorsal and caudal fins spotted. Spiny dorsal fin with trailing flaps. Caudal fin rounded in juveniles, trilobed or double-concave in adults. HABITAT: VA to FL and in the Gulf of Mexico. Found over soft mud and sandy bottoms from about 30 to 560 ft. BIOLOGY: Rock seabass are hermaphroditic. Females mature in one year, then become male in second year. May live to four years of age.

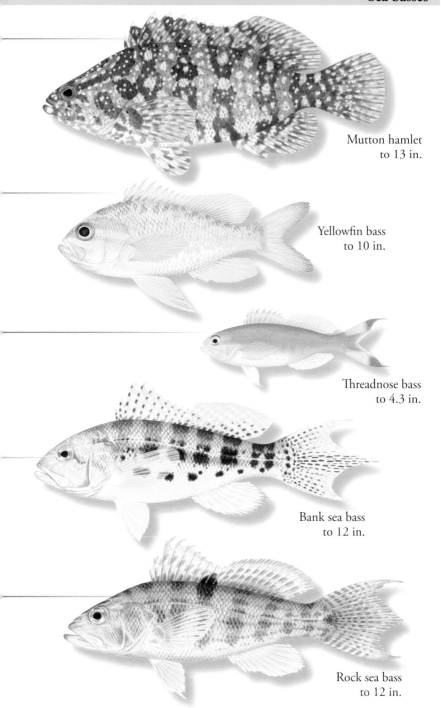

Mutton hamlet
to 13 in.

Yellowfin bass
to 10 in.

Threadnose bass
to 4.3 in.

Bank sea bass
to 12 in.

Rock sea bass
to 12 in.

Black sea bass - *Centropristis striata* (Linnaeus, 1758)

FEATURES: Blackish to grayish on body and fins. Centers of scales whitish. Unpaired fins with whitish streaks and margins. Sides with several obscure bars. May have dark stripe on sides. May pale or darken. Tips of spiny dorsal fin with fleshy tabs. Caudal fin rounded to trilobed, with trailing filaments. Adult males develop hump on nape. HABITAT: MA to FL. In Gulf of Mexico only along coast of FL. Over rocky and soft bottoms, around reefs, pilings, and jetties. From shore to about 250 ft. BIOLOGY: Black sea bass are hermaphroditic. Feed on invertebrates. Sought for sport.

Graysby - *Cephalopholis cruentata* (Lacepède, 1802)

FEATURES: Body and fins grayish, bluish, or brownish with small, evenly-spaced orange brown to reddish spots. Four black or white spots along back under dorsal-fin base. Dorsal, caudal, and anal fins with reddish inner margin. May pale or darken in color. HABITAT: NC to FL, Gulf of Mexico, Bahamas, and Caribbean Sea to Brazil. Also Bermuda. Found over seagrass beds and coral reefs from shore to about 560 ft. BIOLOGY: Graysby feed at dusk and dawn. Secretive during the day.

Coney - *Cephalopholis fulva* (Linnaeus, 1758)

FEATURES: Three color phases: red (deep water); orange brown to bicolored (shallow water); yellow (shallow to deep water). All have small, dark-circled, pale blue spots. Always with two dark spots on lower lip and two dark spots on upper peduncle. Night pattern is usually pale with irregular bars. HABITAT: SC to FL, Gulf of Mexico, Bahamas, and Caribbean Sea to Brazil. Also Bermuda. Around coral reefs and rocky bottoms in clear water to about 150 ft. BIOLOGY: Coney are hermaphroditic. May hide in caves or under ledges during day. Feed on crustaceans and small fishes. Occasionally follow moray eels in search of prey.

Marbled grouper - *Dermatolepis inermis* (Valenciennes, 1833)

FEATURES: Pattern variable. Brownish to grayish with pale blotches and mottling. Dark spots scattered between blotches. Margins of fins usually pale. Juveniles black to dark brown with irregular white blotches and rounded caudal fin. Head profile steep, nearly straight. Body deep. HABITAT: NC to FL, Gulf of Mexico, Bahamas, and Caribbean Sea to Brazil. Around ledges, over reefs, in caves and crevices, and around oil rigs from about 70 to 820 ft. BIOLOGY: Marbled grouper are solitary and shy. Sought as sportfish. Currently listed as Near Threatened.

Dwarf sand perch - *Diplectrum bivittatum* (Valenciennes, 1828)

FEATURES: Pale greenish gray above, pearly below. Two brownish stripes on upper sides. Stripes are continuous in juveniles, become broken and intersected by dark saddles in adults. Snout and cheeks with pale blue lines. Dorsal fins with orange margins and spots. Caudal fin with irregular orange bands and elongate upper ray. Two small dark spots on upper caudal peduncle. HABITAT: E FL, Gulf of Mexico to Brazil. Absent from Bahamas and Antilles. Over muddy silt bottoms from near shore to about 390 ft. Other name: Dwarf seabass.

Graysby
to 13 in.

Black sea bass
to 2 ft.

Coney
to 15.4 in.

Marbled grouper
to 3 ft.

juvenile

Dwarf sand perch
to 10 in.

201

Sand perch - *Diplectrum formosum* (Linnaeus, 1766)

FEATURES: Shades of brown dorsally, fading to white below. Narrow, wavy blue lines on snout and head merge with narrow blue stripes on body. Dark midlateral stripe may run from head to a dark spot at caudal-fin base. Five to seven diffuse dark bars on sides. Dorsal and caudal fins tannish with blue lines. Caudal fin usually forked, with upper lobe longer than lower lobe. HABITAT: VA to FL, Gulf of Mexico, and from Colombia to Brazil. Also Bermuda. Rare in Bahamas and Antilles. Found over sandy to shelly bottoms from shore to about 240 ft. BIOLOGY: Sand perch feed primarily on crustaceans. Spawn from March to September. Other name: Sand seabass.

Rock hind - *Epinephelus adscensionis* (Osbeck, 1765)

FEATURES: Body and fins greenish to buff with numerous reddish brown spots and scattered pale blotches. Three to five dark blotches below dorsal-fin base. Single dark blotch at top of caudal peduncle. Dark spots along margin of caudal fin may form a band. Juveniles with larger and fewer dark spots. HABITAT: MA (rare) to FL, Gulf of Mexico, and Caribbean Sea. Also Bermuda and eastern Atlantic. Occur around rocky bottoms, reefs, jetties, and oil rigs from about 6 to 330 ft. BIOLOGY: Rock hind feed on crabs and fishes. Other name: Calico grouper. Listed as Critically Endangered.

Speckled hind - *Epinephelus drummondhayi* Goode & Bean, 1878

FEATURES: Adults shades of brown with a dense covering of small, pearly speckles on body and fins. Juveniles yellowish with numerous pearly spots on body and fins. Caudal fin with nearly straight margin. Body relatively deep. HABITAT: NC to FL Keys, Gulf of Mexico to Quintana Roo. Also Bermuda. Occur over rocky bottoms from about 80 to 600 ft. BIOLOGY: May live to 25 years. Females carry up to 2 million eggs. Feed on fishes and invertebrates. Other name: Kitty Mitchell. Critically Endangered.

Yellowedge grouper - *Epinephelus flavolimbatus* Poey, 1865

FEATURES: Yellowish tan to grayish brown dorsally, pale below. Thin, pale blue line from eye to preopercular corner. Rows of obscure, pale spots evident in juveniles; usually fade with age. Adults may or may not display spots. Juveniles also with black saddle on caudal peduncle. Margins of dorsal, caudal, and pectoral fins yellow. HABITAT: NC to FL, Gulf of Mexico to Brazil. Over rocky, sandy, or mud bottoms from about 200 to 1,200 ft. BIOLOGY: Yellowedge grouper are hermaphroditic. May live to about 20 years. Feed on invertebrates and fishes. Currently listed as Vulnerable.

Red hind - *Epinephelus guttatus* (Linnaeus, 1758)

FEATURES: Buff, greenish white, to pale reddish brown with reddish brown spots dorsally and bright red spots ventrally. Spiny dorsal fin with yellow tips. Soft dorsal, caudal, and anal fins with a broad blackish inner margin and a narrow white margin. HABITAT: NC to FL, Gulf of Mexico, Bahamas, and Caribbean Sea to Brazil. Also Bermuda. Found over rocky bottoms and reefs from about 6 to 330 ft. BIOLOGY: Red hind are hermaphroditic. Females may carry up to 3 million eggs. Feed on octopods, crustaceans, and fishes. Other name: Strawberry grouper.

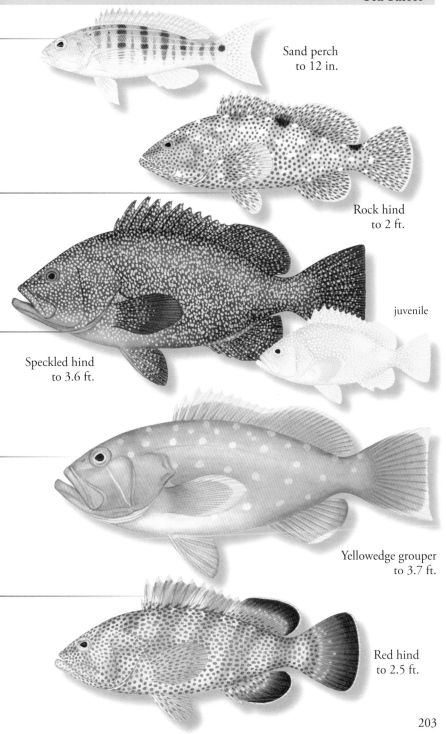

Sand perch
to 12 in.

Rock hind
to 2 ft.

juvenile

Speckled hind
to 3.6 ft.

Yellowedge grouper
to 3.7 ft.

Red hind
to 2.5 ft.

Goliath grouper - *Epinephelus itajara* (Lichtenstein, 1822)

FEATURES: Brownish, brownish yellow to greenish with small dark and pale spots. Smaller specimens with broken bars on body. Bars often radiate from eyes. Fins variably banded and spotted. Head broad, flattened between comparatively small eyes. Spiny dorsal fin low. Caudal fin rounded. Body robust. HABITAT: FL, Gulf of Mexico, Bahamas, and Caribbean Sea to Brazil. Also in eastern Pacific. In shallow waters, around wrecks, and in mangrove swamps, bays, and harbors to about 100 ft. BIOLOGY: Territorial and slow growing. May live to 37 years old. Females may carry up to 5 million eggs. Other name: Jewfish. Protected in US. Listed as Critically Endangered.

Red grouper - *Epinephelus morio* (Valenciennes, 1828)

FEATURES: Reddish brown with diffuse, pale blotches and spots. Bars often radiate from eyes. Dark margins on soft dorsal, caudal and anal fins. Inside of mouth bright reddish orange. Spiny dorsal fin with a tall second spine and nearly straight margin. HABITAT: NC to FL, Gulf of Mexico, Bahamas, Caribbean Sea to Brazil. Also Bermuda. Around reefs, in crevices, under ledges, and over sandy or mud bottoms to about 950 ft. BIOLOGY: Red grouper are hermaphroditic. Females carry up to 5 million eggs. Sought commercially and for sport. Currently listed as Near Threatened.

Misty grouper - *Epinephelus mystacinus* (Poey, 1852)

FEATURES: Brownish with 8–9 darker bars and paler, diffuse spots. Bands radiate from eyes. Last two bands on peduncle are darker than others and fuse to appear as one. Dark 'moustache' just above jaws. Fin margins pale. Juveniles with more contrasting body bars and pale caudal fin. HABITAT: NC to FL, Gulf of Mexico, Bahamas, and Antilles to Trinidad from about 100 to 1,600 ft. Also Bermuda and Honduras.

Warsaw grouper - *Epinephelus nigritus* (Holbrook, 1855)

FEATURES: Dark reddish brown to almost black dorsally, fading to dull reddish gray below. Juveniles with yellowish caudal fin and scattered whitish spots on body. Spiny dorsal fin with ten spines; second spine distinctly elongate. Pelvic-fin origin anterior to lower pectoral-fin base. HABITAT: MA to FL, Gulf of Mexico, Cuba, Haiti, and Trinidad. Also Venezuela to Brazil. Adults over rough, rocky bottoms from about 180 to 1,700 ft. Juveniles may be found around reefs and jetties. BIOLOGY: Some authors place Warsaw grouper in the genus *Hyporthodus*. Critically Endangered.

Snowy grouper - *Epinephelus niveatus* (Valenciennes, 1828)

FEATURES: Large specimens uniformly dark brown. Spiny dorsal-fin margin blackish. Other fin margins may be pale blue. Intermediate specimens with regularly spaced white spots. Small juveniles also with dark saddle on peduncle and yellowish pectoral and caudal fins. Pelvic-fin origin below or posterior to lower pectoral-fin base. Upper opercular margin distinctly arched. HABITAT: MA to Cuba, Gulf of Mexico to Brazil. Bermuda, Great Bahama Bank, and Bimini Islands. Adults over rocky bottoms from about 30 to 1,300 ft. BIOLOGY: Feed on crustaceans and fishes. May live to 27 years. Some authors place Snowy grouper in the genus *Hyporthodus*. Vulnerable.

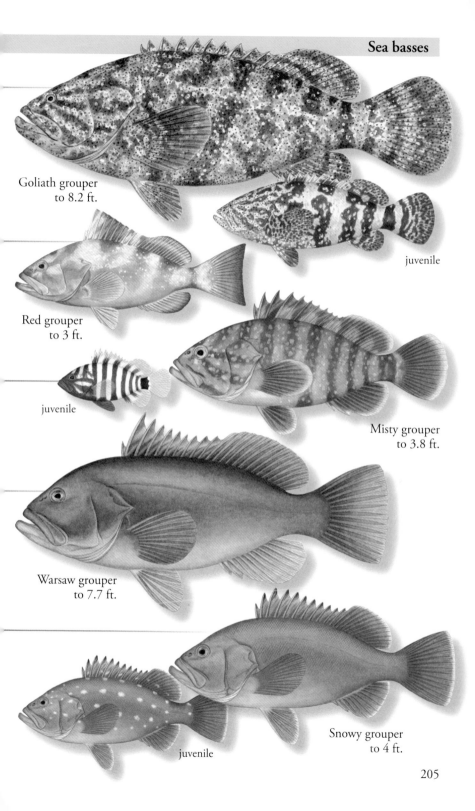

Goliath grouper
to 8.2 ft.

juvenile

Red grouper
to 3 ft.

juvenile

Misty grouper
to 3.8 ft.

Warsaw grouper
to 7.7 ft.

juvenile

Snowy grouper
to 4 ft.

Nassau grouper - *Epinephelus striatus* (Bloch, 1792)

FEATURES: Buff colored with five dark brownish bars on body. May have pale blotches scattered on head and body. Specimens from deep water are pinkish to reddish ventrally. Diagonal band runs from upper jaw through eye to dorsal-fin origin. Small dark spots around eyes. Margin of spiny dorsal fin yellowish. Black saddle on caudal peduncle. Color may change fairly rapidly. HABITAT: FL to Yucatán, Bahamas, Caribbean Sea to Brazil. Northwestern Gulf of Mexico and Bermuda. Occur from shore to about 300 ft. Adults are found over coral reefs, juveniles over seagrass beds. BIOLOGY: Nassau grouper form large, complex spawning aggregations. Spawning correlates to phases of the Moon. Prey on crustaceans and fishes. Currently listed as Endangered.

Spanish flag - *Gonioplectrus hispanus* (Cuvier, 1828)

FEATURES: Body covered in alternating yellow and salmon-colored stripes. Yellow bar runs from snout through eye and along base of dorsal fins. Cheeks with yellow spots and lines. Large, deep red spot on anterior portion of anal fin. White blotch on abdomen. HABITAT: NC to FL, Gulf of Mexico, Cuba, Jamaica. Also Venezuela to Brazil. Occur over rocky bottoms from about 200 to 1,200 ft. BIOLOGY: Spanish flag is the only species within the genus *Gonioplectrus*.

Streamer bass - *Hemanthias aureorubens* (Longley, 1935)

FEATURES: Reddish orange dorsally, silvery on sides and below. Dorsal fins and caudal-fin lobes yellowish. Pectoral fins pinkish. Eyes large, mostly yellow with red. Snout short, mouth oblique. Spiny dorsal fin with small trailing tabs or filaments. Caudal fin deeply forked. HABITAT: NJ to FL, and northern Gulf of Mexico. Also Colombia to Surinam. Reported over semihard bottoms from about 300 to 2,000 ft. BIOLOGY: Streamer bass were previously known as *Prontogrammus aureorubens*.

Longtail bass - *Hemanthias leptus* (Ginsburg, 1952)

FEATURES: Rosy to reddish dorsally. Silvery below. May have indistinct yellowish markings on sides. Yellow stripe below eye from snout to opercular margin. All fins except pectorals yellow with red at base. Larger specimens with an elongate filament on third dorsal-fin spine. Dorsal and anal fins with trailing filaments. Pelvic fins long, trailing. HABITAT: NC to FL, northern and southern Gulf of Mexico, Venezuela to Surinam. Over hard bottoms from about 100 to 2,000 ft. BIOLOGY: Longtail bass larvae hatch as both male and female. Females may eventually transform into males.

Red barbier - *Hemanthias vivanus* (Jordan & Swain, 1885)

FEATURES: Reddish dorsally, pale to silvery below. Two yellow stripes on each side of head. Yellowish areas on sides. Males with long dorsal-fin filaments, yellow anal fin, and red pelvic fins. Females with relatively short dorsal-fin filaments, mottled anal fin, and pink pelvic fins. Caudal fin deeply forked in both. HABITAT: NJ to FL, Bahamas, northeastern and southern Gulf of Mexico. Also western and southern Caribbean Sea to Brazil. Found from about 65 to 1,400 ft. BIOLOGY: Red barbier are hermaphroditic. Form large, fast-moving schools. Feed on crustaceans.

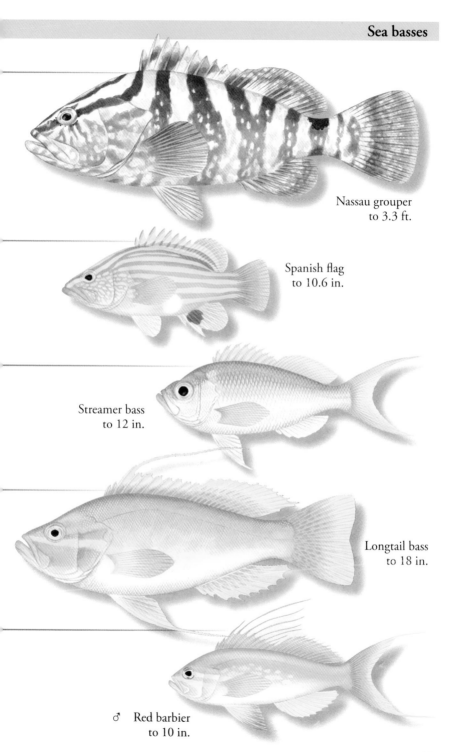

Nassau grouper
to 3.3 ft.

Spanish flag
to 10.6 in.

Streamer bass
to 12 in.

Longtail bass
to 18 in.

♂ Red barbier
to 10 in.

Serranidae - Sea basses, *cont.*

Yellowbelly hamlet - *Hypoplectrus aberrans* Poey, 1868

FEATURES: Two color forms reported. Form 1: brown dorsal color reaches ventrally to abdomen. Form 2 (shown): brown or bluish dorsal color does not reach past body midline. Both forms yellow ventrally and with yellow caudal, anal, pelvic, and pectoral fins. HABITAT: Form 1, Central America. Form 2, S FL (rare) and Antilles. Over shallow reefs and rocky areas. BIOLOGY: Yellowbelly hamlet are hermaphroditic.

Yellowtail hamlet - *Hypoplectrus chlorurus* (Cuvier, 1828)

FEATURES: Body uniformly dark brownish to bluish black. Caudal fin pale to yellow. Pectoral fins usually colorless, occasionally yellowish. HABITAT: S FL (rare; observations may be erroneous), Antilles to Venezuela. Reef associated, near bottom from about 10 to 75 ft. BIOLOGY: Hermaphroditic and solitary. Feed on crustaceans and fishes. Believed to mimic Yellowtail damselfish, *Microspathodon chrysurus*.

Blue hamlet - *Hypoplectrus gemma* Goode & Bean, 1882

FEATURES: Body and fins iridescent blue to bluish black, with upper and lower margins of caudal fin dark blue to black. Color may pale at night. Body relatively deep, compressed. Dorsal fin broad, rounded. Caudal fin concave. HABITAT: S FL, FL Keys, and Bahamas. Reef associated in shallow water. BIOLOGY: Blue hamlet are hermaphroditic. Thought to mimic Blue chromis, *Chromis cyanea*.

Golden hamlet - *Hypoplectrus gummigutta* (Poey, 1851)

FEATURES: Body and fins bright yellow. Blackish blotch on snout and jaws surrounded by bright blue lines. May also have blue lines behind eyes and at base of opercle. Dorsal, anal, and pelvic fins may have thin, blue margins. HABITAT: S FL, Bahamas, Greater Antilles, Nicaragua, Honduras. Reef associated to about 150 ft. BIOLOGY: Observed hybridizing with Yellowbelly hamlet, *Hypoplectrus aberrans*.

Shy hamlet - *Hypoplectrus guttavarius* (Poey, 1852)

FEATURES: Head, abdomen, and all fins yellow. Body from nape to caudal-fin base dark bluish black to purplish black. Blackish blotch with bright blue border on snout. May have blue lines around eyes and to corner of preopercle. Body relatively deep and compressed. Dorsal fin rounded. Caudal fin concave. HABITAT: S FL, Bahamas, and Antilles. Reef associated in shallow water. BIOLOGY: Shy hamlet are hermaphroditic. Thought to mimic Rock beauty, *Holacanthus tricolor*.

Indigo hamlet - *Hypoplectrus indigo* (Poey, 1851)

FEATURES: Head and body with alternating deep blue and whitish bars. Blue bar below spiny dorsal fin widest. Pelvic fins deep blue, others pale blue. Body relatively deep and compressed. Dorsal fin broad, rounded. HABITAT: S FL, Bahamas, Greater Antilles. Also off portions of central and northern South America. Reef associated. Occur in shallow water to about 145 ft. BIOLOGY: Indigo hamlet are hermaphroditic. Found singly or in pairs. May defend feeding territories.

Yellowbelly hamlet
to 5 in.

Yellowtail hamlet
to 5 in.

Blue hamlet
to 5 in.

Golden hamlet
to 5 in.

Shy hamlet
to 5 in.

Indigo hamlet
to 5.5 in.

Serranidae - Sea basses, *cont.*

Black hamlet - *Hypoplectrus nigricans* (Poey, 1852)

FEATURES: Overall brownish black to bluish or purplish black. Bars and blotches absent. Body relatively deep. Dorsal fin broad, rounded. HABITAT: S FL, Bahamas, and Antilles. Also eastern coast of Central America. Reef associated. Occur in shallow, tropical water. BIOLOGY: Hermaphroditic. Found singly or in pairs. May defend feeding territories. Thought to mimic Dusky damselfish, *Stegastes adustus.*

Barred hamlet - *Hypoplectrus puella* (Cuvier, 1828)

FEATURES: Tan to yellowish with six brownish bars on head and body. Bar below spiny dorsal fin is widest. Bright blue lines and spots on head and chest. Color may pale or darken. HABITAT: FL, Gulf of Mexico, Bahamas, and Caribbean Sea to Venezuela. Also Bermuda. Reef associated. Prefer murky water to about 100 ft. BIOLOGY: Barred hamlet are hermaphroditic. Pairs spawn at same site on consecutive nights. May hybridize with other hamlets. Most abundant hamlet in the area.

Butter hamlet - *Hypoplectrus unicolor* (Walbaum, 1792)

FEATURES: Pale yellow to creamy with whitish to bluish highlights. Snout may have black blotch, or blotch may be absent. Blue ring around eyes. May have numerous lines and spots on head. Black blotch on caudal peduncle; blotch may form ring around peduncle. HABITAT: S FL, N and S Gulf of Mexico, Bahamas, Caribbean Sea. Reef associated in shallow water. BIOLOGY: May hybridize with Barred hamlet.

Tan hamlet - *Hypoplectrus sp.*

FEATURES: Pale brown to yellowish brown. Bluish wash around mouth. Some reported with dark smudge on snout and dark spot on pectoral-fin base. Others reported to have spot on caudal peduncle. Still others observed with no markings on body. HABITAT: Reported from S FL, SE Gulf of Mexico, Caribbean Sea, and Bermuda. Reef associated. BIOLOGY: This species is not yet described.

Candy basslet - *Liopropoma carmabi* (Randall, 1963)

FEATURES: Yellow with five blue to purplish stripes on body. Each stripe is bordered above and below by a red stripe. Dorsal fin with a large black spot bordered by a blue ring. Caudal fin with two separate black spots bordered by blue rings. Anal fin lacks black spot. HABITAT: FL Keys, Bahamas to northern South America. Reef associated over coral, sand, and coral rubble bottoms, from about 50 to 230 ft. BIOLOGY: Candy basslet are reclusive and territorial.

Wrasse basslet - *Liopropoma eukrines* (Starck & Courtenay, 1962)

FEATURES: Reddish dorsally. Broad, dark reddish brown stripe runs from snout to caudal fin. Stripe is bordered above and below by a yellowish stripe. Caudal fin with white margin and dark blotch. Dorsal and anal fins lack dark spot. HABITAT: NC to FL Keys, northern Gulf of Mexico. Around rocky or coral reefs from about 98 to 490 ft. BIOLOGY: Wrasse basslet are secretive and occur singly or in pairs.

210

Black hamlet
to 5 in.

Barred hamlet
to 5 in.

Butter hamlet
to 5 in.

Tan hamlet
to 5 in.

Candy basslet
to 2 in.

Wrasse basslet
to 5 in.

211

Cave basslet - *Liopropoma mowbrayi* Woods & Kanazawa, 1951

FEATURES: Body and fins shades of red. Yellowish band from snout tip to eyes. Dorsal and anal fins with a small black spot and bluish white margin at tips. Spot on anal fin may be small to absent. Caudal fin with bluish white margin and black inner band. HABITAT: S FL, Bahamas, to northern South America. Also Bermuda. Reef associated from about 98 to 200 ft. BIOLOGY: Cave basslet are solitary.

Peppermint basslet - *Liopropoma rubre* Poey, 1861

FEATURES: Yellowish with five reddish brown stripes on body. Each stripe bordered above and below by a pinkish to reddish stripe. Dorsal and anal fins with a large black spot and white margin. Caudal fin with white margin and two spots that may merge at center. HABITAT: S FL, Bahamas, Caribbean Sea to Venezuela. Also Bermuda and portions of the Gulf of Mexico. Reef associated in deep recesses from about 10 to 150 ft. BIOLOGY: Solitary and secretive. Other name: Swissguard basslet.

Western comb grouper - *Mycteroperca acutirostris* (Valenciennes, 1828)

FEATURES: Grayish brown with paler blotches and spots on head and body. Largest individuals mostly uniformly colored. Three to four brown stripes radiate from eyes to opercular margin. Another brown stripe radiates from upper jaw to lower opercular margin. Ventral stripes become wavy on chest. Fins dark with irregular pale spots and streaks. Trailing margin of anal fin pointed in adults. Caudal-fin margin rounded in juveniles, becoming concave in adults. HABITAT: NW Gulf of Mexico, Cuba, Jamaica, Lesser Antilles, and northern South America to Brazil. Reported to be common around Port Aransas, TX. Adults occur over rocky bottoms from shore to about 82 ft. Juveniles occur around turtle-grass beds, mangroves, and shallow reefs.

Black grouper - *Mycteroperca bonaci* (Poey, 1860)

FEATURES: Head and body grayish to dark brown with dark grayish, brownish, or reddish spots that blend into streaks and rectangular, chain-like patterns. Soft dorsal, caudal, and anal fins with broad, dark inner margins and narrow, whitish outer margins. Pectoral fins with yellowish to orange margin. Corner of preopercle evenly rounded, lacks notch. HABITAT: FL Keys, Gulf of Mexico, Bahamas, Antilles to Brazil. Found over rocky bottoms and coral reefs and around jetties to about 330 ft. Juveniles found around mangroves. BIOLOGY: Black grouper are hermaphroditic. Near Threatened.

Yellowmouth grouper - *Mycteroperca interstitialis* (Poey, 1860)

FEATURES: Pale brownish gray with close-set, brown spots dorsally and on sides. Some may have faint bars or may be uniformly colored. Mouth and spiny dorsal-fin margin always yellowish. Pectoral fins with dark rays, pale membranes, and whitish margin. Pelvic fins comparatively short. Caudal-fin margin evenly serrated. Juveniles tricolored, becoming blotched with age. HABITAT: FL, Gulf of Mexico, Bahamas, Antilles to Brazil. Over coral reefs and rocky bottoms to about 490 ft. BIOLOGY: Yellowmouth grouper feed primarily on fishes. Sought commercially. Listed as Vulnerable.

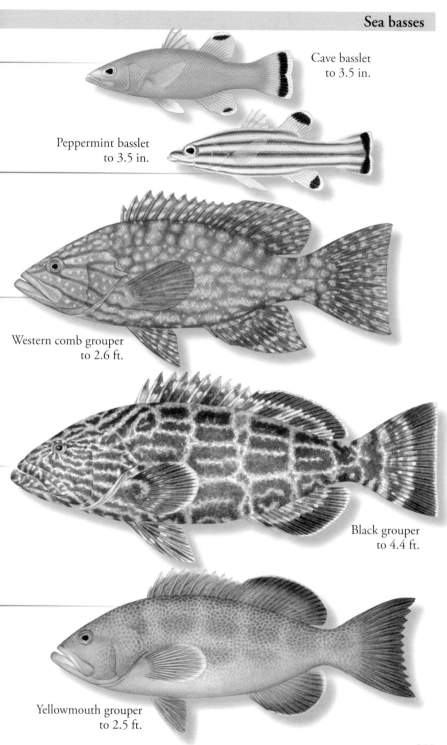

Cave basslet
to 3.5 in.

Peppermint basslet
to 3.5 in.

Western comb grouper
to 2.6 ft.

Black grouper
to 4.4 ft.

Yellowmouth grouper
to 2.5 ft.

Gag - *Mycteroperca microlepis* (Goode & Bean, 1879)

FEATURES: Females and juveniles grayish with darker vermiculations that may blend into saddles or bars. Males with two patterns: pale gray with posterior dark vermiculations and dark ventral area; or with posterior vermiculations, blackish snout, and dark ventral and posterior areas. Soft dorsal- and anal-fin margins always rounded. HABITAT: NC to FL, and Gulf of Mexico to Yucatán. Also Cuba. Juveniles north to MA. Also Bermuda. Adults offshore over rocky or reef bottoms to about 500 ft. Juveniles in estuaries and over seagrass beds. BIOLOGY: Gag are hermaphroditic. Adults solitary or in large groups. Form spawning aggregations.

Scamp - *Mycteroperca phenax* Jordan & Swain, 1884

FEATURES: Usually grayish brown with small, close-set, darker spots. Spots may form clusters or may overlie diffuse vermiculations. Larger specimens pale anteriorly, dark posteriorly, with dark vermiculations. Corners of mouth yellowish. Pectoral fins with dark inner margin, pale outer margin. Caudal fin concave, unevenly serrated. HABITAT: NC to FL, Gulf of Mexico, northern South America. Over rocky and live bottoms, from about 33 to 330 ft. Juveniles occur around jetties and mangroves. BIOLOGY: Feed on octopods, crustaceans, and fishes. May live to 21 years old.

Tiger grouper - *Mycteroperca tigris* (Valenciennes, 1833)

FEATURES: Head and body with close-set brownish, reddish to blackish spots over whitish, grayish to greenish background. Spots cluster or blend to form oblique bars on upper sides. Ventral area marbled. Fins with irregular spots and bands. Pectoral fin usually with yellowish margin. Juveniles yellow with blackish mid-body stripe that develops into oblique bars. HABITAT: S FL, Bahamas, Antilles to Brazil. In Gulf of Mexico from FL; also TX to Yucatán. Also Bermuda. Over reefs and rocky bottoms from about 33 to 130 ft. BIOLOGY: Hermaphroditic. Ambush prey.

Yellowfin grouper - *Mycteroperca venenosa* (Linnaeus, 1758)

FEATURES: Greenish gray in shallow water, reddish in deeper water. Both with oblong, darker blotches, small blackish spots outlined in red dorsally, and small red spots ventrally. Both phases with black inner margins on dorsal, caudal, anal, and pelvic fins. Pectoral fins always with yellow margin. HABITAT: NC, FL, Gulf of Mexico, Bahamas, and Antilles to Brazil. Also Honduras, Nicaragua, and Bermuda. Around reefs and rocky and muddy bottoms from about 6 to 450 ft. Juveniles over seagrass beds. BIOLOGY: Feed on squids and fishes. Flesh may be toxic. Near Threatened.

Atlantic creolefish - *Paranthias furcifer* (Valenciennes, 1828)

FEATURES: Brownish red to purplish dorsally, creamy to salmon below. Dark red blotch at upper pectoral-fin base. Three pale or dark spots below dorsal fin. Sometimes with spots on posterior portion of lateral line. Spiny dorsal-fin membranes with yellowish spots. Caudal fin deeply forked. HABITAT: NC to FL, Gulf of Mexico, Bahamas, and Caribbean Sea to Brazil. Also Bermuda and Eastern Atlantic. Over hard bottoms and coral reefs from about 33 to 210 ft. BIOLOGY: Feed on zooplankton.

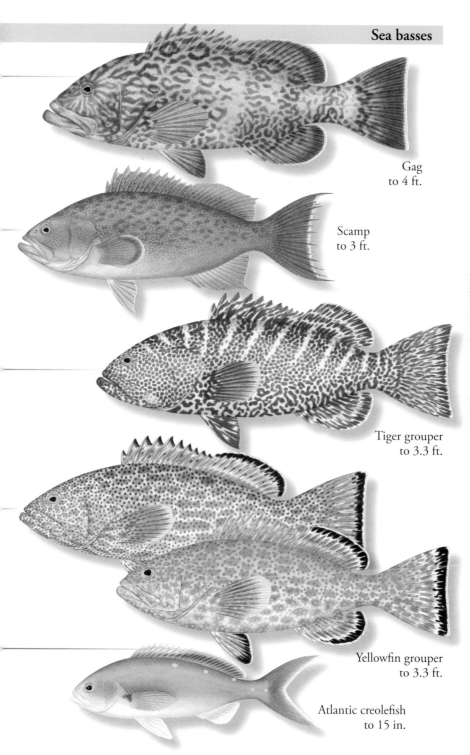

Gag
to 4 ft.

Scamp
to 3 ft.

Tiger grouper
to 3.3 ft.

Yellowfin grouper
to 3.3 ft.

Atlantic creolefish
to 15 in.

Apricot bass - *Plectranthias garrupellus* Robins & Starck, 1961

FEATURES: Reddish dorsally, paler ventrally. Cheeks dull orange. Spiny dorsal fin reddish orange. Soft dorsal, caudal and anal fins yellowish. Spiny dorsal fin tall, angular, with third spine tallest. Lateral line runs close to dorsal-fin base. HABITAT: NC to FL, eastern Gulf of Mexico, Bahamas, eastern and western Carribean Sea. Found near rocky and rubble bottoms from about 42 to 1,200 ft. BIOLOGY: Apricot bass feed on crustaceans. Beleived to be hermaphroditic.

Roughtongue bass - *Pronotogrammus martinicensis* (Guichenot, 1868)

FEATURES: Head and body rosy to orange red. A pale yellow or greenish yellow band runs from snout, under eyes, to opercular margin. May also have two other bars extending from eyes. Irregular yellowish bars and blotches on forebody. Fins rosy to orange red. Third dorsal-fin spine tall. Caudal fin may have elongate upper and lower rays. HABITAT: NC to FL, Gulf of Mexico, Bahamas, and Caribbean Sea to Brazil. Also Bermuda. Found near bottom, often around soft corals, from about 180 to 750 ft. BIOLOGY: Roughtongue bass are hermaphroditic. Feed on crustaceans. Previously known as *Holanthias martinicensis*.

Reef bass - *Pseudogramma gregoryi* (Breder, 1927)

FEATURES: Body and fins brownish red with paler blotches. Head brownish to greenish with dark-edged, whitish bars radiating from eyes. Opercle with large, black, ocellated spot. Each eye with a small, fleshy tentacle. Spiny dorsal fin low. HABITAT: S FL and Bahamas to northern South America. Also reported from Antilles and Bermuda. Occur around hard corals and coral rubble from about 3 to 200 ft. BIOLOGY: Reef bass are secretive. Reported to flare gills at rivals. Feed on crustaceans and worms.

Freckled soapfish - *Rypticus bistrispinus* (Mitchill, 1818)

FEATURES: Brownish above, yellowish to creamy below. Upper portion of head dark with pale stripe along profile. Body and unpaired fins covered with close-set, tiny spots. Spots are more dense dorsally. Pattern may be uniform or broken by underlying, diffuse, pale blotches. Juveniles with a dark stripe on sides. HABITAT: E FL, eastern and southern Gulf of Mexico, Bahamas, and Antilles to Venezuela. Near bottom over sandy areas from near shore to about 260 ft. BIOLOGY: Freckled soapfish are solitary. Hide in shells and burrows during the day, feed on crustaceans at night. Skin is covered in toxic mucus.

Whitespotted soapfish - *Rypticus maculatus* Holbrook, 1855

FEATURES: Brownish with scattered white spots on upper sides of body. Spots may merge. Upper portion of head brownish with pale stripe along profile. HABITAT: NC to FL and Gulf of Mexico. Occur near bottom around rocky areas, coral reefs, jetties, and pilings from near shore to about 300 ft. BIOLOGY: Whitepotted soapfish are secretive and nocturnal. Reported to lie on their sides pressed against rocks or under ledges. Feed on crustaceans and fishes.

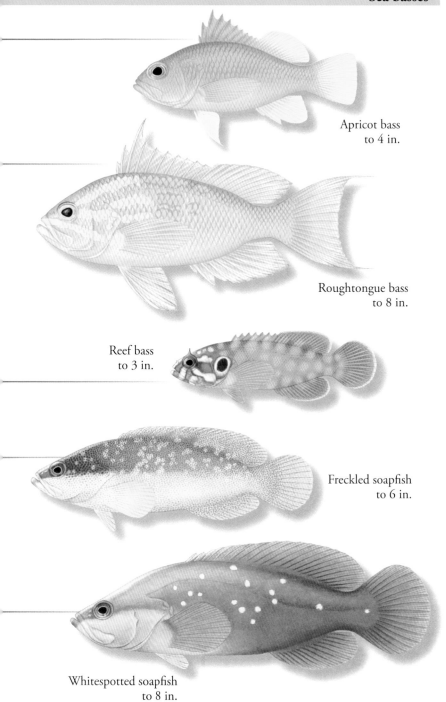

Apricot bass
to 4 in.

Roughtongue bass
to 8 in.

Reef bass
to 3 in.

Freckled soapfish
to 6 in.

Whitespotted soapfish
to 8 in.

Greater soapfish - *Rypticus saponaceus* (Bloch & Schneider, 1801)

FEATURES: Dark bluish black to brownish with dense scattering of small, irregular, pale flecks and blotches on head, body, and fins. Lower jaw and abdomen pale. Profile of head and nape may be pale. Juveniles pale gray with dark markings that form reticulations. HABITAT: E FL to Brazil, and Gulf of Mexico. Also Bermuda and eastern Atlantic. Found around soft bottoms and hard bottoms from about 3,200 ft. BIOLOGY: Greater soapfish are nocturnal and secretive. Often lie motionless on sides. Skin is covered in toxic mucus. Feed on crustaceans and fishes.

Spotted soapfish - *Rypticus subbifrenatus* Gill, 1861

FEATURES: Brownish to olivaceous with fairly evenly spaced dark, ocellated spots on head, body, and fins. Abdomen pale. Older specimens with fewer spots posteriorly. Head profile with pale stripe. HABITAT: S FL, Gulf of Mexico, Bahamas, and Antilles to Venezuela. Also Bermuda. Around rocky areas and reefs to about 70 ft. BIOLOGY: Spotted soapfish are secretive and solitary. Adults prefer clear water. Skin is covered in toxic mucus.

School bass - *Schultzea beta* (Hildebrand, 1940)

FEATURES: Rusty orange dorsally turning orange below with three rows of irregular pale to white blotches that form a chain-like pattern dorsally and on sides. Snout brown. Reddish bar below each eye. Caudal fin yellow; sometimes with a dark, chevron-shaped mark. Mouth very protrusible, lacks teeth. HABITAT: NC to FL, eastern and southern Gulf of Mexico, Bahamas, Antilles to northern South America. Reef associated from about 70 to 550 ft. BIOLOGY: School bass form small schools close to bottom while feeding on plankton. Believed to be hermaphroditic. Previously known as *Serranus beta*.

Pygmy sea bass - *Serraniculus pumilio* Ginsburg, 1952

FEATURES: Buff colored with irregular, brownish bars on sides and peduncle. Abdomen whitish. Reddish bars radiate from eyes. Small, black spots below dorsal-fin base. Dark blotch at rear of spiny dorsal fin. Lateral line with row of dark brown dots. HABITAT: NC to FL, Gulf of Mexico, Greater Antilles to Venezuela. Occur around reefs, seagrass beds, and sandy and shelly bottoms from near shore to about 540 ft. BIOLOGY: Pygmy sea bass are hermaphroditic. Feed on crustaceans.

Orangeback bass - *Serranus annularis* (Günther, 1880)

FEATURES: Color slightly variable. Broad, reddish to blackish saddle below spiny dorsal fin. Yellowish saddles below soft dorsal fin and on peduncle. Lower sides with yellowish bars and spots. Two rectangular-shaped, black-outlined, orange markings behind eyes. HABITAT: S FL, E and N Gulf of Mexico, Bahamas, Antilles, to northern South America. Also Bermuda. Reef associated from about 33 to 230 ft. Also reported over sandy, rocky, and rubble areas. BIOLOGY: Orangeback bass are hermaphroditic. Usually occur in pairs. Reported staying close to sheltering crevices.

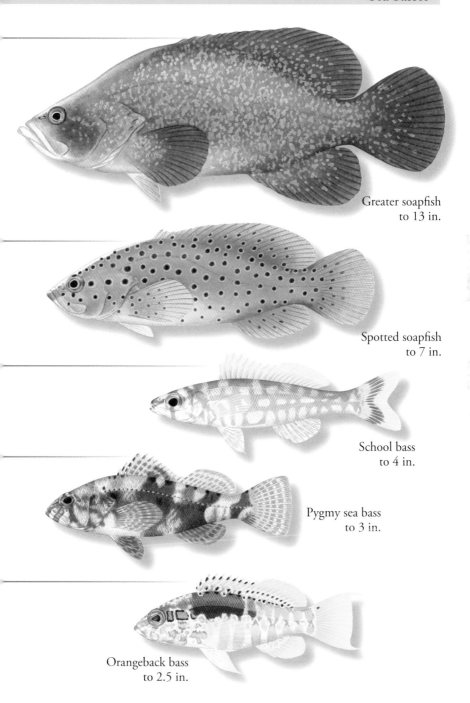

Greater soapfish
to 13 in.

Spotted soapfish
to 7 in.

School bass
to 4 in.

Pygmy sea bass
to 3 in.

Orangeback bass
to 2.5 in.

Blackear bass - *Serranus atrobranchus* (Cuvier, 1829)

FEATURES: Pearly whitish with six to eight tannish and brownish bars dorsally and on sides. Midbody bar extends from spiny dorsal-fin margin to abdomen; bar may be dark to faint. Yellowish patch below midbody bar may be bright to absent. Opercle with a black blotch on inner surface that is visible from the outside. Fins pale to yellowish. HABITAT: FL and northern Gulf of Mexico to Brazil. Occur near bottom from about 33 to 900 ft. BIOLOGY: Blackear bass are hermaphroditic.

Lantern bass - *Serranus baldwini* (Evermann & Marsh, 1899)

FEATURES: Color and pattern varies with depth; colors brighter in deeper waters. Upper sides with rows of reddish to blackish blotches forming stripes. Always with irregular, yellow midbody stripe. Lower sides with four dark, oval blotches, each with a yellowish to reddish bar below. Always four, small black spots at caudal-fin base. HABITAT: S FL, Bahamas, Antilles to Brazil. Occur near bottom over rocky and rubble areas from about 3 to 260 ft. Also over seagrass beds. Juveniles may be found in abandoned conch shells. BIOLOGY: Lantern bass are hermaphroditic and territorial. Feed on shrimps and fishes.

Snow bass - *Serranus chionaraia* Robins & Starck, 1961

FEATURES: Whitish with broad, brown stripes radiating from eyes that merge with rows of brown blotches on body. Lower cheeks and chest with brown blotches. Large, white area on abdomen behind pectoral-fin base. Midsection of caudal peduncle usually lacks blotches. Brownish bar at caudal-fin base. Upper and lower caudal-fin margins with row of brown spots. HABITAT: SE FL, FL Keys, Puerto Rico, Honduras. Near bottom around reefs and rubble bottoms from about 150 to 295 ft.

Tattler - *Serranus phoebe* Poey, 1851

FEATURES: Whitish to tannish with a brownish bar below each eye. A brownish bar from anterior margin of spiny dorsal fin to abdomen is followed by a silvery white bar on abdomen above anus. Dark stripe on sides from mid-body to caudal-fin base. May also have indistinct bars at nape and on posterior sides. All markings except silvery white bar fade with age. HABITAT: SC, FL, Gulf of Mexico, Greater Antilles to northern South America. Also Bermuda. Around reefs and rocky areas from about 88 to 590 ft. BIOLOGY: Hermaphroditic and solitary. Feed primarily on shrimps.

Belted sandfish - *Serranus subligarius* (Cope, 1870)

FEATURES: Head with irregular, reddish brown spots and a dark band through eye. Scales outlined in reddish brown. Large black blotch on anterior portion of soft dorsal fin merges with dark bar below. Abdomen abruptly silvery white. Fins banded. Snout pointed, head profile sloping. HABITAT: NC to FL, and in Gulf of Mexico from FL to Veracruz. Rare in FL Keys. Found near bottom over rocky and mixed bottoms, around jetties and outcroppings from about 3 to 60 ft.; possibly deeper. BIOLOGY: Belted sandfish are hermaphroditic. Territorial. Prefer turbid water.

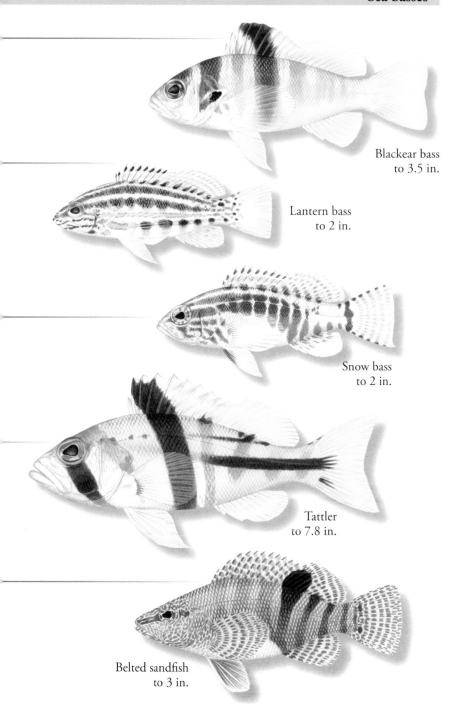

Blackear bass
to 3.5 in.

Lantern bass
to 2 in.

Snow bass
to 2 in.

Tattler
to 7.8 in.

Belted sandfish
to 3 in.

Serranidae - Sea basses, *cont.*

Tobaccofish - *Serranus tabacarius* (Cuvier, 1829)

FEATURES: Whitish with orange brown midbody stripe or with orange brown from midline to abdomen. Alternating brown to grayish saddles on back. Saddles may form U-shapes. Caudal fin with a dark brown stripe on upper and lower lobes, often appearing as a V-shape. HABITAT: GA to FL, Bahamas, Antilles to Brazil. Also E Gulf of Mexico and Bermuda. Reef associated. Occur from about 10 to 200 ft. BIOLOGY: Tobaccofish may follow goatfishes in search of prey. Hermaphroditic.

Harlequin bass - *Serranus tigrinus* (Bloch, 1790)

FEATURES: Whitish with irregular grayish to black bars on head and body that are intersected by horizontal rows of pale to dark spots. Sides may have yellow cast. Dorsal, caudal, and anal fins spotted. Spiny dorsal fin may have black blotch between third and fifth spines. Snout pointed. HABITAT: S FL, Bahamas, Antilles to Venezuela. Also Bermuda. Over reefs and around scattered rock or coral from about 3 to 120 ft.

Chalk bass - *Serranus tortugarum* Longley, 1935

FEATURES: Alternating reddish brown and bluish white bars dorsally. Reddish bars may become pinkish below. Sides with bluish to lavender sheen. Ventral area whitish. Caudal fin may be bluish white, reddish brown, or slightly banded. HABITAT: S FL, SE Gulf of Mexico, Bahamas, Antilles to Venezuela. Reef associated from about 40 to 1,300 ft. BIOLOGY: Often in small groups.

ALSO IN THE AREA: Yellowtail bass, *Bathyanthias cubensis*; Saddle bass, *Serranus notospilus*, p. 431.

Grammatidae - Basslets

Fairy basslet - *Gramma loreto* Poey, 1868

FEATURES: Deep bluish purple anteriorly, yellow posteriorly. Rusty bar from lower jaw tip through eye. Yellowish bar from lower eye to opercle. Black spot on anterior portion of spiny dorsal fin. Caudal-fin margin either rounded or slightly concave. HABITAT: FL Keys, Bahamas, Antilles, Quintana Roo to Venezuela. Reef associated. Around caves and ledges from about 3 to 200 ft. BIOLOGY: Often swim upside down. Males care for nests. Apparently introduced to FL Keys.

Threeline basslet - *Lipogramma trilineatum* Randall, 1963

FEATURES: Yellowish with scales outlined in greenish blue. Bright blue stripe from tip of upper jaw to dorsal-fin origin. Bright blue line from top of eyes to below spiny dorsal fin. May have short blue line below eyes. Cheeks pinkish to purplish. Dorsal fin yellowish with blue margin. Anal and pelvic fins with yellowish spots. HABITAT: SE FL, Dry Tortugas, Bahamas, scattered in Caribbean Sea. Found near bottom around coral and rocky ledges and walls to about 310 ft.

ALSO IN THE AREA: Dusky basslet, *Lipogramma anabantoides*, p. 431.

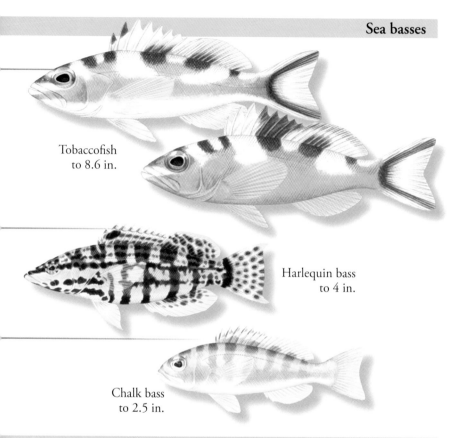

Tobaccofish
to 8.6 in.

Harlequin bass
to 4 in.

Chalk bass
to 2.5 in.

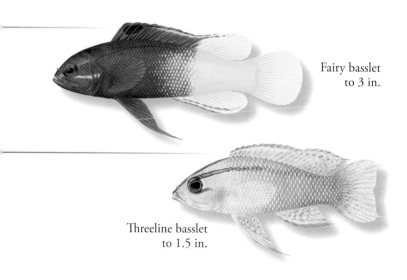

Fairy basslet
to 3 in.

Threeline basslet
to 1.5 in.

Opistognathidae - Jawfishes

Swordtail jawfish - *Lonchopisthus micrognathus* (Poey, 1860)

FEATURES: Tan with narrow, pale blue bars anteriorly and spots posteriorly. Dorsal and anal fins with blue margins. Pelvic fins bluish. Caudal fin with bluish upper and lower margins and elongate, trailing rays. HABITAT: S FL, Gulf of Mexico, to northern South America. Over relatively uniform silty or muddy bottoms to about 280 ft. BIOLOGY: Swordtail jawfish are usually colonial. Build and maintain burrows that may reach about 14 inches in depth. May live commensally with the Smoothwrist soft crab, *Chasmocarcinus cylindricus*. Feed on zooplankton.

Yellowhead jawfish - *Opistognathus aurifrons* (Jordan & Thompson, 1905)

FEATURES: Head bright to pale yellow fading to pearly bluish to tannish posteriorly. Pair of black spots on chin. Dorsal fin with narrow blue margin. Body otherwise unmarked. Caudal fin rounded to slightly pointed. HABITAT: S FL, Bahamas, to Barbados to Central America. In Gulf of Mexico from Dry Tortugas and Flower Garden Banks. Over sand and coral rubble around reefs from about 3 to 200 ft. BIOLOGY: Yellowhead jawfish use their mouths to excavate and maintain burrows. Males mouth-brood eggs. Popular in aquarium trade.

Moustache jawfish - *Opistognathus lonchurus* Jordan & Gilbert, 1882

FEATURES: Olive gray dorsally, tannish on sides. Abdomen whitish. Snout dark. Jaws bluish. Dusky orange moustache above upper jaw. Irregular blue lines on sides. Dorsal and anal fins with blue banding. All fins except pectoral fins with blue margins. HABITAT: SC to FL, northeastern and northwestern Gulf of Mexico, Greater Antilles to Guyana. On sandy, rubble bottoms around reefs to about 300 ft. BIOLOGY: Moustache jawfish build and maintain burrows.

Banded jawfish - *Opistognathus macrognathus* Poey, 1860

FEATURES: Body brownish with irregular, pale spots and mottling. Four to six brown blotches at dorsal-fin base. Blotches may merge with brown bars below. Dorsal fin yellowish with irregular, pale bands. Dark blotch between sixth and ninth dorsal-fin spines may be present. Males with elongate, banded upper jaw. HABITAT: S FL to northern South America. Absent from Jamaica. Over sand, rubble, and rocky bottoms around reefs from near surface to about 140 ft. BIOLOGY: Banded jawfish build and maintain burrows that they line with small stones. Apparently monogamous.

Mottled jawfish - *Opistognathus maxillosus* Poey, 1860

FEATURES: Body brownish to grayish brown with irregular pale spots and mottling. Large pale blotches form rows below dorsal-fin base and along ventral area. Two large, pale blotches at caudal-fin base. Dorsal fin brownish with pale spotting and five to six dark blotches along base; first blotch may be indistinct. Blotch between sixth and eighth spines largest and darkest. HABITAT: S FL to northern South America. Also eastern Gulf of Mexico. Over sandy, rocky, and coral rubble bottoms from about 3 to 25 ft. BIOLOGY: Construct elaborate burrows lined with coral debris.

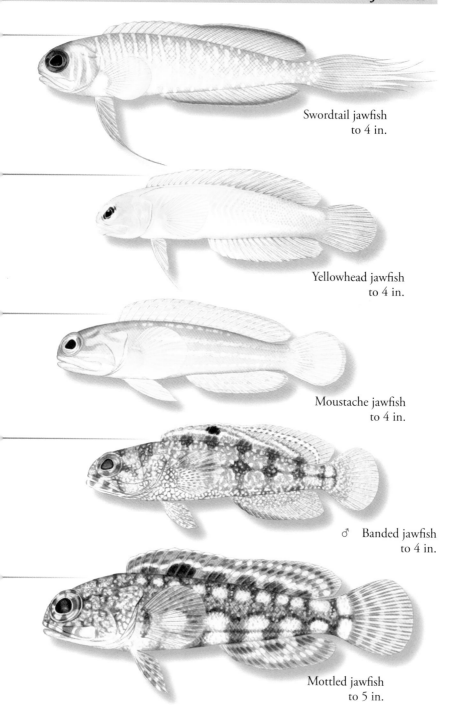

Swordtail jawfish
to 4 in.

Yellowhead jawfish
to 4 in.

Moustache jawfish
to 4 in.

♂ Banded jawfish
to 4 in.

Mottled jawfish
to 5 in.

Opistognathidae - Jawfishes, *cont.*

Spotfin jawfish - *Opistognathus robinsi* Smith-Vaniz, 1997

FEATURES: Head brownish with paler spots and mottling. Body brownish with pale blotches arranged in uneven rows on sides. Dorsal fin brown with bands of pale spots. Prominent ocellated black spot between third and seventh dorsal-fin spines. Males with elongated upper jaw. HABITAT: SC to FL, Bahamas, and northern Gulf of Mexico. On sandy to rubble bottoms from near shore to about 150 ft. Reported in estuaries and lagoons. BIOLOGY: Spotfin jawfish construct and maintain burrows. Males mouth-brood eggs until hatching.

Dusky jawfish - *Opistognathus whitehursti* (Longley, 1927)

FEATURES: Head mottled brown. Body mottled dark brown with obscure dark bars reaching dorsal-fin base. Row of close-set, irregular white blotches along dorsal profile. Usually with pale blue to green blotch between second and fourth dorsal-fin spines; blotch may be obscure to absent. Soft dorsal fin yellowish with pale and dark banding. HABITAT: S FL, eastern Gulf of Mexico, Bahamas to Brazil. From about 3 to 156 ft. Occur over sandy and rubble bottoms, often around edges of turtle-grass beds. BIOLOGY: Construct and maintain burrows. May live in groups.

ALSO IN THE AREA: Megamouth jawfish, *Opistognathus melachasme*; Yellowmouth jawfish, *Opistognathus nothus*, p. 431.

Priacanthidae - Bigeyes

Bulleye - *Cookeolus japonicus* (Cuvier, 1829)

FEATURES: Deep red fading to pinkish or silvery below, or uniformly red. Dorsal, caudal and anal fins red with blackish margins. Pelvic fins with blackish membranes. Eyes large with red irises. Dorsal and anal fins elongate posteriorly. Pelvic fins expanded with inner rays attached to abdomen by membrane - become shorter with age. HABITAT: Circumglobal in tropical to temperate seas. In western Atlantic from VA to Argentina. Juveniles recorded to Nova Scotia. Over hard bottoms from about 200 to 1,300 ft. BIOLOGY: Bulleye feed on crustaceans and fishes. May live to nine years or more. Often confused with Glasseye, *Heteropriacanthus cruentatus*. Other names: Longfin bulleye, Longfinned bullseye.

Glasseye snapper - *Heteropriacanthus cruentatus* (Lacepède, 1801)

FEATURES: Orange red, or red to pink with pale to silvery bars and blotches. May be uniformly colored. Soft dorsal, caudal, and anal fins faintly spotted. Spiny dorsal- and caudal-fin margins sometimes dusky. Eyes large with red irises. Pelvic fins comparatively short, attached to abdomen by membrane. HABITAT: Circumglobal in tropical to temperate seas. In western Atlantic from NJ (rare) to Argentina and Gulf of Mexico. Occur in shallow water to 65 ft. BIOLOGY: Glasseye snapper are secretive and nocturnal. Hide in holes and crevices and under ledges during the day. Feed on a variety of invertebrates. Taken by hook-and-line, by spearing, and in traps. Other name: Glasseye. Previously known as *Priacanthus cruentatus* and *Cookeolus boops*.

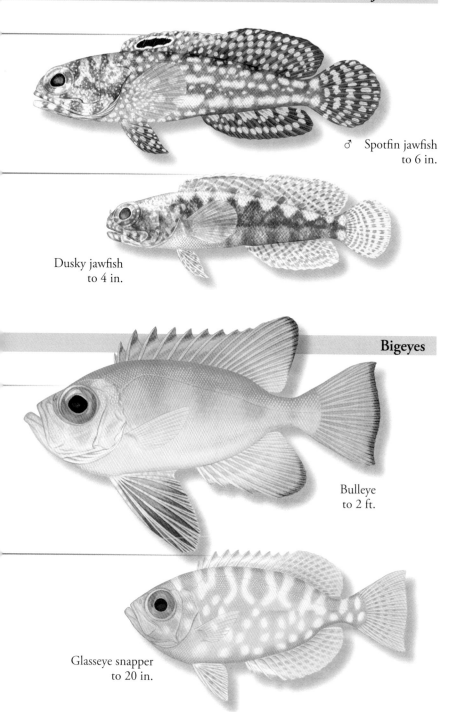

♂ Spotfin jawfish
to 6 in.

Dusky jawfish
to 4 in.

Bulleye
to 2 ft.

Glasseye snapper
to 20 in.

Bigeye - *Priacanthus arenatus* Cuvier, 1829

FEATURES: Uniformly red, but may change to silvery with broad bars. Row of small, dark red spots sometimes along lateral line. Dorsal and anal fins with faint spots. Pelvic fins with blackish membranes and small black spot at base. Spiny dorsal-fin margin comparatively straight. Caudal fin margin concave. HABITAT: On both sides of Atlantic in tropical to subtropical waters. In western Atlantic from NC to FL, Gulf of Mexico, Bahamas, Caribbean Sea to Argentina. Also Bermuda. Over reefs and rocky bottoms from about 65 to 820 ft. BIOLOGY: Bigeye feed on invertebrates.

Short bigeye - *Pristigenys alta* (Gill, 1862)

FEATURES: Body red to salmon-colored, often with diffuse, paler bars dorsally. Soft dorsal, caudal, and anal fins pale with black margins. Pelvic fins red with black margins. Juveniles mottled in red to salmon with banded fins. Eyes large with red irises. Inner rays of pelvic fins attached to abdomen. Scales comparatively large. Body comparatively deep. HABITAT: MA to FL, Gulf of Mexico, Bahamas, and Caribbean Sea to Venezuela. Also Bermuda. Juveniles to ME. Found over hard bottoms from about 16 to 410 ft. BIOLOGY: Short bigeye are secretive, hide in recesses.

Apogonidae - Cardinalfishes

Bigtooth cardinalfish - *Apogon affinis* (Poey, 1875)

FEATURES: Translucent pale pink, salmon to bronze. Always with dusky to blackish stripe from snout tip through eyes. Caudal fin may have dusky margin. Otherwise unmarked. HABITAT: S FL and FL Keys, Gulf of Mexico, Bahamas to Surinam. Also Bermuda. Reported from the eastern central Atlantic. Reef associated around hollow coral heads, caves, and overhangs. Typically from about 65 to 295 ft.; reported to 984 ft. BIOLOGY: Bigtooth cardinalfish spawn in pairs. Males mouth-brood eggs.

Bridle cardinalfish - *Apogon aurolineatus* (Mowbray, 1927)

FEATURES: Translucent salmon, pinkish to golden, with enlarged melanophores on body. Body and fins otherwise unmarked. May have two streaks behind eyes. HABITAT: S FL, northern and southern Gulf of Mexico, Bahamas, to northern South America. Over seagrass beds and around coral reefs. From about 3 to 245 ft. BIOLOGY: Bridle cardinalfish form congregations and seek shelter among sea anemones. Feed on zooplankton. Form pairs during spawning. Males mouth-brood eggs.

Barred cardinalfish - *Apogon binotatus* (Poey, 1867)

FEATURES: Pinkish to pale red and somewhat translucent. May pale or darken or have iridescent highlights. Always with one blackish bar from rear second dorsal-fin base to rear anal-fin base and one blackish bar on caudal peduncle near caudal-fin base. HABITAT: SE FL, eastern Gulf of Mexico, Bahamas, Antilles to Venezuela. Reef associated from shore to about 200 ft. BIOLOGY: Barred cardinalfish are nocturnal. Reported to form aggregations in hiding places during the day.

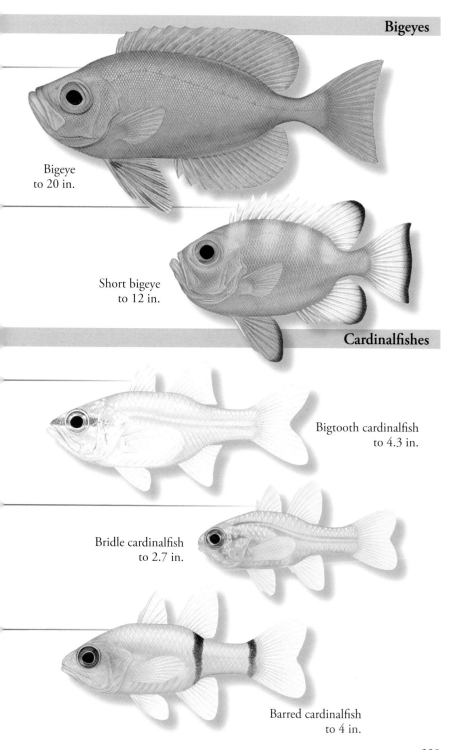

Bigeyes

Bigeye
to 20 in.

Short bigeye
to 12 in.

Cardinalfishes

Bigtooth cardinalfish
to 4.3 in.

Bridle cardinalfish
to 2.7 in.

Barred cardinalfish
to 4 in.

Whitestar cardinalfish - *Apogon lachneri* Böhlke, 1959

FEATURES: Reddish with small black spot followed by small white spot on upper caudal peduncle just behind second dorsal-fin base. Black speckles cluster on outer margins of dorsal and anal fins. Scales may appear to have iridescent flecks. HABITAT: S FL, southern Gulf of Mexico, Bahamas, Antilles to Belize. Reef associated in clear waters from about 17 to 230 ft. BIOLOGY: Whitestar cardinalfish are secretive and nocturnal. Remain hidden during the day.

Slendertail cardinalfish - *Apogon leptocaulus* Gilbert, 1972

FEATURES: Deep red with three dark bars on body. First bar under first dorsal fin; second bar from second dorsal-fin base to anal-fin base; third bar circles end of caudal peduncle. Anterior portions of second dorsal and anal fins dark. Upper and lower lobes of caudal fin dark. Snout pointed. Caudal peduncle long and slender. HABITAT: Boca Raton, FL, Bahamas, Isla de Providencia, and Belize. Occur in caves and crevices over coral reefs and rocky slopes from about 60 to 100 ft. May be more common in deeper water. BIOLOGY: Slendertail cardinalfish are rare in the area. Little is known of their habits.

Flamefish - *Apogon maculatus* (Poey, 1860)

FEATURES: Deep orange red to dusky red with black flecks that cluster to form bar through eye to opercular margin. Eyes with a white stripe above and below pupil. Black blotch below second dorsal-fin base. Usually with faint, dark, bar-like saddle on caudal peduncle. HABITAT: MA to FL, Gulf of Mexico, Bahamas to Venezuela. Also Bermuda. Around reefs, sea walls, and pilings from near shore to about 330 ft. BIOLOGY: Flamefish hide in or near reef caves and crevices during the day. Feed at night. Males mouth-brood eggs.

Mimic cardinalfish - *Apogon phenax* Böhlke & Randall, 1968

FEATURES: Deep orange red to red with iridescent highlights. At night, highlights become turquoise blue. Diffuse, triangular-shaped blackish bar below rear second dorsal-fin base. Wide, blackish saddle or bar on caudal peduncle near caudal-fin base. HABITAT: FL Keys, Bahamas, islands off coast of Venezuela. Occur around reefs and rocky areas with little sand from about 10 to 165 ft. BIOLOGY: Male Mimic cardinalfish mouth-brood eggs. Solitary, secretive. Reported hiding in crevices during the day, foraging at night.

Broadsaddle cardinalfish - *Apogon pillionatus* Böhlke & Randall, 1968

FEATURES: Reddish and somewhat transparent with some luster. Diffuse, blackish bar below rear of second dorsal fin is followed by a white bar and a diffuse blackish saddle. A small, white bar at caudal-fin base may be present or absent. HABITAT: S FL, Bahamas to Venezuela. Reef associated from about 50 to 295 ft. Reported around reef faces, slopes, and walls. BIOLOGY: Broadsaddle cardinalfish are secretive. Hide in reef caves and crevices by day, emerge at night to feed.

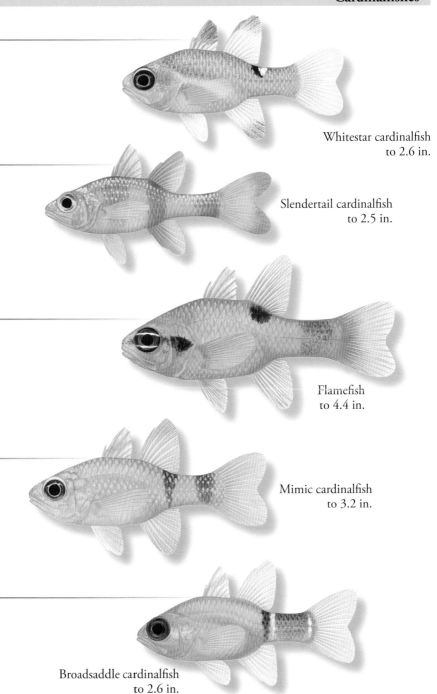

Whitestar cardinalfish
to 2.6 in.

Slendertail cardinalfish
to 2.5 in.

Flamefish
to 4.4 in.

Mimic cardinalfish
to 3.2 in.

Broadsaddle cardinalfish
to 2.6 in.

Apogonidae - Cardinalfishes, *cont.*

Pale cardinalfish - *Apogon planifrons* Longley & Hildebrand, 1940

FEATURES: Iridescent, pale pink. Narrow reddish to blackish bar from rear of second dorsal-fin base to rear of anal-fin base. Large, reddish to blackish spot or wide bar on caudal peduncle near caudal-fin base. Body color becomes pearly at night. HABITAT: S FL, eastern Gulf of Mexico, Bahamas to Brazil. Also Bermuda. Reef associated from about 10 to 100 ft. BIOLOGY: Pale cardinalfish are reclusive and solitary. Reported leaving crevices and caves at night to forage on zooplankton. Form pairs during spawning. Males mouth-brood eggs.

Twospot cardinalfish - *Apogon pseudomaculatus* Longley, 1932

FEATURES: Deep orange red with faint to dark blotch on opercle. Black spot on back below rear of second dorsal fin. Black spot on upper caudal peduncle near caudal-fin base. Eyes with a whitish stripe above and below pupil. HABITAT: MA (rare) to FL, Gulf of Mexico, Bahamas to Brazil. Also Bermuda. Over reefs and hard bottoms, around harbors, pilings, and sea walls from near shore to about 1,320 ft. BIOLOGY: Twospot cardinalfish are nocturnal. Males brood eggs in mouth. Feed on plankton.

Sawcheek cardinalfish - *Apogon quadrisquamatus* Longley, 1934

FEATURES: Translucent reddish orange, or reddish to brownish orange, with darker melanaphores covering body. Small to large diffuse, dark spot on caudal peduncle at caudal-fin base. Fins yellowish. HABITAT: S FL, Bahamas to southern Brazil. Around reefs and over seagrass flats, sand, and rubble bottoms from about 4 to 245 ft. BIOLOGY: Sawcheek cardinalfish have been reported seeking shelter around sea urchins, anemones, and tube sponges. Feed at night.

Belted cardinalfish - *Apogon townsendi* (Breder, 1927)

FEATURES: Reddish yellow to pinkish and somewhat translucent. Narrow dark bar from posterior base of second dorsal fin to anal-fin base. Narrow dark bar on caudal peduncle, and a broken bar at caudal-fin base. Area between bars on peduncle darkens at night to form a ringed blotch. HABITAT: S FL, eastern and western Gulf of Mexico, Bahamas, Antilles to northern South America. Also Bermuda. Around coral and rocky areas from about 10 to 180 ft. BIOLOGY: Belted cardinalfish are nocturnal. Seek shelter among urchin spines.

Bronze cardinalfish - *Astrapogon alutus* (Jordan & Gilbert, 1882)

FEATURES: Head and abdomen silvery, body brownish. Dark bars radiate from eyes. Brownish melanophores concentrate to form flecks and obscure blotches on scales. Fins brownish to bronze and densely covered in darker flecks. Pelvic fins with bronze hue; may have blackish margins. Pelvic fins do not reach past anterior third of pelvic-fin base. HABITAT: NC to Venezuela and eastern Gulf of Mexico. Reported from Bahamas. Occur around reefs and seagrass beds in shallow water. BIOLOGY: Bronze cardinalfish have been reported to occur in the mantle cavity of the West Indian fighting conch, *Strombus pugilis.*

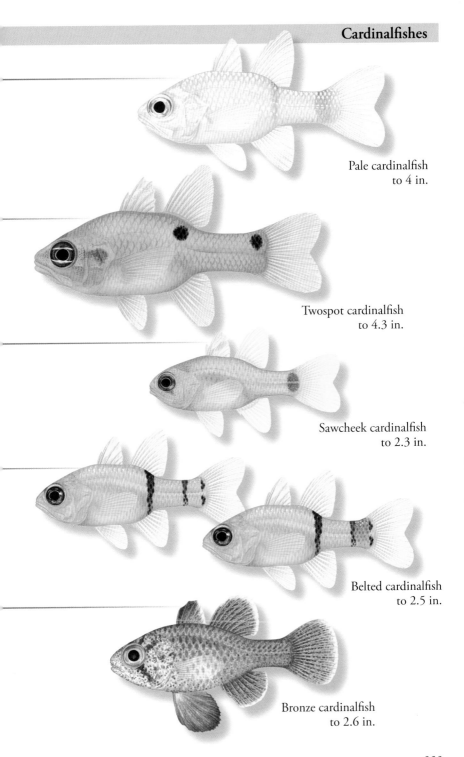

Pale cardinalfish
to 4 in.

Twospot cardinalfish
to 4.3 in.

Sawcheek cardinalfish
to 2.3 in.

Belted cardinalfish
to 2.5 in.

Bronze cardinalfish
to 2.6 in.

233

Blackfin cardinalfish - *Astrapogon puncticulatus* (Poey, 1867)

FEATURES: Head and abdomen silvery. Dark bars radiate from eyes. Brown melanophores concentrate to form a dense covering of flecks and spots on head and body. Fins finely to densely peppered with brown flecks. Margins of second dorsal and anal fins lack pigment. Pelvic fins reach between anterior and middle third of anal-fin base. HABITAT: S FL, eastern Gulf of Mexico, Bahamas to Brazil. Also Bermuda. Found around reefs, over seagrass beds, and in empty shells to about 26 ft. BIOLOGY: Blackfin cardinalfish form pairs during spawning. Males mouth-brood eggs.

Conchfish - *Astrapogon stellatus* (Cope, 1867)

FEATURES: Head and abdomen silvery. Dark bars radiate from eyes. Brown melanophores concentrate to form small brown spots and flecks on head and body. Fins brownish and densely peppered with small brown flecks. Pelvic fins greatly expanded, reach to rear third of pelvic-fin base. HABITAT: S FL, eastern and southern Gulf of Mexico, Bahamas, Antilles to Venezuela. Also Bermuda. Reef associated from about 3 to 130 ft. BIOLOGY: Conchfish live commensally in the mantle cavity of the Queen conch, *Strombus giga*s, and in the Stiff penshell, *Atrina rigida*. Nocturnal.

Freckled cardinalfish - *Phaeoptyx conklini* (Silvester, 1915)

FEATURES: Translucent, usually with pinkish cast. Body densely covered with dark flecks that concentrate on scales. Flecks form a bar below eyes and a blotch near caudal-fin base. Second dorsal and anal fins always with dark band just above base. Second and third dorsal-fin spines about equal in length. Eyes very large. HABITAT: S FL, northern and southern Gulf of Mexico, Bahamas to Venezuela. Also Bermuda. Over shallow coral, rocky, and rubble bottoms. Often in empty shells or containers.

Dusky cardinalfish - *Phaeoptyx pigmentaria* (Poey, 1860)

FEATURES: Translucent with brownish to reddish cast. Body with dark flecks and spots concentrated on scales. Flecks form a bar below eyes and a blotch near caudal-fin base. Second dorsal and anal fins unmarked. Third dorsal-fin spine usually taller than second spine. HABITAT: S FL, northern and southern Gulf of Mexico, Antilles to Brazil. Also Bermuda. On coral, seagrass, or shelly bottoms from near shore to about 140 ft. BIOLOGY: Dusky cardinalfish court and pair during spawning. Males mouth-brood eggs.

Sponge cardinalfish - *Phaeoptyx xenus* (Böhlke & Randall, 1968)

FEATURES: Translucent orangish to lavender brown. Head with yellow cast. Small brownish spots and flecks on head and body. Flecks form a bar below eyes and a blotch near caudal-fin base. Second dorsal and anal fins yellowish, with or without narrow, faint to dark band at base. Second dorsal-fin spine slightly taller than third spine. HABITAT: S FL, eastern and northern Gulf of Mexico, Bahamas to Venezuela. Associated with cylindrical sponges on rocky and coral bottoms. BIOLOGY: Share tube-shaped sponge cavities with gobies and brittle stars. Feed at night.

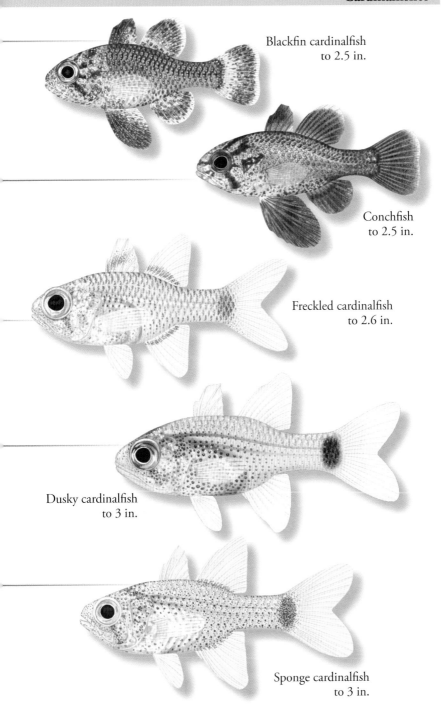

Blackfin cardinalfish
to 2.5 in.

Conchfish
to 2.5 in.

Freckled cardinalfish
to 2.6 in.

Dusky cardinalfish
to 3 in.

Sponge cardinalfish
to 3 in.

235

Goldface tilefish - *Caulolatilus chrysops* (Valenciennes, 1833)

FEATURES: Pale bluish violet with yellow cast dorsally. Silvery to pearly white below. Bright yellow streak from nostril to below eye. Faint blue streak below yellow streak. Irises golden. Black spot just above pectoral-fin base. Membranes of soft dorsal and caudal fins with pale yellow spots. HABITAT: NC to FL, Gulf of Mexico, and Greater Antilles to Brazil. Bottom-dwelling, usually over rubble bottoms from 295 to about 630 ft. BIOLOGY: Feed on invertebrates and small fishes.

Blackline tilefish - *Caulolatilus cyanops* Poey, 1866

FEATURES: Violet to blue with yellowish sheen and reticulations dorsally. Silvery to white below. Predorsal ridge yellow. May have dark stripe below dorsal-fin base. Spiny portion of dorsal fin yellowish. Upper and lower caudal-fin lobes yellowish. Black spot above pectoral-fin base. HABITAT: NC to FL, Gulf of Mexico, and Antilles to northern South America. Bottom-dwelling from about 148 to 1,624 ft.

Anchor tilefish - *Caulolatilus intermedius* Howell Rivero, 1936

FEATURES: Pale violet brown dorsally. White below. Black bar runs from jaws around anterior margin of eyes and up to black predorsal ridge. Black spot above pectoral-fin base. Dorsal fin with a series of dark blotches that form an irregular band. HABITAT: In Gulf of Mexico from NW FL to Yucatán. Also Cuba. Over muddy bottoms from about 147 to 951 ft. Other name: Gulf bareye.

Blueline tilefish - *Caulolatilus microps* Goode & Bean, 1878

FEATURES: Dark brown gray dorsally, buff white ventrally. Pale yellow streak from tip of snout to eyes, underlined by bright blue streak. Irises golden. Predorsal ridge black. Dorsal fin with yellowish margin. Anal fin with dusky inner margin. Caudal fin rays yellowish at base. HABITAT: VA to FL and Gulf of Mexico to Yucatán. Demersal on mud and rubble bottoms from 98 to about 775 ft.

Tilefish - *Lopholatilus chamaeleonticeps* Goode & Bean, 1879

FEATURES: Bluish green to bluish gray dorsally with small, close-set, irregular yellow spots. Fading to milky white ventrally. Juveniles with fewer, larger yellow spots. Yellowish predorsal flap present. Caudal fin with irregular yellow spots and lines. HABITAT: Nova Scotia to FL, Gulf of Mexico to Surinam. Near bottom on soft bottoms, usually from 265 to 1,770 ft. BIOLOGY: Construct burrows in bottom substrate. Feed on a variety of invertebrates and fishes. Sought commercially.

Sand tilefish - *Malacanthus plumieri* (Bloch, 1786)

FEATURES: Blue gray dorsally, pearly white below. Irregular pale blue lines on head. Dorsal and anal fins with yellow margins. Caudal fin with yellow upper and lower margins and a dusky patch on upper lobe. Body elongate. HABITAT: Ocracoke Island Inlet, NC, to FL, Gulf of Mexico, Bahamas, and Antilles to Uruguay. Also Bermuda. Bottom-dwelling over sand and rubble, from about 33 to 500 ft.

Goldface tilefish
to 21 in.

Blackline tilefish
to 14 in.

Anchor tilefish
to 12 in.

Blueline tilefish
to 12 in.

Tilefish
to 3.5 ft.

Sand tilefish
to 2 ft.

Pomatomidae - Bluefish

Bluefish - *Pomatomus saltatrix* (Linnaeus, 1766)

FEATURES: Greenish to greenish blue dorsally, silvery below. Dark blotch at pectoral-fin base. Dorsal and caudal fins olivaceous. Body moderately elongate and compressed. Dorsal profile moderately convex. Lower jaw extends beyond upper jaw. Teeth numerous. HABITAT: Worldwide in eight major populations. In western Atlantic from Nova Scotia to FL, Gulf of Mexico to Yucatán and Cuba. Also Bermuda and Colombia to Argentina. Occur coastally over continental shelves. BIOLOGY: Swift and voracious. Adults hunt in loose groups, juveniles in schools, often mangling prey. Migrate north in summer, south in winter. Sought commercially and for sport.

Coryphaenidae - Dolphinfishes

Pompano dolphinfish - *Coryphaena equiselis* Linnaeus, 1758

FEATURES: Bright green blue dorsally, silvery below with golden highlights. Small dark spots scattered on sides. Dorsal fin tall, long-based. Caudal fin deeply forked. Body comparatively deep, dorsal and ventral profiles comparatively convex. Dorsal fin with 52–59 rays. Pectoral fins about half of head length. Anal fin convex, lacks deep anterior notch. HABITAT: Worldwide in tropical and warm temperate seas. Usually offshore and oceanic, may enter coastal waters. Associated with flotsam and *Sargassum* seaweed. BIOLOGY: School with and are often misidentified as Dolphinfish.

Dolphinfish - *Coryphaena hippurus* Linnaeus, 1758

FEATURES: Bright blue green above, golden to silvery below. Small dark spots on sides. Pectoral fins yellowish. Juveniles with bars that extend into dorsal and anal fins. Dorsal fin tall, long-based. Caudal fin deeply forked. Body comparatively shallow, dorsal and anal profiles comparatively straight. Dorsal fin with 58–66 rays. Pectoral fins more than half of head length. Anal fin notched anteriorly. Males with angular head profile, females with rounded head profile. HABITAT: Worldwide in tropical to warm temperate seas. Usually offshore and oceanic, may enter coastal waters. Associated with flotsam and *Sargassum* seaweed. BIOLOGY: Dolphinfish form small schools. Highly migratory. Feed on fishes, crustaceans, and squids. Other name: Mahi-mahi.

Rachycentridae - Cobia

Cobia - *Rachycentron canadum* (Linnaeus, 1766)

FEATURES: Dark brown dorsally and on sides, whitish below. May have two whitish stripes on sides. Dorsal, caudal, and pectoral fins dark brown. Pelvic and anal fins dusky. Snout broad, head compressed. First dorsal fin with seven to nine separate spines. Caudal fin tall, forked. HABITAT: Worldwide in tropical to warm temperate seas, except eastern Pacific. MA to FL, Gulf of Mexico, Bahamas, and Caribbean Sea to Brazil. Also Bermuda. Pelagic around reefs, over rocky bottoms, and in estuaries from near surface to about 4,000 ft. BIOLOGY: Cobia grow rapidly and may reach eight years of age. Feed on invertebrates and fishes. Sought commercially and for sport.

Bluefish

Bluefish
to 3.8 ft.

Dolphinfishes

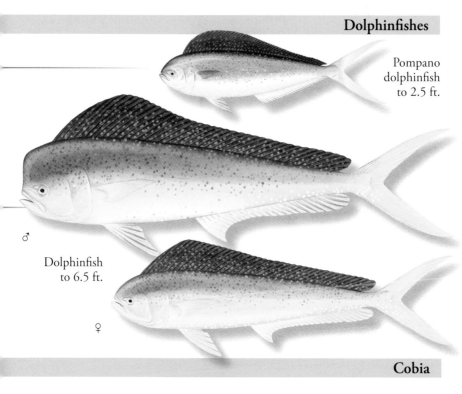

Pompano
dolphinfish
to 2.5 ft.

♂

Dolphinfish
to 6.5 ft.

♀

Cobia

Cobia
to 6.5 ft.

Sharksucker - *Echeneis naucrates* Linnaeus, 1758

FEATURES: Dark bluish to brownish gray above. Dark stripe from mouth to caudal-fin base is bordered above and below by whitish stripe. White stripes may be obscure in larger specimens. Vertical fins dark with thin, pale to white margins. Disk with 21 to 28 laminae (usually 23). Caudal fin with concave margin. HABITAT: Worldwide in tropical to warm temperate seas. From Nova Scotia to Uruguay.

Whitefin sharksucker - *Echeneis neucratoides* Zuiew, 1786

FEATURES: Brownish gray to blackish above. Dark stripe from tip of snout to caudal-fin base is bordered above and below by whitish stripe. Stripes may be obscure in some specimens. Vertical fins with broad, pale to whitish margin. Disk with 18 to 23 laminae (usually 21). Caudal-fin margin is truncate to slightly trilobed. HABITAT: Western Atlantic only. MA to northern South America.

Spearfish remora - *Remora brachyptera* (Lowe, 1839)

FEATURES: Color corresponds to host color. May be brownish, bluish brown to whitish. Dark specimens with pale dorsal- and anal-fin margins. Disk with 15 to 18 laminae, does not reach past pectoral-fin tips. Pectoral fins rounded. Caudal-fin margin slightly concave. HABITAT: Worldwide in tropical to warm temperate seas. Associated with billfishes, also with sharks and molas.

Marlinsucker - *Remora osteochir* (Cuvier, 1829)

FEATURES: Brownish gray to grayish or blackish. Cephalic disk with 15 to 19 laminae, reaches well past pectoral fin tips. Pectoral fins rounded. Caudal peduncle comparatively narrow. Caudal fin with concave margin. HABITAT: Worldwide in tropical to warm temperate seas. BIOLOGY: Associated with billfishes, particularly White marlin and Sailfish. Feed on host's parasites and prey remains.

Remora - *Remora remora* (Linnaeus, 1758)

FEATURES: Overall brownish gray to sooty or blackish. Mottled or uniformly colored. Cephalic disk with 16 to 20 laminae, does not reach past pectoral-fin margin. Pectoral-fin margin blunt. Caudal-fin margin deeply concave. HABITAT: Worldwide in tropical to warm temperate seas. Nova Scotia to Argentina.

White suckerfish - *Remorina albescens* (Temminck & Schlegel, 1850)

FEATURES: Usually whitish. May be pale tannish or grayish. Cephalic disk reaches to pectoral-fin tips. Dorsal and anal fins rounded. HABITAT: Worldwide in tropical to warm temperate seas. FL to Brazil; may occur farther north. BIOLOGY: Usually found attached to Giant manta and Black marlin, with a few accounts of sharks as hosts. Previously known as *Remora albescens*.

ALSO IN THE AREA: Slender suckerfish, *Phtheirichthys lineatus*; Whalesucker, *Remora australis*, p. 431.

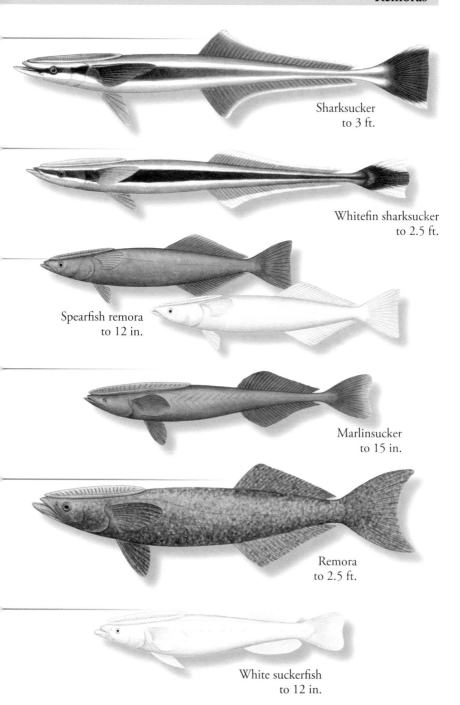

Sharksucker
to 3 ft.

Whitefin sharksucker
to 2.5 ft.

Spearfish remora
to 12 in.

Marlinsucker
to 15 in.

Remora
to 2.5 ft.

White suckerfish
to 12 in.

Carangidae - Jacks and Pompanos

African pompano - *Alectis ciliaris* (Bloch, 1787)

FEATURES: Bluish dorsally, silvery below. May have obscure, dark blotch on upper opercular margin. Grayish blue chevrons on body fade with age. Head profile strongly arched over eyes. Profile below dorsal fin and above anal fin relatively straight. Dorsal and anal fins with pointed lobes in adults, filamentous in juveniles. HABITAT: Worldwide in tropical to warm temperate seas. MA to FL, Gulf of Mexico, Bahamas, and Caribbean Sea to Brazil. Also Bermuda. Adults near bottom, juveniles pelagic. BIOLOGY: Solitary, strong swimmers. Feed mainly on fishes and squid.

Yellow jack - *Caranx bartholomaei* Cuvier, 1833

FEATURES: Body silvery greenish blue with yellowish sheen. Dorsal profile yellowish. All fins with yellowish cast. Body otherwise unmarked. May display pale blotches. Upper and lower profiles almost identical in shape. Straight portion of lateral line begins posterior to anal-fin origin. HABITAT: MA to FL, Gulf of Mexico, Bahamas, Caribbean Sea to Brazil. Also Bermuda. Reef associated over continental shelves and slopes from surface to about 164 ft. Juveniles pelagic. BIOLOGY: Solitary or in small schools.

Blue runner - *Caranx crysos* (Mitchill, 1815)

FEATURES: Shades of blue to green dorsally with metallic sheen. Silvery to golden below. Breeding males blackish. Small black spot on opercular margin. Tips of upper and lower caudal-fin lobes black. Upper and lower body profiles almost identical in shape. Straight portion of lateral line begins anterior to anal-fin origin. HABITAT: Nova Scotia to FL, Gulf of Mexico, Bahamas, and Caribbean Sea to Brazil. Also Bermuda. Over outer continental shelves from surface to about 330 ft. Juveniles associated with *Sargassum* seaweed. BIOLOGY: Schooling. Feed on fishes and invertebrates.

Crevalle jack - *Caranx hippos* (Linnaeus, 1766)

FEATURES: Greenish, bluish, to bluish black above. Silvery to golden below. Small black spot on opercular margin. Black blotch on lower pectoral-fin rays. Jaws extend to below or just beyond rear margin of eyes. HABITAT: Nova Scotia to FL, Gulf of Mexico, Bahamas, and Greater Antilles to Uruguay. Also Bermuda and eastern Atlantic. In brackish to marine waters from surface to about 1,100 ft. Larger specimens may be solitary and in deeper water. May ascend rivers. Juveniles pelagic, around *Sargassum* seaweed. BIOLOGY: Form moderate to large schools. Prized gamefish.

Horse-eye jack - *Caranx latus* Agassiz, 1831

FEATURES: Metallic bluish to bluish gray dorsally, silvery to golden below. Small black spot at upper opercular margin. Posterior scutes silvery to blackish. Upper dorsal-fin lobe may be blackish. Caudal fin yellowish, upper margin may be blackish. Eyes comparatively large. Jaws extend to rear margin of eyes. HABITAT: NJ to FL, Gulf of Mexico, Bahamas, and Caribbean Sea. Also Bermuda. Pelagic. Inshore, offshore, and along sandy beaches. Also enter brackish water and rivers. From surface to about 460 ft. BIOLOGY: Form small schools. Feed on fishes and invertebrates.

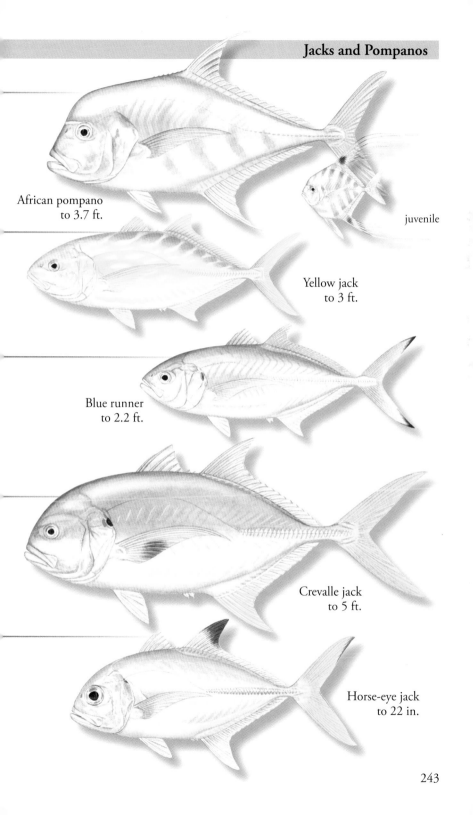

African pompano
to 3.7 ft.

juvenile

Yellow jack
to 3 ft.

Blue runner
to 2.2 ft.

Crevalle jack
to 5 ft.

Horse-eye jack
to 22 in.

243

Black jack - *Caranx lugubris* Poey, 1860

FEATURES: Olive gray to brownish gray or blackish. May be paler below or uniformly dark. Vertical fins, scutes, and outer margin of caudal fin blackish. Dorsal profile of head angular. Jaws extend to middle of eyes. HABITAT: Worldwide in tropical marine waters. In western Atlantic from Gulf of Mexico, Bahamas, Antilles to Brazil. Also Bermuda. Found offshore from about 78 to 213 ft. Pelagic in clear ocean water. BIOLOGY: Black jack occasionally form schools. Feed at night.

Bar jack - *Caranx ruber* (Bloch, 1793)

FEATURES: Pale grayish blue above, silvery below. Blackish band overlying iridescent bluish band runs along dorsal profile, through caudal peduncle, and into lower caudal-fin lobe. Body and fins may turn entirely blackish. HABITAT: NJ to FL, southern Gulf of Mexico, Bahamas, and Caribbean Sea to Venezuela. Also Bermuda. In clear, shallow waters around coastal reefs. Juveniles associate with *Sargassum* seaweed rafts. BIOLOGY: Bar jack are Schooling; occasionally solitary. Make grunting sounds when distressed. Flesh is toxic.

Atlantic bumper - *Chloroscombrus chrysurus* (Linnaeus, 1766)

FEATURES: Metallic blue dorsally, silvery below. Black blotch on upper caudal peduncle. Caudal fin yellowish, others with yellowish tint. Body compressed and with ventral profile more convex than dorsal profile. HABITAT: MA to FL, Gulf of Mexico, and Great Antilles to Uruguay. Also Bermuda and eastern Atlantic. Occur in shallow coastal marine and estuarine waters. Also around mangrove-lined lagoons. Juveniles may occur well offshore and are associated with jellyfish. BIOLOGY: Atlantic bumper are schooling. Feed on fishes, cephalopods, zooplankton, and detritus. Grunt when in distress.

Mackerel scad - *Decapterus macarellus* (Cuvier, 1833)

FEATURES: Bluish black to metallic blue dorsally; may have greenish cast posteriorly. Silvery to white ventrally. Black spot on upper opercular margin. Silvery to bluish stripe present on midline. Caudal fin yellowish green to yellowish. Single, separate finlet behind dorsal and anal fins. HABITAT: Worldwide in clear, open tropical and warm temperate seas. Gulf of Maine to Brazil. Also Bermuda. Absent from Gulf of Mexico. BIOLOGY: Mackerel scad are schooling fishes that feed on planktonic invertebrates.

Round scad - *Decapterus punctatus* (Cuvier, 1829)

FEATURES: Bluish to greenish dorsally, silvery to whitish below. Small black spot on opercular margin. Yellowish to olivaceous stripe on body midline. Up to 14 small, black spots on curved portion of lateral line. Single, separate finlet behind dorsal and anal fins. HABITAT: MA to FL, Gulf of Mexico, Bahamas, and Caribbean Sea to Brazil. Also Bermuda and eastern Atlantic. Occur primarily from mid-water to near bottom to about 295 ft. Juveniles pelagic and near surface. BIOLOGY: Round scad are schooling. Feed on planktonic invertebrates. Other name: Cigarfish.

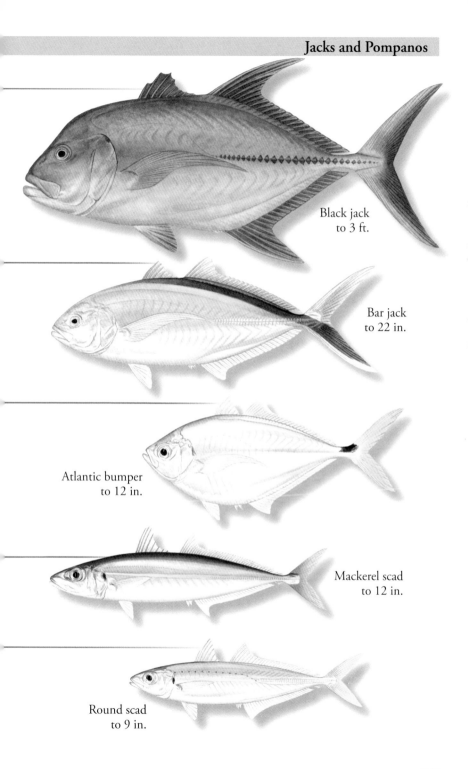

Black jack
to 3 ft.

Bar jack
to 22 in.

Atlantic bumper
to 12 in.

Mackerel scad
to 12 in.

Round scad
to 9 in.

Redtail scad - *Decapterus tabl* Berry 1968

FEATURES: Bluish black to metallic blue dorsally, white ventrally. Small black spot on upper opercular margin. Caudal fin red. Tips of anterior second dorsal-fin rays reddish. Single, separate finlet behind dorsal and anal fins. HABITAT: In tropical to warm temperate seas. NC, S FL, Bermuda, Colombia, and Venezuela. In Gulf of Mexico off LA and Tamaulipas, Mexico. Also Indian and western and central Pacific oceans. Found in mid-water to near bottom from about 490 to 650 ft. BIOLOGY: Redtail scad form schools and feed on planktonic invertebrates. Used as bait.

Rainbow runner - *Elagatis bipinnulata* (Quoy & Gaimard, 1825)

FEATURES: Dark blue, olive blue to dark green dorsally. White ventrally. Greenish to yellowish midbody stripe from snout to caudal fin bordered above and below by narrow bluish stripes. Two separate rays form finlets behind dorsal and anal fins. Dorsal and ventral profiles similarly shaped. Mouth small. HABITAT: Worldwide in tropical to warm temperate seas. MA to FL, Gulf of Mexico, Bahamas, and Caribbean Sea to Brazil. Pelagic. Found near surface over reefs and offshore. BIOLOGY: May form large schools. Feed on fishes, invertebrates. Sought as game and food.

Bluntnose jack - *Hemicaranx amblyrhynchus* (Cuvier, 1833)

FEATURES: Bluish green dorsally, silvery below. Large, blackish blotch on opercle. Dorsal-fin margins blackish. Caudal fin yellowish with blackish tip on upper lobe. Body deep and strongly compressed. Mouth small, snout blunt. HABITAT: NC to FL, Gulf of Mexico to northern South America. Also Cuba. Reported rare along eastern coast of US. Coastal, near bottom of marine and brackish waters to about 160 ft. BIOLOGY: Bluntnose jack are usually solitary. Juveniles associated with jellyfish.

Pilotfish - *Naucrates ductor* (Linnaeus, 1758)

FEATURES: Bluish silver with bluish black bars on body. Three bars extend into second dorsal fin. Caudal-fin lobes with white tips. Band from eyes to dorsal-fin origin absent. Eyes large. First dorsal fin with small, separate spines. Caudal peduncle with lateral keels. HABITAT: Worldwide in tropical to warm temperate seas. Nova Scotia to FL, Gulf of Mexico, Bahamas, and Caribbean Sea to Argentina. Pelagic. BIOLOGY: Adults follow large sharks, rays, and sea turtles. Juveniles associate with seaweed and jellyfish. Feed on ectoparasites, host's prey scraps, and fishes.

Leatherjack - *Oligoplites saurus* (Bloch & Schneider, 1801)

FEATURES: Metallic bluish to greenish dorsally, silvery to white below. Caudal fin yellowish to colorless. Snout bluntly pointed. Jaws narrow, extend to posterior margin of eyes. First dorsal-fin spines separate. Posterior rays of second dorsal and anal fins semidetached. Pectoral fins short. HABITAT: MA to FL, Gulf of Mexico, and Greater Antilles to Brazil. Also in eastern Pacific. Inshore along sandy beaches, bays, and inlets. More often in turbid water. BIOLOGY: Leatherjack form large schools. Juveniles feed on ectoparasites of other fishes. Dorsal- and anal-fin spines venomous.

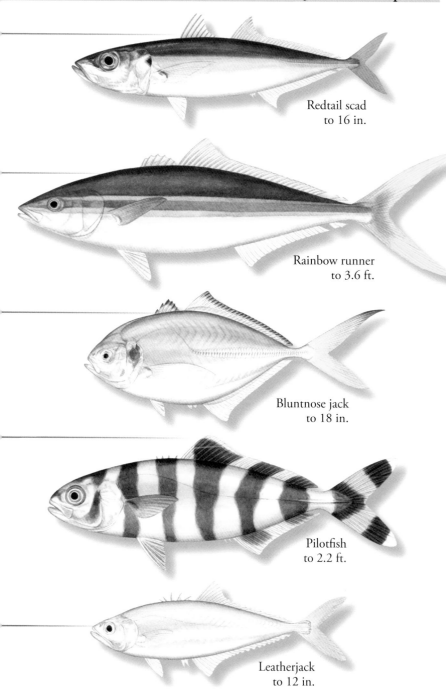

Redtail scad
to 16 in.

Rainbow runner
to 3.6 ft.

Bluntnose jack
to 18 in.

Pilotfish
to 2.2 ft.

Leatherjack
to 12 in.

Carangidae - Jacks and Pompanos, *cont.*

White trevally - *Pseudocaranx dentex* (Bloch & Schneider, 1801)

FEATURES: Pale greenish blue dorsally, fading to silvery below. Yellowish stripe on sides along midline and at bases of dorsal and anal fins. Black spot at opercle margin. Fins yellowish. Dorsal and ventral body profiles similar. Jaws protrusible, upper jaw not reaching anterior margin of eyes. HABITAT: In subtropical waters of Atlantic, Pacific, and Indian oceans from about 260 to 650 ft. In western Atlantic from NC to GA, Bermuda. BIOLOGY: White trevally are schooling fishes that filter and suck invertebrates from the sea bed. Sometimes placed in the genus *Caranx*.

Bigeye scad - *Selar crumenophthalmus* (Bloch, 1793)

FEATURES: Metallic bluish to bluish green dorsally. Silvery below. May have yellowish stripe from opercular margin to upper caudal peduncle. Small black blotch on opercular margin. Eyes very large. Shoulder notched over opercle. Lateral line with pored scales anteriorly, scutes posteriorly. HABITAT: Worldwide in tropical to warm temperate seas. Nova Scotia to FL, Gulf of Mexico, Bahamas, and Caribbean Sea to Brazil. In shallow coastal waters to about 550 ft. BIOLOGY: Schooling. Planktivorous.

Atlantic moonfish - *Selene setapinnis* (Mitchill, 1815)

FEATURES: Silvery white with metallic bluish sheen. Faint dark blotch on opercle and on upper caudal peduncle. Caudal fin may have dusky to yellowish cast; other fins colorless. Juveniles with dark blotch over straight portion of lateral line. Body very deep and compressed. Anterior profile of head steeply sloping, rounded at top. Lower jaw protrudes. HABITAT: Nova Scotia to FL, Gulf of Mexico to Argentina. Absent from Bahamas. Adults near bottom from inshore to about 180 ft. Juveniles pelagic, near surface. BIOLOGY: Schooling. Feed on fishes and crustaceans.

Lookdown - *Selene vomer* (Linnaeus, 1758)

FEATURES: Silvery white with metallic bluish to yellowish sheen. May have silvery bars on body. Body very deep and compressed. Anterior profile of head very steep, angular at top. First dorsal-fin spines elongated in juveniles, reduced in adults. Second dorsal- and anal-fin lobes elongate. Pelvic fins elongate in juveniles, very small in adults. HABITAT: ME to FL, northern and southern Gulf of Mexico, Greater Antilles to Uruguay. Also Bermuda. Occur near bottom of shallow, coastal waters to about 170 ft. Juveniles in estuaries and off beaches. BIOLOGY: Lookdown feed in schools.

Greater amberjack - *Seriola dumerili* (Risso, 1810)

FEATURES: Bluish brown to olivaceous dorsally, often with pinkish luster on sides. Silvery below. Usually with dark band from eyes to first dorsal-fin origin. Faint amber stripe on sides from eyes to caudal fin. Upper jaw broad and rounded posteriorly; reaches to about middle of pupils. Body elongate, comparatively shallow. HABITAT: Worldwide in tropical to warm temperate seas. Nova Scotia to FL, Gulf of Mexico, Bahamas, and Caribbean Sea to Brazil. Also Bermuda. Near bottom over continental shelves and slopes and around reefs and rocky ledges from about 60 to 235 ft.

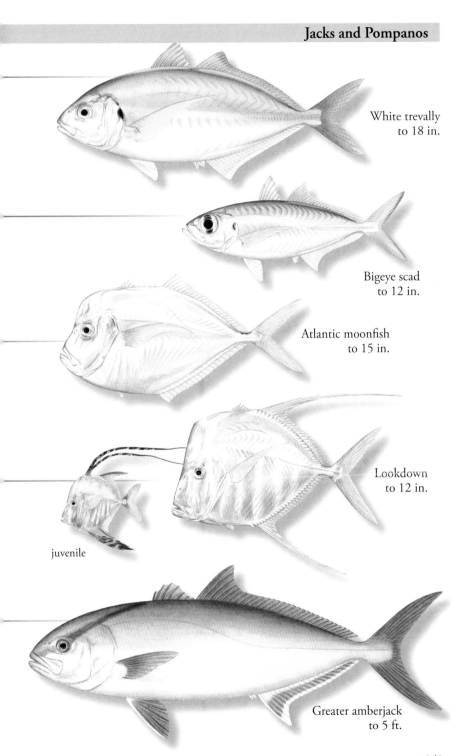

White trevally
to 18 in.

Bigeye scad
to 12 in.

Atlantic moonfish
to 15 in.

Lookdown
to 12 in.

juvenile

Greater amberjack
to 5 ft.

249

Carangidae - Jacks and Pompanos, *cont.*

Lesser amberjack - *Seriola fasciata* (Bloch, 1793)

FEATURES: Dark dusky pinkish to dusky violet dorsally, fading to silvery below. Faint band from eyes to nape. Often with faint amber stripe on sides from eyes to caudal fin. Upper jaw moderately broad posteriorly, reaching to anterior margin of pupil. Eyes comparatively large. HABITAT: MA to FL, Gulf of Mexico, Greater Antilles, Bermuda, and Venezuela. On or near bottoms from about 180 to 490 ft. BIOLOGY: Lesser amberjack are schooling. Feed on squids.

Almaco jack - *Seriola rivoliana* Valenciennes, 1833

FEATURES: Brown, olivaceous, to bluish green dorsally, paler below. Dark band runs from eyes to first dorsal-fin origin. Faint amber stripe on sides from eyes to caudal fin. Upper jaw angular, very broad posteriorly; reaches below anterior margin of pupil. First dorsal- and anal-fin lobes long, pointed. Body comparatively deep. HABITAT: Worldwide in tropical to warm temperate seas. MA to FL, Gulf of Mexico, Bahamas, and Caribbean Sea to Argentina. Also Bermuda. Pelagic in the water column.

Banded rudderfish - *Seriola zonata* (Mitchill, 1815)

FEATURES: Dark grayish to greenish blue dorsally, silvery white below. May have a dark band from eyes to first dorsal fin and an amber stripe on sides from eyes to caudal peduncle. Fins dusky. Second dorsal and caudal fins with pale margins. Juveniles with six bars on sides and dark band from eyes to first dorsal-fin origin. Upper jaw reaches to about rear margin of eyes in adults. Body elongate, comparatively shallow. Snout pointed. HABITAT: ME to FL, Gulf of Mexico to Brazil. Pelagic or near bottom over continental shelves. Juveniles with drifting plants, jellyfish, and larger fishes.

Florida pompano - *Trachinotus carolinus* (Linnaeus, 1766)

FEATURES: Silvery with metallic bluish to greenish sheen. Abdomen silvery to golden. Anal- and caudal-fin lobes yellowish. First dorsal-fin spines small, separate. Second dorsal fin with 22–27 rays. Head profile somewhat rounded, snout blunt. Body deep, compressed. HABITAT: MA to FL, Gulf of Mexico to Brazil. Scattered in Antilles. Along sandy beaches and in brackish bays and inlets from shore to about 130 ft. Adults are pelagic, juveniles occur in beach surf zone. BIOLOGY: Juveniles form large schools, chase prey up onto beach faces. Sought commercially and for sport.

Permit - *Trachinotus falcatus* (Linnaeus, 1758)

FEATURES: Silvery with metallic bluish to greenish sheen. Abdomen silvery, often with golden areas. May also have large gray to black smudge on sides. Anal- and caudal-fin lobes blackish. First dorsal-fin spines small, separate. Second dorsal fin with 17–21 rays. Head profile and snout rounded. Body deep, compressed. HABITAT: MA to Brazil, Gulf of Mexico, Bahamas, and Antilles. Also Bermuda. In coastal waters to about 118 ft. Adults pelagic or near bottom in channels and over seagrass flats, reefs, or mud bottoms. Juveniles occur in beach surf zone. BIOLOGY: Occur singly or in small schools. Juvenile Permit may change color from silver to black. Sought for sport.

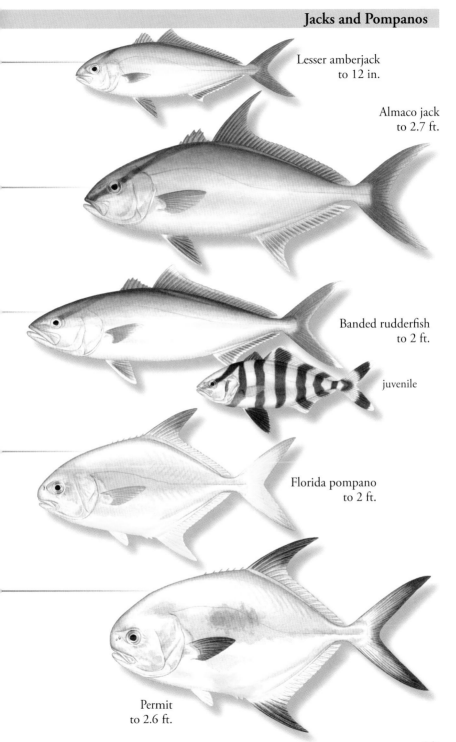

Lesser amberjack
to 12 in.

Almaco jack
to 2.7 ft.

Banded rudderfish
to 2 ft.

juvenile

Florida pompano
to 2 ft.

Permit
to 2.6 ft.

Carangidae - Jacks and Pompanos, *cont.*

Palometa - *Trachinotus goodei* Jordan & Evermann, 1896

FEATURES: Silvery with metallic blue sheen dorsally. Abdomen may have golden tint. Usually with narrow bars on upper sides. Second dorsal and anal-fin lobes elongate, blackish. Upper and lower margins of caudal fin blackish. Head comparatively small. HABITAT: MA to FL, Gulf of Mexico, Bahamas, and Antilles to Brazil. Also Bermuda. Coastal. In beach surf zone and around reefs and rocky areas to about 118 ft.

Rough scad - *Trachurus lathami* Nichols, 1920

FEATURES: Pale to dark blue or bluish green dorsally. Silvery to whitish below. Black spot on opercular margin. First dorsal-fin spines and anterior portion of second dorsal fin dusky. Caudal fin colorless to dusky at margin. Well-developed scutes along entire lateral line; anterior scutes may be somewhat overgrown by body scales. HABITAT: Gulf of Maine south to Argentina. Absent from Antilles. Usually occur coastally and near bottom to about 295 ft. BIOLOGY: Schooling. Caught in trawls.

Cottonmouth jack - *Uraspis secunda* (Poey, 1860)

FEATURES: Grayish, brownish gray, to bluish black or dusky. Juveniles and specimens up to about 12 inches with six or seven bars on body. Bars fade with age. Vertical fins dark. Tongue, floor, and roof of mouth white to creamy. Body oval in shape, laterally compressed. HABITAT: Found scattered from NJ to Brazil. Also Bermuda and Pacific Ocean. Oceanic, in water column from surface to bottoms of about 120 ft. BIOLOGY: Solitary or in small schools. May grunt when in distress. Some authors recognize this species as a junior synonym of *Uraspis helvola*.

Bramidae - Pomfrets

Atlantic pomfret - *Brama brama* (Bonnaterre, 1788)

FEATURES: Grayish silver. Margins of dorsal, caudal, and anal fins blackish. Pectoral-fin margins transparent. Lower jaw protrudes. Area between eyes prominently arched. Vertical fins scaled, rigid. Dorsal fin lobe moderately tall. Caudal-fin lobes similarly shaped. Anal fin usually with 30 rays. HABITAT: Worldwide in tropical to warm temperate seas. Nova Scotia to Belize. Also Bermuda. Not reported from Gulf of Mexico. Oceanic from surface to about 3,300 ft. BIOLOGY: Form small schools. Migrations follow water temperatures. Feed on small fishes and invertebrates.

Caribbean pomfret - *Brama caribbea* Mead, 1972

FEATURES: Dark brown to coppery. Membranes of dorsal and anal fins black. Inside of mouth black. Lower jaw protrudes. Vertical fins scaled, rigid. Upper lobe of caudal fin considerably longer than lower lobe. Anal fin low. Body comparatively deep. HABITAT: NC to FL, northern and southern Gulf of Mexico, Antilles to Brazil. Also Bermuda. Pelagic. From surface to about 1,300 ft. BIOLOGY: Caribbean pomfret likely feed on invertebrates and small fishes.

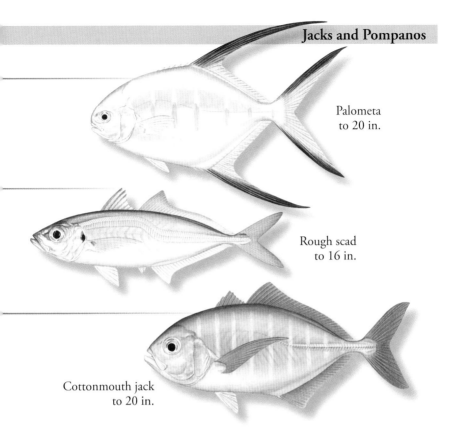

Palometa
to 20 in.

Rough scad
to 16 in.

Cottonmouth jack
to 20 in.

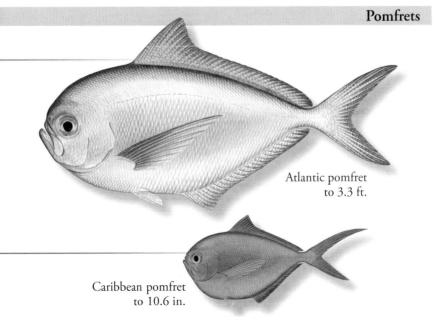

Atlantic pomfret
to 3.3 ft.

Caribbean pomfret
to 10.6 in.

253

Bramidae - Pomfrets, *cont.*

Lowfin pomfret - *Brama dussumieri* Cuvier, 1831

FEATURES: Grayish silvery to dark gray with metallic sheen. Dorsal fin dark. Pectoral and pelvic fins translucent. Vertical fins scaled, relatively rigid. Upper dorsal-fin lobe comparatively low. Upper caudal-fin lobe longer than lower lobe in adults; considerably longer than lower in juveniles. Area between eyes arched. Anal fin usually with 26–28 rays; 27–32 rays reported in Atlantic specimens. HABITAT: Worldwide in tropical seas near edges of continental shelves. In western Atlantic from NC to FL, northern and southwestern Gulf of Mexico and Antilles to Brazil. Pelagic. From surface to about 650 ft. BIOLOGY: Lowfin pomfret are oceanodromous. Reported to spawn year-round. Caught as bycatch. Other name: Lesser bream.

Atlantic fanfish - *Pterycombus brama* Fries, 1837

FEATURES: Dark gray with silvery sheen. Dorsal and anal fins dark. Eyes large. Lower jaw protrudes. Dorsal fin originates over posterior margin of eyes in adults and over pectoral fins in juveniles. Dorsal and anal fins are tall, sail-like, and scaleless; they are able to depress into grooves formed by elongated scales along dorsal and ventral profiles. Dorsal and anal fin lengths vary with age and conditions. Scales on body may or may not be spined. Lateral line absent in adults. HABITAT: Newfoundland to FL, Bahamas, Greater Antilles, northern and southern Gulf of Mexico. Also eastern Atlantic. Offshore, pelagic. Occur from about 82 to 985 ft. BIOLOGY: Atlantic fanfish are reported to spawn year-round over continental shelves of Florida. Seasonally migratory.

Bigscale pomfret - *Taratichthys longipinnis* (Lowe, 1843)

FEATURES: Body and fins blackish with silvery to coppery sheen. Caudal-fin margin, pelvic-fin tips, and lower pectoral-fin margins pale. Lower jaw protrudes. Dorsal- and anal-fin lobes scaled, stiff, long, and pointed. Caudal fin cresent-shaped. Body deep, somewhat compressed. Posterior scales keeled, forming horizontal rows. Lateral line indistinct. HABITAT: Nova Scotia to northern South America. Northern and southern Gulf of Mexico and Puerto Rico. Also Bermuda and eastern Atlantic. Oceanic, pelagic. BIOLOGY: Bigscale pomfret are highly migratory.

Emmelichthyidae - Rovers

Crimson rover - *Erythrocles monodi* Poll & Cadenat, 1954

FEATURES: Dark crimson red dorsally to about midline, pinkish to silvery below. Pectoral and caudal fins red, other fins pinkish. Eyes large. Upper jaw highly protrusible. Mouth toothless or with a few minute teeth. Lower corner of preopercle bluntly rounded. Second dorsal and anal fins with a scaly sheath at base. Caudal fin deeply forked. HABITAT: SC, northern Gulf of Mexico, Bahamas, St. Lucia, Colombia, and Venezuela. Also eastern Atlantic. Found near sand and mud bottoms from about 330 to 985 ft. BIOLOGY: Crimson rover are schooling. Little is known of their life history. Taken as bycatch. Other name: Atlantic rubyfish.

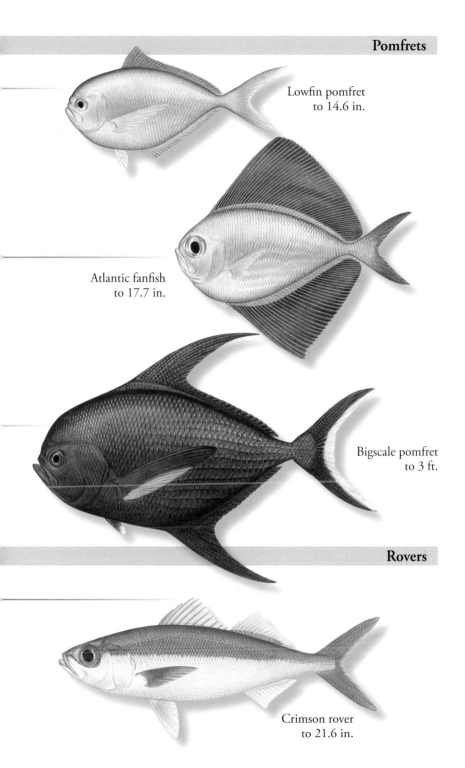

Lowfin pomfret
to 14.6 in.

Atlantic fanfish
to 17.7 in.

Bigscale pomfret
to 3 ft.

Rovers

Crimson rover
to 21.6 in.

Lutjanidae - Snappers

Black snapper - *Apsilus dentatus* Guichenot, 1853

FEATURES: Body and fins brownish violet to brownish black. Outer margin of caudal fin pale. Irises bronze with dark outer ring. Small juveniles deep blue. Last rays of dorsal and anal fins shorter than next-to-last ray. Caudal fin forked to almost straight. HABITAT: FL Keys, Bahamas, Antilles, off Belize, Venezuela, and Galveston, TX. Occur over rocky bottoms and along steep drop-offs from about 40 to 790 ft. BIOLOGY: Black snapper feed on fishes, cephalopods, and tunicates. Juveniles believed to mimic Blue chromis, *Chromis cyanea*.

Queen snapper - *Etelis oculatus* (Valenciennes, 1828)

FEATURES: Dark red to pinkish red above midline. Pinkish to silvery below midline. Spiny dorsal and caudal fins reddish, other fins pinkish. Dorsal fin deeply notched. Last rays of soft dorsal and anal fins long. Lengths of caudal-fin lobes about equal in smaller specimens; upper lobe considerably longer than lower in larger specimens. HABITAT: NC to FL, Gulf of Mexico, Bahamas, and Caribbean Sea to Brazil. Also Bermuda. Over rocky bottoms between about 440 and 1,500 ft. BIOLOGY: Feed on squids, crustaceans, and small fishes. Caught with hook-and-line and trawls.

Mutton snapper - *Lutjanus analis* (Cuvier, 1828)

FEATURES: Olive with reddish tinge dorsally. Sides reddish, ventral area whitish. Sides may be uniformly colored or with pale bars. Blue lines below and behind eyes. Small, black spot present on upper sides. Caudal fin with thin, black margin. Anal fin angular. HABITAT: MA to FL, Gulf of Mexico, Bahamas, Caribbean Sea to Brazil. Found over sandy bottoms, in bays and estuaries, around mangroves, and over coral reefs from about 82 to 312 ft. BIOLOGY: Mutton snapper are solitary. Form large spawning aggregations. Feed on invertebrates and fishes. Currently listed as Vulnerable.

Schoolmaster - *Lutjanus apodus* (Walbaum, 1792)

FEATURES: Reddish brown to olive gray dorsally. Sides and ventral area with reddish tinge. Eight pale bars on sides that fade with age. Fins yellow, or yellow green to pale orange. Usually with solid or broken blue line under eyes. Large canine teeth in upper jaw. Anal fin rounded. HABITAT: MA (rare) to FL, Gulf of Mexico, Bahamas, and Caribbean Sea to Brazil. Also Bermuda. Found in shallow coastal waters over a variety of bottoms. Juveniles enter brackish waters. BIOLOGY: Schoolmaster form aggregations during the day. Feed nocturnally on invertebrates and fishes.

Blackfin snapper - *Lutjanus buccanella* (Cuvier, 1828)

FEATURES: Deep red dorsally. Pale reddish to silvery ventrally. Irises yellow to orange. Fins orangish to yellowish. Always with blackish blotch at pectoral-fin base. Base of soft dorsal fin dark. Smaller specimens with yellowish upper caudal peduncle and caudal fin. Anal fin rounded. HABITAT: NC to FL, Gulf of Mexico, Bahamas, and Caribbean Sea to Brazil. Also Bermuda. Adults over sand and rocky bottoms, drop-offs, and ledges from about 263 to 755 ft. Juveniles from about 115 to 164 ft.

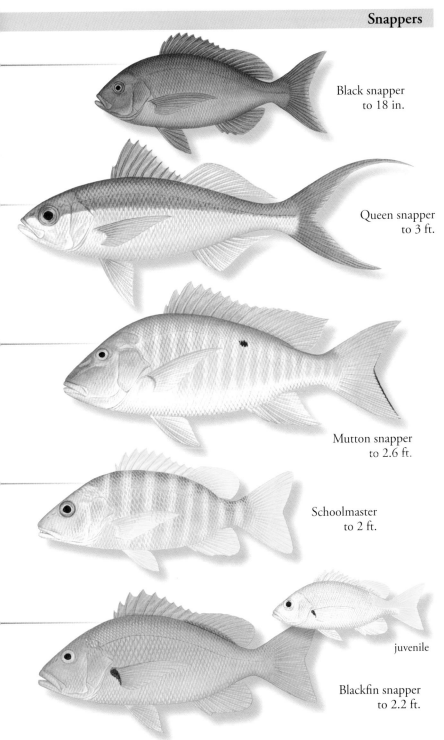

Black snapper
to 18 in.

Queen snapper
to 3 ft.

Mutton snapper
to 2.6 ft.

Schoolmaster
to 2 ft.

juvenile

Blackfin snapper
to 2.2 ft.

257

Red snapper - *Lutjanus campechanus* (Poey, 1860)

FEATURES: Scarlet red to brick red dorsally and on sides. Pinkish below. Irises red. Caudal fin usually with narrow black margin. Small specimens have black blotch on posterior upper sides that fades with age. Anal fin angular with elongate inner rays. HABITAT: MA (rare) to FL and Gulf of Mexico. Adults over rocky bottoms and reefs from about 33 to 624 ft. Juveniles in shallow waters over sandy and muddy bottoms. BIOLOGY: Red snapper feed on invertebrates and fishes. Heavily fished in the US. Protected. SIMILAR SPECIES: Caribbean red snapper, *Lutjanus purpureus*, p. 431.

Cubera snapper - *Lutjanus cyanopterus* (Cuvier, 1828)

FEATURES: Dark to pale reddish gray with silvery reflections dorsally, paler below. Pale bars on upper body fade with age. Pectoral fins grayish to translucent. Body comparatively shallow. Teeth in upper and lower jaws equally developed. Anal fin rounded. HABITAT: Nova Scotia to FL, Gulf of Mexico, Bahamas, and Caribbean Sea to Brazil. Also Bermuda. Adults over rocky bottoms, ledges, and reefs to about 130 ft. Juveniles around mangroves. BIOLOGY: Largest snapper in the area. Form spawning aggregations. Feed on crustaceans and fishes. Currently listed as Vulnerable.

Gray snapper - *Lutjanus griseus* (Linnaeus, 1758)

FEATURES: Dark olive, grayish green, to grayish dorsally. Sides and ventral area paler, with a reddish cast. Centers of scales on sides reddish. Juveniles with a blue line below eyes, a dark bar through eyes, and pale bars on sides that fade with age. HABITAT: MA to FL, Gulf of Mexico, Bahamas, and Caribbean Sea to Brazil. Also Bermuda. From shore to about 590 ft. Around mangroves, rocky areas, coral reefs, estuaries, tidal creeks, and river mouths. Young may enter fresh water. BIOLOGY: Gray snapper feed primarily at night on a variety of invertebrates and fishes. Spawning takes place during full-moon phases. Other name: Mangrove snapper.

Dog snapper - *Lutjanus jocu* (Bloch & Schneider, 1801)

FEATURES: Olive brown with bronze cast dorsally. Reddish with coppery cast below. Upper sides may have pale bars. Usually with a whitish bar below eyes to corner of jaws. May have a series of blue spots or a blue line below eyes to opercular margin. Anal fin rounded. HABITAT: MA (rare) to FL, Gulf of Mexico, Bahamas, and Caribbean Sea to Brazil. Adults over coral reefs. Juveniles in coastal waters, estuaries, and occasionally fresh water. BIOLOGY: Dog snapper are solitary and territorial.

Mahogany snapper - *Lutjanus mahogoni* (Cuvier, 1828)

FEATURES: Olive or grayish dorsally. Sides silvery with reddish cast. Abdomen silvery. Dorsal and caudal fins reddish to yellowish with reddish to dusky margins. Dark blotch may be present on lateral line below soft dorsal fin. HABITAT: NC to FL, Gulf of Mexico, Bahamas, Caribbean Sea to Venezuela. In clear, shallow waters over rocky, sandy, and grassy bottoms. Also near coral reefs. BIOLOGY: Mahogany snapper are schooling. Feed at night on cephalopods, crustaceans, and fishes.

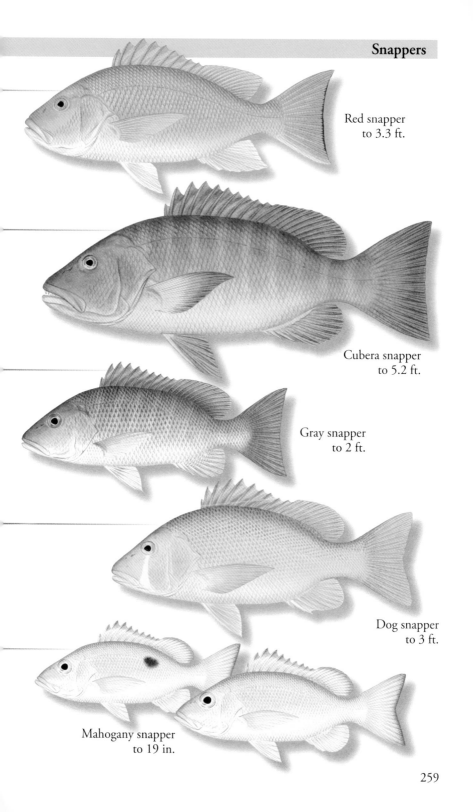

Red snapper
to 3.3 ft.

Cubera snapper
to 5.2 ft.

Gray snapper
to 2 ft.

Dog snapper
to 3 ft.

Mahogany snapper
to 19 in.

Lane snapper - *Lutjanus synagris* (Linnaeus, 1758)

FEATURES: Silvery pink to reddish dorsally. Silvery below. Narrow, yellow stripes on head and sides. May display pale bars on sides. Dark spot below anterior portion of soft dorsal fin; spot may be faint or absent. Dorsal fin with yellow margin. Caudal fin rosy with narrow dark margin. Anal fin rounded. HABITAT: NC to FL, Gulf of Mexico, Bahamas, and Caribbean Sea to Brazil. Also Bermuda. Occur in shallow coastal waters over a variety of bottoms. Primarily around coral reefs and vegetated sandy bottoms. BIOLOGY: Lane snapper feed on fishes and a variety of invertebrates.

Silk snapper - *Lutjanus vivanus* (Cuvier, 1828)

FEATURES: Rosy to pinkish dorsally. Pinkish to silvery below. Irises bright yellow. Juveniles with dark spot below soft dorsal fin. Indistinct, narrow, yellowish stripes on sides. Fins rosy to yellowish. Caudal fin with narrow, dark reddish margin. Anal fin angular. HABITAT: NC to FL, Gulf of Mexico, Bahamas, Caribbean Sea to Brazil. Also Bermuda. Over sandy, rocky, and coral bottoms from about 295 to 650 ft.

Yellowtail snapper - *Ocyurus chrysurus* (Bloch, 1791)

FEATURES: Grayish, bluish, or reddish dorsally with yellowish spots. Bright to dusky yellow stripe from snout to caudal fin. Stripe becomes broader posteriorly, merges with yellow caudal fin. Ventral area silvery white to pinkish. Dorsal fin yellowish. Anal fin broadly rounded. HABITAT: MA to FL, Gulf of Mexico, Bahamas, and Caribbean Sea to Brazil. Also Bermuda. Coastal and around reefs, in water column from shore to about 540 ft. BIOLOGY: Feed on fishes and invertebrates, primarily at night.

Wenchman - *Pristipomoides aquilonaris* (Goode & Bean, 1896)

FEATURES: Reddish to pinkish dorsally. Pinkish to silvery below. Dorsal and caudal fins pinkish to translucent with yellow outer portions. Eyes large. Body comparatively deep. HABITAT: NC to FL, Gulf of Mexico, Caribbean Sea to Brazil. Over hard bottoms, including natural and artificial reefs. BIOLOGY: Feed on fishes.

Slender wenchman - *Pristipomoides freemani* Anderson, 1966

FEATURES: Orange to dark red dorsally. Pinkish to silvery below. Dorsal fin pinkish to translucent with yellow outer portion. Upper caudal-fin lobe yellow, lower lobe pinkish to translucent. Eyes large. Body comparatively shallow. HABITAT: Scattered records from NC to FL, Caribbean Sea to Brazil. Occur from midwater to bottom at depths of about 285 to 720 ft.

Vermilion snapper - *Rhomboplites aurorubens* (Cuvier, 1829)

FEATURES: Deep red above lateral line, pale pinkish to silver below. Fine, bluish lines follow scales above lateral line. Sides with fine, oblique yellowish lines. Irises silvery red. Dorsal fin reddish with orange margin. Caudal fin red. Anal fin broadly rounded. HABITAT: NC to FL, Gulf of Mexico, Bahamas, Caribbean Sea to Brazil. Over rocky bottoms of edges of continental shelves from about 80 to 1,300 ft.

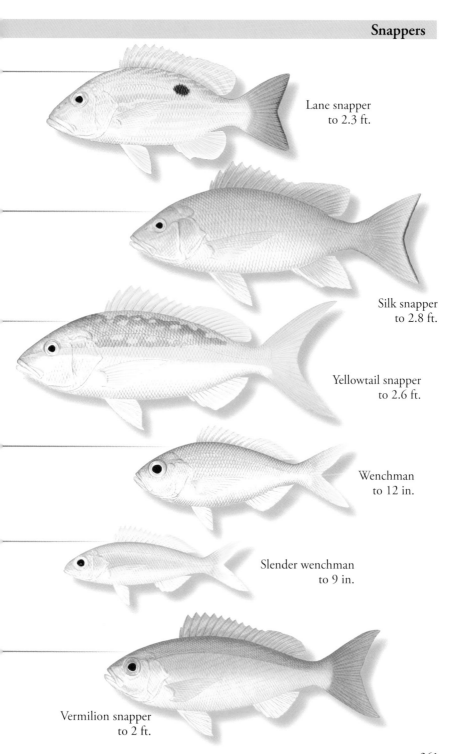

Lane snapper
to 2.3 ft.

Silk snapper
to 2.8 ft.

Yellowtail snapper
to 2.6 ft.

Wenchman
to 12 in.

Slender wenchman
to 9 in.

Vermilion snapper
to 2 ft.

Lobotidae - Tripletails

Atlantic tripletail - *Lobotes surinamensis* (Bloch, 1790)

FEATURES: Variably mottled, flecked, and blotched in shades of brown to olive. Juveniles often yellowish. Head profile steep, concave above eyes. Snout short, jaws large. Spiny dorsal fin low. Soft dorsal, caudal, and anal fins broad, rounded, and overlapping. HABITAT: Worldwide in tropical to warm temperate seas. Nova Scotia to FL, Gulf of Mexico, Bahamas, and Caribbean Sea to Argentina. Offshore, near surface. Often associated with flotsam and *Sargassum* seaweed. BIOLOGY: Sluggish, often floating on sides at the surface. Juveniles may mimic fallen mangrove leaves.

Gerreidae - Mojarras

Irish pompano - *Diapterus auratus* Ranzani, 1842

FEATURES: Silvery, somewhat darker dorsally. Smaller specimens with three dark bars on sides. Spiny dorsal fin with dusky margin. Anal and pelvic fins yellowish. Lower preopercular margin serrated. Anal fin lobed, with strong second spine. Body rhomboid in shape, deep, and compressed. HABITAT: E FL, W Gulf of Mexico, Greater Antilles, northern South America to Brazil. Occur in shallow coastal waters, including estuaries, seagrass beds, mangroves, lagoons, and bays. Commonly enter fresh water. SIMILAR SPECIES: Silver mojarra, *Diapterus rhombeus*, p. 431.

Spotfin mojarra - *Eucinostomus argenteus* Baird & Girard, 1855

FEATURES: Silvery. Smaller specimens with dark, oblique bars and spots that fade with age. Spiny dorsal fin with dusky margin. Tips of caudal fin may be dusky. Anal fin low, fairly straight-edged. Scaleless area between eyes surrounded by scales. Body moderately slender. HABITAT: NJ (rare), NC to FL, Gulf of Mexico, Bahamas, Antilles, northern South America to Brazil. Also Bermuda. Over sandy and shelly bottoms of continental shelves. Occasionally in inlets and estuaries. BIOLOGY: Spotfin mojarra feed on benthic invertebrates.

Silver jenny - *Eucinostomus gula* (Quoy & Gaimard, 1824)

FEATURES: Silvery, somewhat darker dorsally. Smaller specimens with about seven oblique bars and blotches on upper sides that fade with age. Spiny dorsal fin with dusky margin. Scaleless area between eyes surrounded by scales. Body moderately deep. HABITAT: MA (rare), NC to FL, Gulf of Mexico, Bahamas, and Caribbean Sea to Argentina. Also Bermuda. Primarily over shallow seagrass beds, but also over open sandy bottoms. Rarely in fresh water. BIOLOGY: Feed on benthic invertebrates.

Tidewater mojarra - *Eucinostomus harengulus* Goode & Bean, 1879

FEATURES: Silvery with dark bars and spots on upper sides, markings more evident in smaller specimens. Second and third bars form Y-shape. Spiny dorsal fin with dusky margin. Scaleless area between eyes open anteriorly. Body moderately slender. HABITAT: VA to FL, Gulf of Mexico, Bahamas, Antilles, and northern South America to Brazil. Also Bermuda. In protected estuaries, over seagrass beds and sandy and muddy bottoms, and around mangroves. Commonly enter freshwater tributaries.

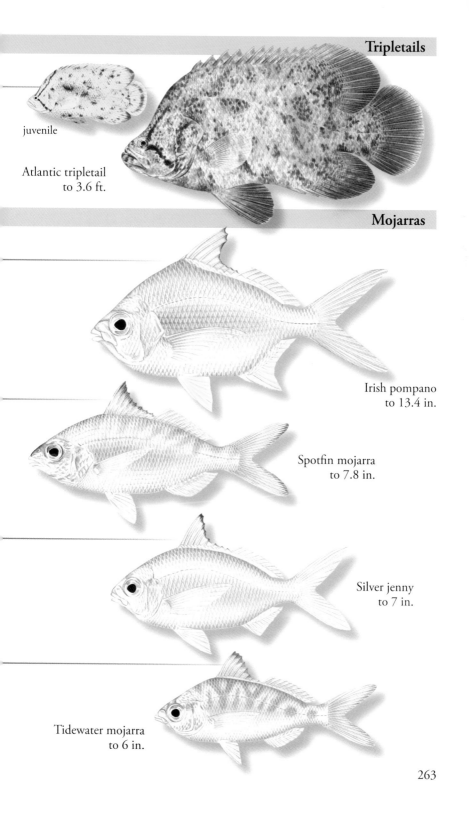

Tripletails

juvenile

Atlantic tripletail
to 3.6 ft.

Mojarras

Irish pompano
to 13.4 in.

Spotfin mojarra
to 7.8 in.

Silver jenny
to 7 in.

Tidewater mojarra
to 6 in.

Bigeye mojarra - *Eucinostomus havana* (Nichols, 1912)

FEATURES: Silvery with darker, oblique bars on upper sides. Black blotch on upper portion of spiny dorsal fin. Eyes very large. Body moderately slender. HABITAT: E FL, Bahamas, Antilles, and northern coast of South America to Venezuela. Coastal. Over shallow sandy, muddy, and vegetated bottoms. Also around mangroves and sandy beaches. Do not enter estuaries. BIOLOGY: Feed on benthic invertebrates.

Slender mojarra - *Eucinostomus jonesii* (Günther, 1879)

FEATURES: Silvery greenish dorsally, silvery below. Smaller specimens with darker, oblique bars that fade with age. Distinct V-shaped mark on snout. Spiny dorsal fin with dusky margin. Scaleless area between eyes open anteriorly. Body slender. HABITAT: E FL, Bahamas, and Antilles. Also Bermuda, S TX, Tamaulipas, Veracruz, and Campeche. Strays as far as VA. Over high-energy sandy and grassy bottoms, in inlets, and along beaches. Do not enter estuaries.

Mottled mojarra - *Eucinostomus lefroyi* (Goode, 1874)

FEATURES: Silvery with about seven dusky, oblique, wavy bars on upper sides. Spiny dorsal fin may be colorless or may have dusky margin. Scaleless area between eyes open anteriorly. Anal fin with two spines. HABITAT: NC (rare), E FL, Bahamas, all Antilles, and southern Gulf of Mexico. Also Bermuda. Along high-energy sandy beaches and inlets. Do not enter estuaries.

Flagfin mojarra - *Eucinostomus melanopterus* (Bleeker, 1863)

FEATURES: Silvery, somewhat darker dorsally. Body unmarked. Dorsal fin with a black outer margin and a white inner band. Scaleless area between eyes open anteriorly. HABITAT: NJ (rare), E FL, Greater and Lesser Antilles, Western Gulf of Mexico to Brazil. Also eastern Atlantic. In shallow coastal waters, inlets, and lagoons over sandy, muddy, and shelly bottoms. Occasionally in fresh water.

Striped mojarra - *Eugerres plumieri* (Cuvier, 1830)

FEATURES: Silvery with brownish, greenish, or bluish tinge dorsally. Silvery below. Dark stripes follow scales along sides. Spiny dorsal fin tall. Second spine of anal fin strong, elongate. Body rhomboid-shaped, moderately deep. HABITAT: SC to W FL, Greater Antilles, western Gulf of Mexico to Brazil. Occur in shallow coastal waters, mangrove-lined creeks, and lagoons. Enter fresh water.

Yellowfin mojarra - *Gerres cinereus* (Walbaum, 1792)

FEATURES: Silvery to tannish with darker bars and spots on upper body. Pelvic fin yellow. Anal fin may have yellowish tint. Dorsal fin unmarked. Dorsal profile over eyes slightly concave. Body moderately deep. HABITAT: FL, Bermuda, Bahamas, Antilles, and western Gulf of Mexico to Brazil. Also in eastern Pacific from Baja CA to Peru. In shallow coastal waters along sandy beaches, around mangroves, over seagrass beds and coral reefs, and in bays. Enter brackish and fresh water.

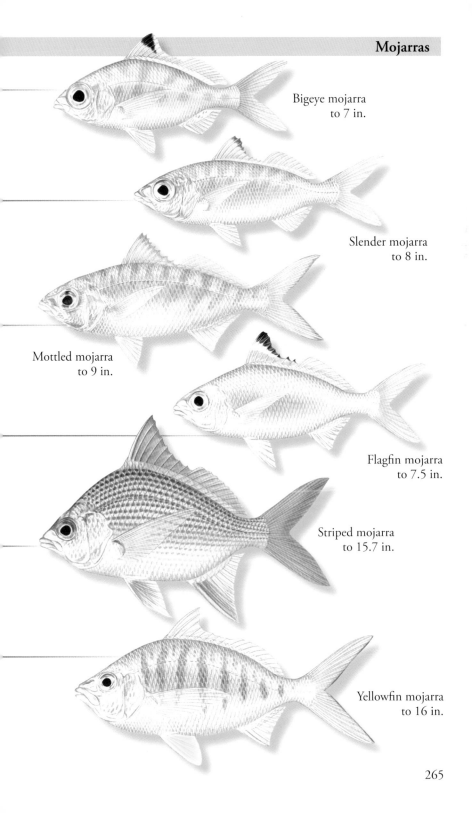

Bigeye mojarra
to 7 in.

Slender mojarra
to 8 in.

Mottled mojarra
to 9 in.

Flagfin mojarra
to 7.5 in.

Striped mojarra
to 15.7 in.

Yellowfin mojarra
to 16 in.

Haemulidae - Grunts

Black margate - *Anisotremus surinamensis* (Bloch, 1791)

FEATURES: Silvery. Each scale on back and sides with black, central spot. Broad, black bar on lower midbody. All fins dark gray to blackish. Body deep, compressed. Head profile sloping. HABITAT: S FL, E Texas, southern Gulf of Mexico, Bahamas, and Caribbean Sea to Brazil. Over shallow coral reefs and rocky areas from shore to about 130 ft. BIOLOGY: Black margate feed on echinoderms, crustaceans, and small fishes. Caught in traps and by hook-and-line. Other name: Thicklip.

Porkfish - *Anisotremus virginicus* (Linnaeus, 1758)

FEATURES: Body with alternating silvery white and yellow stripes. Two black bands on head: the first extends from mouth through eye to nape; the second from rear gill margin to anterior base of dorsal fin. All fins yellow. Body deep, compressed. Head profile steeply sloping. HABITAT: FL, Gulf of Mexico, Caribbean Sea to Brazil. Rare in northern Gulf of Mexico and Bahamas. Over coral reefs and hard bottoms from shore to about to 165 ft. BIOLOGY: Porkfish feed mainly at night on a variety of invertebrates. Juveniles feed on ectoparasites of other fishes.

Barred grunt - *Conodon nobilis* (Linnaeus, 1758)

FEATURES: Brownish dorsally, becoming pale below. Upper body with brown bars and tan stripes. Bars and stripes form checkerboard pattern. All fins with some yellow. Preopercle serrated, with two enlarged spines at lower corner. HABITAT: E FL, Gulf of Mexico, to Brazil. Also Jamaica, Puerto Rico, and Lesser Antilles. In shallow, turbid waters over sandy shores and soft bottoms to about 330 ft. Also in bays and estuaries. BIOLOGY: Barred grunt feed primarily at night on crustaceans and small fishes.

Margate - *Haemulon album* Cuvier, 1830

FEATURES: Pearly gray dorsally, silvery below. Scales on upper body with dark margins. Soft dorsal and caudal fins dark gray. Inside of mouth pale red. Body oblong, compressed. Profile of head almost straight. HABITAT: S FL, Bahamas, and Caribbean Sea to Brazil. Also Bermuda. Reported from northeastern Gulf of Mexico. Occur over coral reefs and hard bottoms from about 65 to 200 ft. BIOLOGY: Feed on a variety of invertebrates and small fishes. Other name: White margate.

Tomtate - *Haemulon aurolineatum* Cuvier, 1830

FEATURES: Silvery white to tannish with two dark yellowish stripes on body. First stripe runs from nape to rear base of soft dorsal fin; second stripe runs from snout to caudal peduncle. May also have narrow, faint yellow stripes on body. Black blotch often present at base of caudal fin. Body oblong, compressed. Head profile slightly convex. HABITAT: Chesapeake Bay to FL, Gulf of Mexico, Bahamas, and Caribbean Sea to Brazil. Also Bermuda. Found over a variety of natural and artificial bottoms from shore to about 130 ft. BIOLOGY: Tomtate may form large schools. Feed on a variety of invertebrates, small fishes, and algae. Other name: Tomtate grunt.

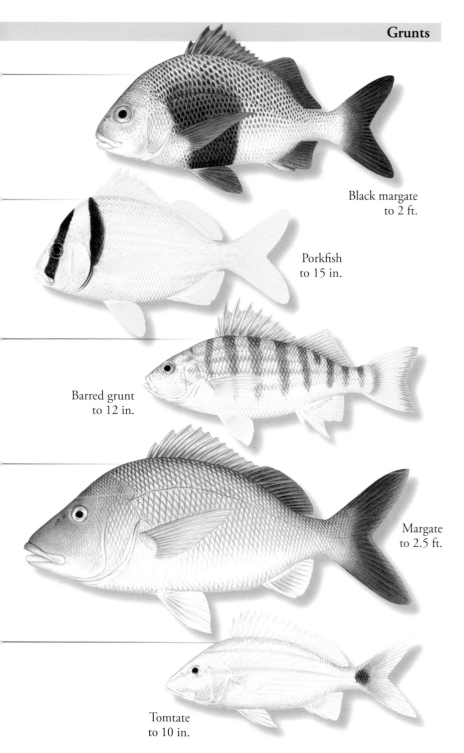

Black margate
to 2 ft.

Porkfish
to 15 in.

Barred grunt
to 12 in.

Margate
to 2.5 ft.

Tomtate
to 10 in.

Haemulidae - Grunts, *cont.*

Caesar grunt - *Haemulon carbonarium* Poey, 1860

FEATURES: Silvery gray with bronze stripes on head and body. Abdomen may be dusky gray to black. Spiny dorsal fin with bronze membranes. Soft dorsal, caudal, and anal fins dark gray to blackish. Inside of mouth red. Body oblong, compressed. Head profile evenly convex. HABITAT: S FL, southern Gulf of Mexico, Bahamas, and Caribbean Sea to Brazil. Also Bermuda. Found over coral reefs and hard bottoms from shore to about 100 ft. BIOLOGY: Caesar grunt feed at night on a variety of invertebrates. Taken by hook-and-line, traps, and seines.

Smallmouth grunt - *Haemulon chrysargyreum* Günther, 1859

FEATURES: Silvery white with bright yellow stripes on head and sides. All fins except pectoral fins yellow. Scales below lateral line form uniformly horizontal rows. Body oblong, somewhat elongate, and compressed. Head profile evenly convex. HABITAT: S FL, southern Gulf of Mexico, Bahamas, and Caribbean Sea to Brazil. Adults over coral reefs. Juveniles over coral reefs, hard bottoms, and seagrass beds to about 130 ft. BIOLOGY: Smallmouth grunt feed on small crustaceans and plankton.

French grunt - *Haemulon flavolineatum* (Desmarest, 1823)

FEATURES: Head and body covered in alternating silvery white and yellowish to bronze stripes. Stripes follow scales on sides. Fins yellowish to bronze. Inside of mouth is reddish. Eyes are large. Profile of head almost straight. Body oblong, compressed. HABITAT: SC to FL, southern Gulf of Mexico, Bahamas, and Caribbean Sea to Brazil. Also Bermuda. Occur around a variety of soft and hard bottoms from shore to about 130 ft. BIOLOGY: Schooling. Feed at night on crustaceans.

Spanish grunt - *Haemulon macrostomum* Günther, 1859

FEATURES: Body silvery gray with dark grayish stripes on sides. Upper portion of back and caudal peduncle yellowish. Ventral area may be grayish to blackish. Outer margins of dorsal, caudal and anal fins yellowish. Pectoral fin yellowish. Body oblong, compressed. Profile from nape to snout almost straight. HABITAT: FL, Bahamas, and Antilles and northern South America to Brazil. Also reported from TX. Occur in clear water over coral reefs and hard bottoms to about 130 ft. BIOLOGY: Spanish grunt feed on crustaceans and echinoderms. Rarely in schools.

Cottonwick - *Haemulon melanurum* (Linnaeus, 1758)

FEATURES: Body pearly white to silvery white with bronze to yellow stripes on head and sides. Black band runs from nape along dorsal profile to caudal fin, where it forms a V-shape along the inner rays. Lower portion of dorsal fin black, outer portion pearly to colorless. Inside of mouth pale red. Body oblong, compressed. Head profile evenly convex. HABITAT: NC to FL, northern Gulf of Mexico, Bahamas, and Caribbean Sea to Brazil. Also Bermuda. Found in clear water over coral reefs, hard bottoms, and adjacent seagrass beds from shore to about 130 ft. BIOLOGY: Cottonwick feed on crustaceans and echinoderms. Often in schools.

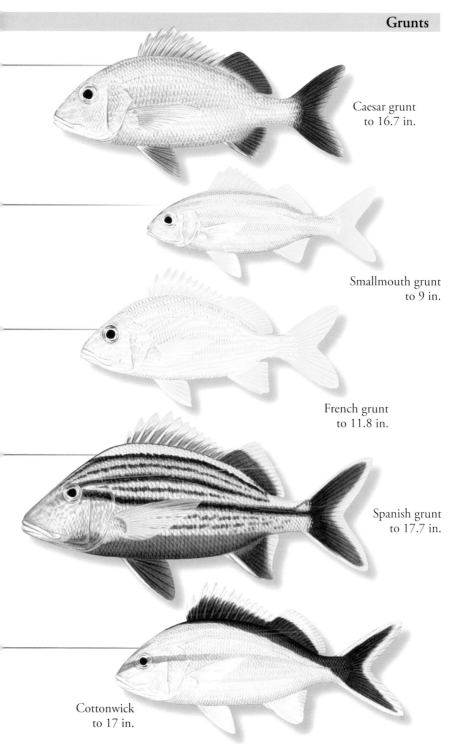

Caesar grunt
to 16.7 in.

Smallmouth grunt
to 9 in.

French grunt
to 11.8 in.

Spanish grunt
to 17.7 in.

Cottonwick
to 17 in.

Sailor's choice - *Haemulon parra* (Desmarest, 1823)

FEATURES: Body pearly gray with large blackish spots on scales that form oblique lines. Spiny dorsal, anal, pelvic, and pectoral fins grayish. Soft dorsal and caudal fins dark gray to blackish. Outer margins of irises yellowish. Head profile slightly convex to nearly straight. HABITAT: S FL, Bahamas, and Caribbean Sea to Brazil. Occur from shore to about 130 ft. Juveniles over seagrass beds, adults over open areas of reefs and other structures. BIOLOGY: Sailor's choice are schooling. Feed at night on invertebrates and small fishes.

White grunt - *Haemulon plumierii* (Lacepède, 1801)

FEATURES: Head and area above pectoral fins with narrow, alternating, blue and yellow wavy lines. Scales on upper body with silvery blue centers and yellow to bronze margins. Abdomen silvery to white. Spiny dorsal fin chalky to yellow white. Soft dorsal, caudal, and anal fins pale bronze. Profile of head nearly straight. HABITAT: Chesapeake Bay to FL, Gulf of Mexico, Bahamas, and Caribbean Sea to Brazil. Occur over a variety of bottoms and reefs from shore to about 130 ft. BIOLOGY: White grunt school during the day. Feed at night on invertebrates and small fishes.

Bluestriped grunt - *Haemulon sciurus* (Shaw, 1803)

FEATURES: Head and body yellow to bronze with blue wavy lines and spots. Blue lines and spots bordered above and below by darker blue. Inner portions of soft dorsal and caudal fins blackish to brownish black. Anal, pelvic, and pectoral fins yellow to pale. Inside of mouth red. Profile of head slightly convex to almost straight. HABITAT: SC to FL, southern Gulf of Mexico, Bahamas, Caribbean Sea to Brazil. Also Bermuda. Found over a variety of bottoms, structures, and reefs from shore to about 130 ft. BIOLOGY: Bluestriped grunt feed on crustaceans and small fishes.

Striped grunt - *Haemulon striatum* (Linnaeus, 1758)

FEATURES: Grayish white dorsally, shading to silvery below, with five distinct brownish to yellowish stripes on head and upper sides. Snout yellowish. Head profile slightly convex. Body oblong, elongate. HABITAT: S FL, southern Gulf of Mexico, Bahamas, Caribbean Sea to Brazil. Occur over outer reefs from about 40 to 330 ft. BIOLOGY: Striped grunt feed on plankton and small crustaceans.

Pigfish - *Orthopristis chrysoptera* (Linnaeus, 1766)

FEATURES: Greenish gray dorsally with iridescent highlights. Silvery on sides, pearly below. Dark orange spots and irregular lines on head and body. Upper body may have pale to dark grayish bars. Fins dusky to yellowish; may have dusky margins. Upper caudal-fin lobe slightly longer than lower lobe. HABITAT: NY to FL, Gulf of Mexico, Cuba, and Bermuda. Near shore in shallow waters over soft bottoms. Often in bays, estuaries. BIOLOGY: Feed on crustaceans and fishes. Grunt when distressed.

ALSO IN THE AREA: Burro grunt, *Pomadasys crocro*, p. 431.

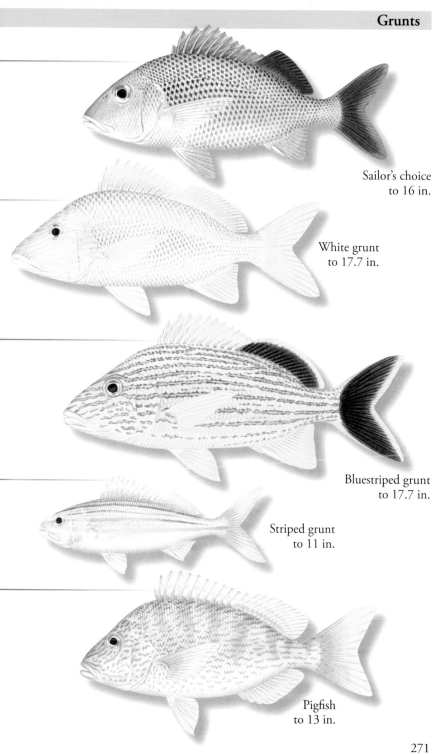

Sailor's choice
to 16 in.

White grunt
to 17.7 in.

Bluestriped grunt
to 17.7 in.

Striped grunt
to 11 in.

Pigfish
to 13 in.

Inermiidae - Bonnetmouths

Bonnetmouth - *Emmelichthyops atlanticus* Schultz, 1945

FEATURES: Metallic greenish to yellowish gray dorsally. Silvery on sides, white along abdomen. Three brownish stripes on upper sides. Mouth highly protrusible. First dorsal fin widely separated from second dorsal fin. HABITAT: FL Keys, Bahamas, Cuba, Virgin Islands, Barbados, and Tobago to northern South America. Found around coral and patch reefs and over sandy bottoms to about 300 ft. BIOLOGY: Bonnetmouth are swift and schooling. Feed on zooplankton and small fishes.

Boga - *Inermia vittata* Poey, 1860

FEATURES: Metallic bluish to greenish dorsally. Silvery on sides, white along abdomen. A broad yellowish green stripe runs from snout to caudal-fin base. Several broken greenish brown stripes on upper sides. Caudal fin with purplish cast. Mouth highly protrusible. Spiny and soft dorsal fins continuous, deeply notched. HABITAT: S FL, northern Gulf of Mexico, Bahamas, and Lesser Antilles to northern South America. Also Bermuda. Occur in open water and along coast to about 300 ft.

Sparidae - Porgies

Sheepshead - *Archosargus probatocephalus* (Walbaum, 1792)

FEATURES: Grayish to brownish dorsally, fading to silvery below. May have golden cast. Five to six blackish to dark brown bars on body. Head and fins grayish to dusky. Anterior teeth are incisor-like. Upper and lower lips grooved. Pectoral fins long. Caudal fin forked. HABITAT: Nova Scotia to FL, Gulf of Mexico to Brazil. Occur along coast, in estuaries around muddy and rocky bottoms, and over hard substrates to about 40 ft. Also around pilings. BIOLOGY: *Archosargus probatocephalus* is subdivided into three subspecies by some authors. Feed on mollusks and crustaceans.

Sea bream - *Archosargus rhomboidalis* (Linnaeus, 1758)

FEATURES: Silvery gray to olivaceous with golden yellow stripes on head and body. Blackish spot about as large as eyes on lateral line near origin. Dorsal-fin margins blackish. Anal and pelvic fins yellowish. Anterior teeth incisor-like. Upper and lower lips grooved. HABITAT: NJ to FL, Gulf of Mexico, Greater and Lesser Antilles to Brazil. Over muddy and vegetated bottoms and around mangroves. BIOLOGY: Feed on mollusks and crustaceans. Other name: Western Atlantic seabream.

Grass porgy - *Calamus arctifrons* Goode & Bean, 1882

FEATURES: Variably pale olive brown with about seven darker bars on body that extend into fins. Silvery-centered scales pepper body. Pale blue streaks below and around eyes. Dark spot on anterior portion of lateral line often blends with dark bar. Caudal fin often with dark V-shaped mark at base. May pale or darken. Head profile fairly evenly convex. HABITAT: S FL and in Gulf of Mexico from FL Keys to LA. Over seagrass beds from near shore to about 72 ft. BIOLOGY: Grass porgy feed primarily on invertebrates.

272

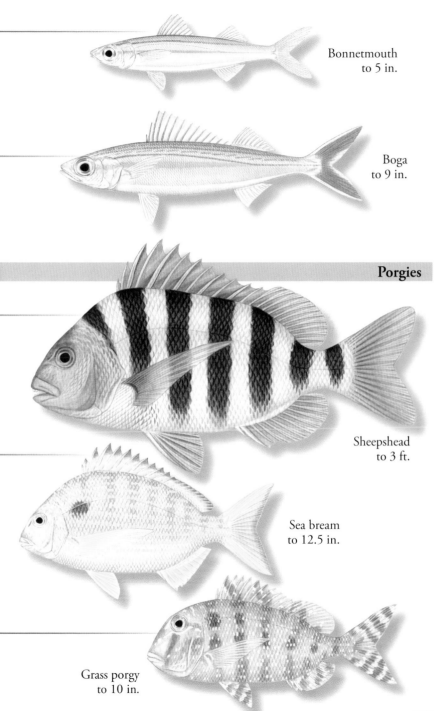

Bonnetmouth
to 5 in.

Boga
to 9 in.

Sheepshead
to 3 ft.

Sea bream
to 12.5 in.

Grass porgy
to 10 in.

273

Sparidae - Porgies, *cont.*

Jolthead porgy - *Calamus bajonado* (Bloch & Schneider, 1801)

FEATURES: Silvery to pale brownish. Scale centers silvery to pale with brassy edges. Scaleless portion of head brownish. Snout with horizontal silvery stripe. Corners of mouth always dark orange. Silvery blue lines above and below eyes. Juveniles with dark bars on body and caudal fin. Profile of snout forms between 43° to 55° angle with midline. HABITAT: RI (rare) to FL, northern and southern Gulf of Mexico, Bahamas, and Antilles to Brazil. Also Bermuda. Occur in coastal waters to about 150 ft. BIOLOGY: Jolthead porgy feed on sea urchins, mollusks, and crabs.

Saucereye porgy - *Calamus calamus* (Valenciennes, 1830)

FEATURES: Silvery with brassy to golden highlights. May display dark blotches. Scaleless portion of head silvery with golden to brassy areas. Corners of mouth silvery to yellowish. Blue line along lower margin of eyes. Small blue spot at upper pectoral-fin base. Profile of snout forms between 60° to 65° angle with midline. HABITAT: NC, FL, Gulf of Mexico, Bahamas, and Antilles to Brazil. Also Bermuda. Reported to be rare to absent from W FL and in western Gulf of Mexico. Found along coastal waters to about 250 ft. Adults occur around color reefs, juveniles over seagrass beds. BIOLOGY: Saucereye porgy may undergo rapid color and pattern changes.

Whitebone porgy - *Calamus leucosteus* Jordan & Gilbert, 1885

FEATURES: Bluish silver on head and body. May show dark blotches or about five dark bars on sides. Snout purplish gray. Narrow blue lines above and below eyes. Fins dusky. Spot at upper pectoral-fin base always absent. Pectoral fins comparatively long, reaching beyond anal-fin origin. HABITAT: NC to FL Keys, Gulf of Mexico to Yucatán. Occur over soft bottoms from about 33 to 330 ft. BIOLOGY: Whitebone porgy are sought for sport and as food.

Knobbed porgy - *Calamus nodosus* Randall & Caldwell, 1966

FEATURES: Rosy silver. Scale centers iridescent bluish. Snout purplish with yellowish to bronze spots. Narrow, blue line under eyes. A diffuse, blue spot often present on upper pectoral-fin base. Snout steeply sloping, forms between 57° to 65° angle with midline. Bony knob present above nostrils; well developed in large adults. Nape humped in large adults. HABITAT: NC to FL Keys. In Gulf of Mexico from S FL to Pensacola, FL, and from Port Aransas, TX, to Campeche Bank. Over hard bottoms from about 30 to 290 ft.

Sheepshead porgy - *Calamus penna* (Valenciennes, 1830)

FEATURES: Silvery with iridescent reflections. May display about seven dark bars on sides. Cheeks silvery with a wash of yellowish brown. A brownish bar runs from below eyes to corner of mouth. Sometimes with a faint blue line below eye. Always with a small black spot on upper pectoral-fin base. Upper lip evenly divided by a lengthwise groove. Head profile evenly convex. Pectoral fins reach to above anal fin origin. HABITAT: S and W FL, Bahamas, Antilles to northern South America. Found over hard to semihard bottoms from about 10 to 285 ft.

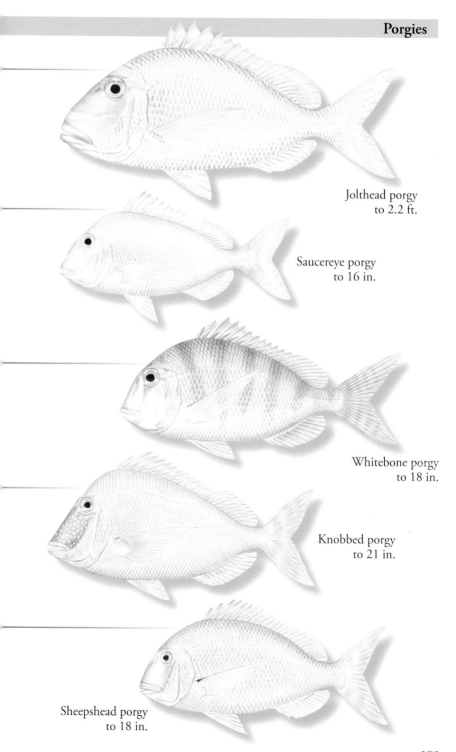

Jolthead porgy
to 2.2 ft.

Saucereye porgy
to 16 in.

Whitebone porgy
to 18 in.

Knobbed porgy
to 21 in.

Sheepshead porgy
to 18 in.

Littlehead porgy - *Calamus proridens* Jordan & Gilbert, 1884

FEATURES: Iridescent silvery with yellowish highlights. Scale centers blue. Snout yellowish with narrow, blue wavy lines. Blue line below and behind each eye. Small blue patch at upper opercular margin. Corners of mouth pale yellow. Head profile steeply sloping, especially in adults. Profile of snout forms between 57° to 64° angle with midline. HABITAT: FL, western and southern Gulf of Mexico, Cuba, Haiti, and Dominican Republic. Found near bottom in shallow coastal waters to about 195 ft. BIOLOGY: Littlehead porgy feed primarily on invertebrates.

Silver porgy - *Diplodus argenteus* (Valenciennes, 1830)

FEATURES: Silvery with pale yellow reflections. Opercular membrane blackish. About nine faint, narrow bars on upper body. Bars fade with age. Black blotch on upper caudal peduncle. Fins with yellowish cast. Teeth in front of jaws protrude. Body oval in profile and laterally compressed. HABITAT: S FL, Bahamas, all Antilles to northern South America. Occur in clear, shallow coastal waters over rocky and coral bottoms. BIOLOGY: Feed on algae, mollusks, and crabs.

Spottail pinfish - *Diplodus holbrookii* (Bean, 1878)

FEATURES: Iridescent golden dorsally and on sides. Silvery below. Opercular membrane black. About nine faint, narrow bars on upper sides. Bars fade with age. Large black saddle on upper caudal peduncle. Dorsal fin membranes with brownish cast. Pectoral, caudal, and anal fins with golden cast. Teeth in front of jaws protrude. Body oval in profile and laterally compressed. HABITAT: Chesapeake Bay to northwestern Gulf of Mexico. Occur in shallow coastal waters over reefs and in bays and harbors. Prefer vegetated bottoms. BIOLOGY: Feed on a variety of invertebrates.

Pinfish - *Lagodon rhomboides* (Linnaeus, 1766)

FEATURES: Body with alternating iridescent bluish and yellowish to bronze stripes that intersect grayish bars. Stripes and bars form checkerboard pattern. Black spot on lateral-line origin. Dorsal fin with yellowish bands. Anal fin with yellowish inner band. Body oval and laterally compressed, with upper and lower profiles similarly shaped. HABITAT: MA (rare) to FL, Gulf of Mexico, northern Cuba, and Bermuda. Found near bottom in a variety of shallow coastal habitats, including bays, estuaries, and canals. Often around vegetated and hard bottoms. May enter fresh water. BIOLOGY: Feed on a variety of plants, small fishes, and invertebrates. Often used as bait.

Red porgy - *Pagrus pagrus* (Linnaeus, 1758)

FEATURES: Iridescent pinkish to reddish dorsally, becoming silvery below. Dorsal, caudal, and pectoral fins with pinkish to reddish cast. May display reddish bars. Opercular margin dark. Body oblong and laterally compressed. Head profile gently sloping. HABITAT: NY to FL, Gulf of Mexico to Argentina. Also eastern Atlantic and Mediterranean Sea. Absent from Bahamas and Antilles. Found near bottom over rocky and hard sand bottoms from about 32 to 260 ft.

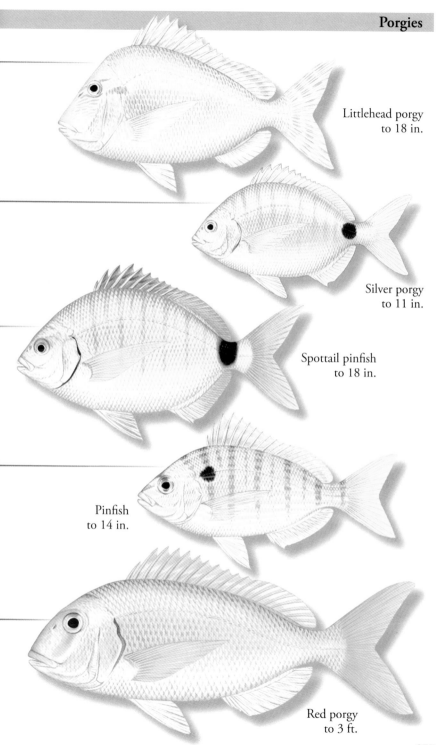

Littlehead porgy
to 18 in.

Silver porgy
to 11 in.

Spottail pinfish
to 18 in.

Pinfish
to 14 in.

Red porgy
to 3 ft.

277

Sparidae - Porgies, *cont.*

Longspine porgy - *Stenotomus caprinus* Jordan & Gilbert, 1882

FEATURES: Silvery with olivaceous cast dorsally. Juveniles may have faint, dark, narrow bars. Body otherwise unmarked. Teeth in front of jaws incisor-like. Head profile almost straight, steeply sloping. First two dorsal-fin spines very short. Third through fifth spines very tall. Body deeply oblong in profile. HABITAT: Eastern coast of FL and Gulf of Mexico to Yucatán Peninsula. Also possibly from NC. Found over muddy bottoms from about 16 to 390 ft. BIOLOGY: Longspine porgy feed on a variety of invertebrates, detritus, and small fishes.

Scup - *Stenotomus chrysops* (Linnaeus, 1766)

FEATURES: Dull silvery with iridescent blue reflections. May display dark bars. Blue line between eyes. Blue patch above each eye. Blue band along dorsal profile. Fins dusky with blue spots. Dorsal profile of head slightly concave above eye. Body oval in profile. HABITAT: Nova Scotia to FL (rare south of NC). Found near bottom in coastal waters, primarily over hard bottoms. Occur inshore in summer, offshore in winter. BIOLOGY: Feed on a variety of invertebrates.

Polynemidae - Threadfins

Atlantic threadfin - *Polydactylus octonemus* (Girard, 1858)

FEATURES: Pale silvery olive, silvery yellow, to dusky silver dorsally. Whitish ventrally. Dorsal, caudal, and anal fins dusky yellow with blackish margins. Pectoral fins black. Lower eight (sometimes nine) pectoral-fin rays separate, long, and filamentous. Scales comparatively large. HABITAT: NY to FL, Gulf of Mexico to Yucatán. Rare along Atlantic coast. Occur along coast in surf and over muddy bottoms from about 16 to 72 ft. Commonly enter estuaries. BIOLOGY: Atlantic threadfin are seasonally migratory.

Littlescale threadfin - *Polydactylus oligodon* (Günther, 1860)

FEATURES: Dusky silver dorsally, whitish ventrally. Dorsal and caudal fins dark, blackish at margins. Anal and pelvic fins with dark inner portions. Pectoral fins black. Lower seven (sometimes eight) pectoral-fin rays separate, long, and filamentous. Scales comparatively small. HABITAT: SE FL, Bahamas, Jamaica, Lesser Antilles, northeastern South America to Brazil. Occur near bottom and close to shore along exposed sandy beaches. BIOLOGY: Utilize free pectoral-fin rays to locate prey.

Barbu - *Polydactylus virginicus* (Linnaeus, 1758)

FEATURES: Silvery olive to blue gray dorsally, whitish ventrally. Dorsal and caudal fins dusky to yellowish, with blackish margins. Pelvic and anal fins with pale margins. Pectoral fins pale, with dark inner area. Lower seven pectoral-fin rays separate, long, and filamentous. HABITAT: NJ (rare) to FL, Greater and Lesser Antilles, Yucatán to Brazil. Also Bermuda. Occur near bottom along sandy and muddy flats and beaches and around mangroves. Also found in estuaries and river mouths. BIOLOGY: Barbu feed primarily at night on a variety of invertebrates, fishes, and plants.

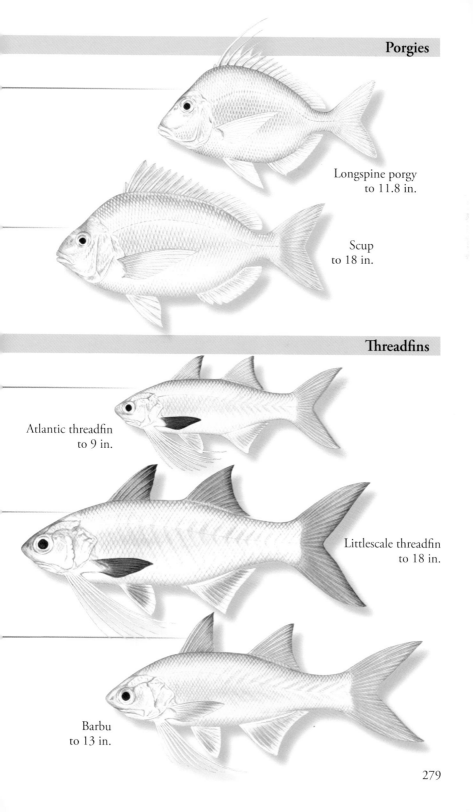

Longspine porgy
to 11.8 in.

Scup
to 18 in.

Atlantic threadfin
to 9 in.

Littlescale threadfin
to 18 in.

Barbu
to 13 in.

Silver perch - *Bairdiella chrysoura* (Lacepède, 1802)

FEATURES: Silvery, greenish, to bluish dorsally. Silvery to yellowish below. Obscure stripes follow scales on sides. Fins dusky to pale yellow. Preopercle with several small spines at corner. Dorsal fin deeply notched. Caudal fin with straight to angular margin. HABITAT: Cape Cod to FL. In Gulf of Mexico from FL to northern Mexico. Found along coast over shallow sandy and muddy bottoms. Also in estuaries. Occasionally in fresh water. BIOLOGY: Silver perch migrate into estuarine nurseries during summer. Feed on crustaceans, worms, and sometimes fishes. Other name: Silver croaker.

Blue croaker - *Corvula batabana* (Poey, 1860)

FEATURES: Bluish gray with silvery to brassy highlights. Dark spots scattered on upper sides; form rows along scales on sides. Preopercle finely serrated at corner. Soft dorsal and anal fins scaled at base. Caudal-fin margin rounded. HABITAT: S FL, Bay of Campeche, and Greater Antilles. Occur in clear, highly saline waters over vegetated mud flats and over coral reefs from about 33 to 100 ft. BIOLOGY: Blue croaker feed mainly on crustaceans. Previously placed in the genus *Bairdiella*.

Striped croaker - *Corvula sanctaeluciae* (Jordan, 1890)

FEATURES: Grayish with silvery highlights dorsally. Silvery below. Brownish spots follow scales along back and on sides. Rows on back are straight anteriorly, oblique under dorsal fins, and straight posteriorly. Fins pale yellowish with small dark flecks. Soft dorsal and anal fins scaled at base. HABITAT: E FL (rare), Greater and Lesser Antilles, and Costa Rica to Guyana. Over muddy and sandy bottoms of inshore waters. Juveniles around rocky areas. BIOLOGY: Feed mainly on shrimps.

Sand seatrout - *Cynoscion arenarius* Ginsburg, 1930

FEATURES: Pale to dark grayish green dorsally, iridescent on sides. Silvery below. Pelvic and anal fins pale to yellowish. Dark blotch at pectoral-fin base. Body otherwise unmarked. Mouth large, lower jaw protrudes. Pair of large canine teeth in upper jaw tip. Anal fin with 10–11 rays. Caudal fin angular. HABITAT: In Gulf of Mexico from FL to TX. Rare in Bay of Campeche. Possibly in E FL. In shallow coastal areas over sandy bottoms, in surf, and in estuaries. BIOLOGY: Sand seatrout move into estuaries during summer. Feed on crustaceans and fishes. Other name: Sand weakfish.

Spotted seatrout - *Cynoscion nebulosus* (Cuvier, 1830)

FEATURES: Silvery gray with iridescent reflections dorsally. Silvery on sides and below. Posterior upper sides with round black spots. Dorsal and caudal fins spotted. Mouth large, with lower jaw protruding. Pair of large canine teeth in upper jaw at tip. Juveniles with angular caudal fin, adults with straight to slightly concave caudal fin. HABITAT: NY to FL. In Gulf of Mexico from FL to Laguna Madre, Mexico. In coastal waters over sandy bottoms and seagrass beds, around rocks, and in marshes, tide pools, and estuaries. BIOLOGY: Spotted seatrout spawn in bays. Young spend first year on seagrass flats. Populations declining. Other name: Spotted weakfish.

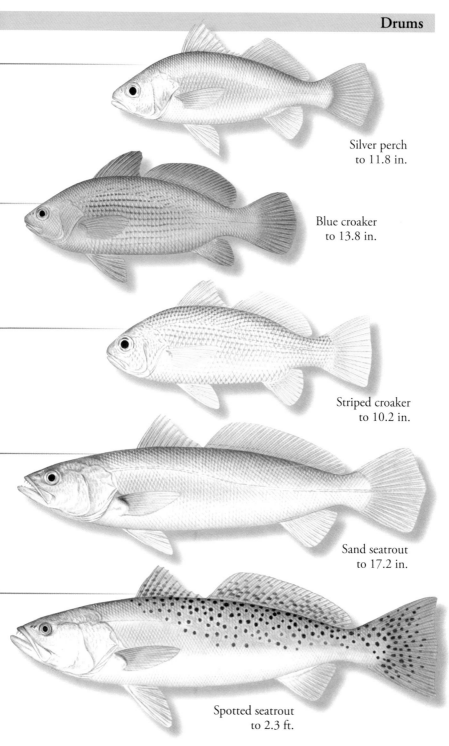

Silver perch
to 11.8 in.

Blue croaker
to 13.8 in.

Striped croaker
to 10.2 in.

Sand seatrout
to 17.2 in.

Spotted seatrout
to 2.3 ft.

Silver seatrout - *Cynoscion nothus* (Holbrook, 1848)

FEATURES: Shades of silver dorsally, silvery below. Upper sides may have faint spots along scales. Fins dusky to yellowish. Body otherwise unmarked. Mouth large, lower jaw protrudes. Pair of canine teeth in upper jaw. Anal fin with 8–11 rays. Caudal fin angular to straight. HABITAT: VA to FL, in Gulf of Mexico from FL to TX. Inshore over sandy bottoms, along beaches, and in inlets and river mouths. BIOLOGY: Feed on crustaceans and fishes. Sought as foodfish. Other name: Silver weakfish.

Weakfish - *Cynoscion regalis* (Bloch & Schneider, 1801)

FEATURES: Greenish gray with iridescent highlights dorsally. Silvery on sides and below. Upper sides with dark flecks and spots that follow scales and form oblique rows. Inside of opercle dark, visible externally. Pelvic and anal fins yellowish. Mouth large, lower jaw protrudes. Pair of large canine teeth in upper jaw at tip. HABITAT: Nova Scotia to SW FL, where it is rare. In shallow coastal waters over sandy and muddy bottoms, in sounds, and in surf. Move into estuaries during summer. BIOLOGY: Weakfish spawn in deep-water inlets and estuaries. Feed mainly on crustaceans and fishes. Sought as gamefish. Other names: Gray weakfish, Gray trout.

Jackknife-fish - *Equetus lanceolatus* (Linnaeus, 1758)

FEATURES: Pearly tan with iridescent reflections. Three white-edged, blackish bands on body: one through eyes; a second from nape through pelvic fins; the third from first dorsal fin along body to tip of caudal fin. Pectoral fins pale to colorless. Head profile nearly straight. Caudal fin pointed. Juveniles with black streak on snout. HABITAT: SC to FL, W Gulf of Mexico, and Caribbean Sea to Brazil. Also Bermuda. Occur over coastal sandy and muddy bottoms. Also around reefs and hard bottoms to about 196 ft. BIOLOGY: Feed on invertebrates. Popular in aquarium trade.

Spotted drum - *Equetus punctatus* (Bloch & Schneider, 1801)

FEATURES: Whitish with wide, dark brown bands on head and body: one through eyes; a second from nape to pelvic fins; the third from first dorsal fin to caudal-fin base with narrow bands above and below. Dorsal, caudal and anal fins dark brown with white spots. Pectoral fins dark brown. Head profile almost straight. Caudal fin bluntly pointed. Juveniles with dark spot on snout. HABITAT: S FL, Bahamas, all Antilles, and Panama to Brazil. Found primarily over coral reefs. BIOLOGY: Secretive and solitary. Feed on soft corals and invertebrates. Popular in aquarium trade.

Banded drum - *Larimus fasciatus* Holbrook, 1855

FEATURES: Silvery grayish olive dorsally, silvery with gold reflections on sides, and silvery below. Seven to nine brownish bars on upper body. Fins dusky brownish to coppery. Snout short, blunt. Mouth large, oblique. Lower jaw protrudes. Caudal fin margin angular. HABITAT: MA to FL, in Gulf of Mexico from FL to Veracruz. Rare in southern FL. Over muddy and sandy bottoms from near shore to about 320 ft. BIOLOGY: Banded drum feed mainly on shrimps. Caught as trawl bycatch.

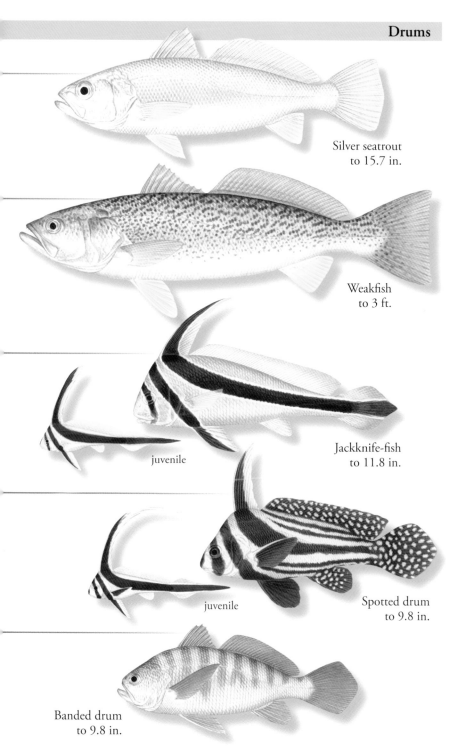

Silver seatrout
to 15.7 in.

Weakfish
to 3 ft.

juvenile

Jackknife-fish
to 11.8 in.

juvenile

Spotted drum
to 9.8 in.

Banded drum
to 9.8 in.

Spot - *Leiostomus xanthurus* Lacepède, 1802

FEATURES: Silvery gray with coppery to pinkish highlights dorsally. Silvery on sides and below. Upper body with 11–15 oblique, brownish bars. Distinct black spot behind upper opercular margin. Snout short, rounded. Mouth small, nearly horizontal. Dorsal profile arched. HABITAT: MA to FL, in Gulf of Mexico from FL to Veracruz. Found in coastal waters over sandy and muddy bottoms from near shore to about 197 ft. BIOLOGY: Spot feed mainly on bottom-dwelling worms, crustaceans, and detritus. Migrate into estuaries in summer and fall.

Southern kingfish - *Menticirrhus americanus* (Linnaeus, 1758)

FEATURES: Silvery gray to golden brown dorsally and on sides. Ventral area white. Intensity of color varies. Often with 7–8 faint to dark brownish bands on upper body; when present, first two bands form V-shape. Mouth small, nearly horizontal. Ventral profile nearly straight. Abdomen flattened. HABITAT: MA to S FL, Gulf of Mexico, western Caribbean Sea to Brazil. Rare in S FL and Venezuela. Over sandy mud to hard sand bottoms in shallow coastal waters. Also in surf and estuaries. BIOLOGY: Feed on bottom-dwelling invertebrates. Other name: Southern kingcroaker.

Gulf kingfish - *Menticirrhus littoralis* (Holbrook, 1847)

FEATURES: Pale silvery gray dorsally and on sides. Ventral area white. First dorsal-fin tip dusky. Upper caudal-fin tip usually black. Body otherwise unmarked. Mouth small, nearly horizontal. Ventral profile nearly straight. Abdomen flattened. HABITAT: VA to FL, Gulf of Mexico, western Caribbean Sea to Brazil. In surf and over sandy and sandy mud bottoms. Occasionally in estuaries. BIOLOGY: Feed on bottom-dwelling invertebrates. Sought as foodfish. Other names: Gulf whiting, Sea mullet.

Northern kingfish - *Menticirrhus saxatilis* (Bloch & Schneider, 1801)

FEATURES: Silvery gray to golden brown dorsally and on sides. Always with at least five broad to narrow bands on upper body. First and third bands form V-shape. Horizontal stripe from about mid-body to caudal fin. Mouth small, nearly horizontal. Second dorsal-fin spine elongate. Ventral profile nearly straight. Abdomen flattened. HABITAT: ME to FL, in Gulf of Mexico from FL to Campeche Bay. In shallow coastal waters over sandy mud bottoms, also in surf and in estuaries.

Atlantic croaker - *Micropogonias undulatus* (Linnaeus, 1766)

FEATURES: Silvery gray dorsally with pinkish to golden reflections. Sides silvery to golden. Ventral area silvery. Brownish spots on sides form numerous oblique, broken bars. Dorsal fins with dusky spots along base. Dark blotch at pectoral-fin base. Preopercle serrated, spiny at corner. Caudal fin trilobed in adults. HABITAT: MA to FL, in Gulf of Mexico from FL to Campeche Bay. Over sandy and muddy bottoms in coastal waters from near shore to about 197 ft. Also in estuaries. BIOLOGY: Atlantic croaker feed on bottom-dwelling worms, crustaceans, and small fishes. Caught by hook-and-line and in trawl bycatch. Other name: Croaker.

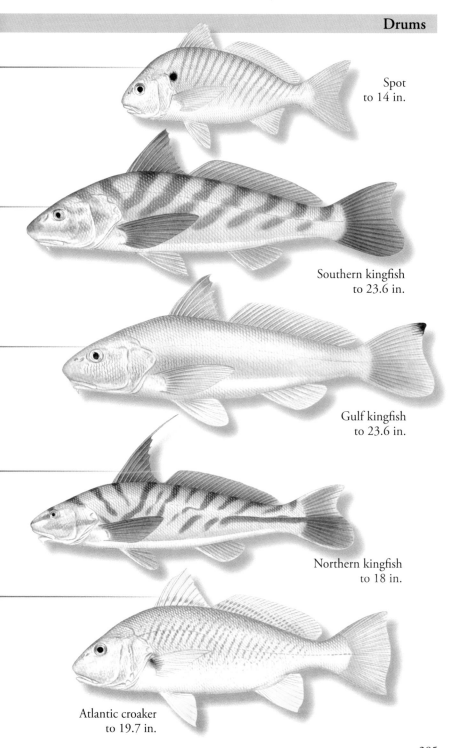

Spot
to 14 in.

Southern kingfish
to 23.6 in.

Gulf kingfish
to 23.6 in.

Northern kingfish
to 18 in.

Atlantic croaker
to 19.7 in.

Reef croaker - *Odontoscion dentex* (Cuvier, 1830)

FEATURES: Silvery gray to brownish gray with brownish spots on scales. Large black blotch at pectoral-fin base. Eyes large, dark. Snout short, blunt. Mouth large with pair of canine teeth in upper jaw at tip. Dorsal profile evenly convex, similar to ventral profile. HABITAT: S FL, FL Keys, Greater and Lesser Antilles, and Costa Rica to Brazil. In shallow coastal waters over sandy mud bottoms and coral reefs. BIOLOGY: Feed on shrimps, small fishes. Caught by hook-an-line and in trawls.

High-hat - *Pareques acuminatus* (Bloch & Schneider, 1801)

FEATURES: Head and body with alternating, irregular, pearly white and dark brown to grayish stripes. First dorsal fin with dark brown inner band, other fins dark brown. Mouth small, nearly horizontal. Dorsal profile arched, ventral profile nearly straight. Juveniles with two spots on snout, dark upper pectoral fins, and tall, trailing dorsal fin. HABITAT: VA to FL, Gulf of Mexico, and Caribbean Sea to Brazil. In clear, coastal waters over sandy and muddy bottoms and coral reefs to about 196 ft. BIOLOGY: High-hat feed on bottom-dwelling invertebrates. Popular in aquarium trade.

Blackbar drum - *Pareques iwamotoi* Miller & Woods, 1988

FEATURES: Pearly white to grayish with one dark brown bar through eyes. A second, broad bar below dorsal fin merges with a horizontal stripe that extends to caudal-fin margin. Broad pale brown area below soft dorsal fin. Mouth small, nearly horizontal. Dorsal profile arched, ventral profile nearly straight. Juveniles similarly patterned, but may lack pale brown area. HABITAT: NC to FL, Gulf of Mexico, western Caribbean Sea to Brazil. Also Bermuda. Over sandy mud, reefs, and hard bottoms from about 120 to 600 ft. BIOLOGY: Feed on gastropods. Other name: Gulf cubbyu.

Cubbyu - *Pareques umbrosus* (Jordan & Eigenmann, 1889)

FEATURES: Brown to grayish brown with alternating narrow and very thin dark stripes on sides. Stripes merge on head to form V-shapes. All fins dark. Mouth small, nearly horizontal. Ventral profile nearly straight. Juveniles with oval-shaped mark between eyes, colorless pectoral fins, and tall, but not trailing, dorsal fin. HABITAT: NC to FL, Gulf of Mexico, western Caribbean Sea to Brazil. Found in shallow coastal waters over sandy mud, hard bottoms, and reefs from about 120 to 600 ft.

Black drum - *Pogonias cromis* (Linnaeus, 1766)

FEATURES: Silvery bronze to almost black dorsally and on sides. Ventral area silvery. Juveniles silvery gray with four to five black bars on sides. Bars fade with age. Mouth small, nearly horizontal. Chin with 10–13 pairs of barbels. Dorsal profile arched. Ventral profile nearly straight. Dorsal fin deeply notched. Caudal-fin margin nearly straight. HABITAT: Nova Scotia to FL, in Gulf of Mexico from FL to TX. Uncommon in Antilles and Atlantic coast of South America. Occur over coastal sandy and mud bottoms, near river mouths, and in estuaries. Also in surf. BIOLOGY: Feed on bottom-dwelling organisms. Sought commercially and for sport. Listed as Endangered.

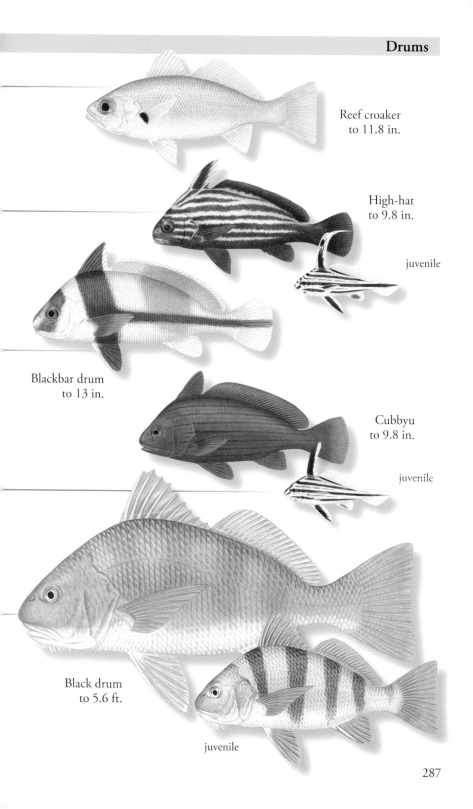

Reef croaker
to 11.8 in.

High-hat
to 9.8 in.

juvenile

Blackbar drum
to 13 in.

Cubbyu
to 9.8 in.

juvenile

Black drum
to 5.6 ft.

juvenile

Sciaenidae - Drums, *cont.*

Red drum - *Sciaenops ocellatus* (Linnaeus, 1766)

FEATURES: Color varies. Iridescent coppery, brassy, to gray dorsally, silvery below. Scales darker at centers, forming oblique bands that fade with age. At least one black, ocellated spot at upper caudal-fin base. Snout bluntly rounded. Mouth small, nearly horizontal. Dorsal profile evenly convex. Body robust. HABITAT: MA to FL, in Gulf of Mexico from FL to northern Mexico. Along coastal waters over sandy and sandy mud bottoms and in estuaries and surf zone. BIOLOGY: Red drum are migratory with seasons and tides. Feed on crustaceans, mollusks, and small fishes. Sought commercially and for sport. Other names: Redfish, Channel bass, Puppy drum.

Star drum - *Stellifer lanceolatus* (Holbrook, 1855)

FEATURES: Silvery grayish to silvery olive dorsally. Fins dusky to pale. Spiny dorsal fin with dark margin. Snout bluntly rounded. Mouth moderately large, oblique. Profile of nape slightly concave. Dorsal fin notched. Caudal fin somewhat pointed. HABITAT: VA to FL, in Gulf of Mexico from FL to TX. Found in coastal waters over hard sandy mud bottoms from shore to about 65 ft. Also in estuaries. BIOLOGY: Feed on small crustaceans. Common trawl bycatch. Other name: American star drum.

Sand drum - *Umbrina coroides* Cuvier, 1830

FEATURES: Silvery gray dorsally and on sides. Silvery below. Centers of scales dark, forming faint stripes. Upper sides with 8–9 dark to faint bars. Fins dusky. Snout bluntly rounded. Mouth small. Small barbel on chin. HABITAT: VA to FL, western Gulf of Mexico, Greater and Lesser Antilles, and northern South America to Brazil. Along shallow beaches, over muddy bottoms, and in estuaries. Sometimes over reefs.

Mullidae - Goatfishes

Yellow goatfish - *Mulloidichthys martinicus* (Cuvier, 1829)

FEATURES: Silvery olive gray to silvery tan dorsally. Ventral area slivery white. A bright yellow stripe on sides merges with yellow caudal fin. Dorsal fins and tips of anal and pelvic fins yellow. Dark red blotches appear on body at night. Snout gently sloping. Two long barbels on chin. Ventral profile nearly straight. HABITAT: S FL, Gulf of Mexico, Bahamas, and Caribbean Sea to Brazil. Also Bermuda. Found along shallow coasts over and around reefs and seagrass beds. BIOLOGY: Yellow goatfish usually feed at night. Form large schools.

Red goatfish - *Mullus auratus* Jordan & Gilbert, 1882

FEATURES: Reddish dorsally, silvery below. Two to five yellowish to reddish stripes on sides. First dorsal fin pale, with blackish red stripe above orangish stripe. Second dorsal fin banded. Caudal fin reddish, lacks distinct bands. Snout steeply sloping. Two long barbels on chin. HABITAT: MA to FL, Gulf of Mexico, Caribbean Sea to Guyana. Absent from Bahamas. Over coastal muddy and silty bottoms from about 30 to 300 ft. BIOLOGY: Use barbels to locate bottom-dwelling invertebrates.

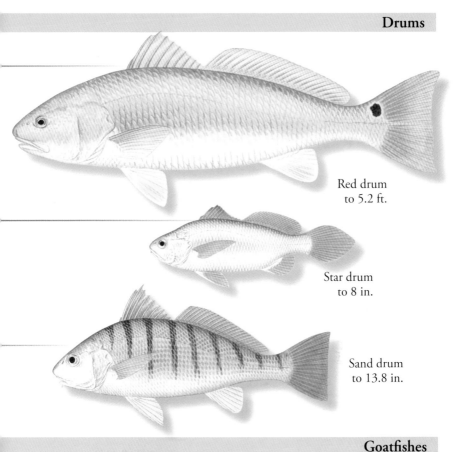

Red drum
to 5.2 ft.

Star drum
to 8 in.

Sand drum
to 13.8 in.

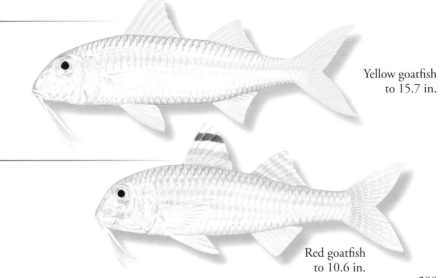

Yellow goatfish
to 15.7 in.

Red goatfish
to 10.6 in.

Mullidae - Goatfishes, *cont.*

Spotted goatfish - *Pseudupeneus maculatus* (Bloch, 1793)

FEATURES: Pearly pink dorsally. Silvery below. Three large, blackish red blotches on sides. Night pattern reddish with interconnected blotches. Daytime resting pattern similar to night pattern. Snout profile almost straight. Two barbels on chin. HABITAT: NJ to FL, Gulf of Mexico, Bahamas, Caribbean Sea to Brazil. Also Bermuda. Along shallow coastal waters over sandy and rocky bottoms near reefs. Juveniles found on seagrass beds. BIOLOGY: Feed on bottom-dwelling invertebrates. Form groups.

Dwarf goatfish - *Upeneus parvus* Poey, 1852

FEATURES: Salmon pink to reddish dorsally with a broken yellowish stripe on each side. Ventral area pearly white. Dorsal and caudal fins distinctly banded. Snout gently sloping. Two barbels on chin. HABITAT: NC to FL, Gulf of Mexico, Greater and Lesser Antilles, and northern South America. Over sandy, silty, and muddy bottoms from about 130 to 330 ft. BIOLOGY: Feed on bottom-dwelling invertebrates.

Pempheridae - Sweepers

Glassy sweeper - *Pempheris schomburgkii* Müller & Troschel, 1848

FEATURES: Head and body dark coppery with iridescent and greenish highlights. Dark band at anal-fin base. Mouth large, oblique. Eyes very large. Single dorsal fin at mid-body line. Ventral profile deeply convex. HABITAT: S FL and Dry Tortugas, Bahamas, and Antilles to Brazil. Also Bermuda. In crevices and caves of coral reefs during the day. In water column at night. BIOLOGY: Feed on zooplankton.

Kyphosidae - Sea chubs

Yellow chub - *Kyphosus incisor* (Cuvier, 1831)

FEATURES: Silvery bluish gray with brassy stripes following scales on sides. Two brassy stripes on sides of head. Opercular membrane gray. Dorsal profile in front of eyes slightly convex. Soft dorsal fin with 13–15 (usually 14) rays. Anal fin with 12–13 (usually 13) rays. HABITAT: MA to FL, Gulf of Mexico, Bahamas, Caribbean Sea to Brazil. Also Bermuda and eastern Atlantic. Usually occur offshore and associated with flotsam and *Sargassum* seaweed. Also in shallow water over hard bottoms and reefs.

Bermuda chub - *Kyphosus sectator* (Linnaeus, 1758)

FEATURES: Silvery bluish gray with yellowish stripes following scales on sides. Two yellowish stripes on sides of head. May display pale spots on head and body. Opercular membrane blackish. Dorsal profile in front of eyes humped. Soft dorsal fin with 11–13 (usually 11) rays. Anal fin with 10–11 (usually 11) rays. HABITAT: MA to FL, Gulf of Mexico, Bahamas, Caribbean Sea to Brazil. Also Bermuda, eastern Atlantic, and Mediterranean. In shallow water over seagrass beds and sandy and rocky bottoms and around coral reefs. Sometimes offshore. Young often associated with *Sargassum* seaweed.

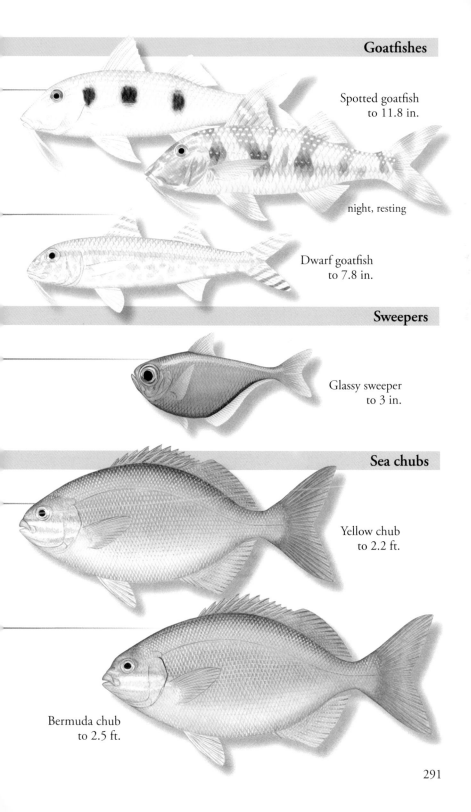

Goatfishes

Spotted goatfish
to 11.8 in.

night, resting

Dwarf goatfish
to 7.8 in.

Sweepers

Glassy sweeper
to 3 in.

Sea chubs

Yellow chub
to 2.2 ft.

Bermuda chub
to 2.5 ft.

291

Chaetodontidae - Butterflyfishes

Foureye butterflyfish - *Chaetodon capistratus* Linnaeus, 1758

FEATURES: Head and ventral area yellowish. Grayish band through each eye from nape to lower opercle. Body pale gray to pale yellow with dark, oblique dashes that follow scales. Large, black, ocellated spot on posterior body. Juveniles with black band through eyes, two broad bars on body, and two ocellated spots posteriorly; upper spot fades with age. HABITAT: MA (rare) to FL, Gulf of Mexico, Bahamas, and Caribbean Sea to Brazil. Also Bermuda. Found over shallow coastal reefs and rocky areas. Juveniles occur over seagrass beds. BIOLOGY: Fourspot butterflyfish feed on small, bottom-dwelling gorgonians, tunicates, and worms. Found singly or in pairs. Popular in the aquarium trade.

Spotfin butterflyfish - *Chaetodon ocellatus* Bloch, 1787

FEATURES: Body and most of head silvery white. Black band through eyes. Yellow areas on snout and yellow band from opercular margin to pectoral-fin base. Fins and posterior portion of body yellow. A black blotch at middle base of soft dorsal fin appears during night, becomes faint or disappears during day. A small black spot at rear tip of soft dorsal fin always present. Juveniles silvery white on head and body; black band through each eye; black blotch at middle base of soft dorsal fin blends with black band on posterior body. HABITAT: MA (rare) to FL, Gulf of Mexico, Bahamas, and Caribbean Sea to Brazil. Also Bermuda. Occur over shallow coastal reefs from shore to about 100 ft. BIOLOGY: Spotfin butterflyfish usually occur in pairs. Feed on bottom-dwelling invertebrates. Popular in aquarium trade.

Reef butterflyfish - *Chaetodon sedentarius* Poey, 1860

FEATURES: Yellowish dorsally, fading to silvery white below. Broad black band from nape, through eyes, to lower opercular margin. Several faint, oblique, brownish bars on sides. A broad blackish band extends from rear of soft dorsal fin, through caudal peduncle, to rear of anal fin; upper portion may be very dark to faint. Caudal fin yellow, lacks banding. Juveniles similar to adults, but with diffuse, dark spot on rear soft dorsal fin. HABITAT: NC to FL, Gulf of Mexico, Bahamas, and Caribbean Sea to Brazil. Also Bermuda. Found over reefs from about 50 to 130 ft. Recorded to 330 ft. BIOLOGY: Reef butterflyfish feed on bottom-dwelling invertebrates.

Banded butterflyfish - *Chaetodon striatus* Linnaeus, 1758

FEATURES: Whitish with broad dark brown to grayish bars on body. The first bar is from nape through eye to breast. The second is from spiny dorsal fin to abdomen. The third is from anterior portion of soft dorsal fin to middle base of anal fin. Sides with dark, oblique lines that follow scale rows. Soft dorsal and anal fins with two broad dark bands. Caudal fin with dark inner band. Juveniles similar to adults, but with large, ocellated spot on anterior soft dorsal fin; sides lack oblique lines. HABITAT: MA (rare) to FL, Gulf of Mexico, Bahamas, and Caribbean Sea to Brazil. Also Bermuda and eastern Atlantic. Found over shallow coastal reefs and rocky areas. Juveniles occur over seagrass beds. BIOLOGY: Banded butterflyfish feed on coral polyps, worms, crustaceans, and mollusk eggs. Popular in aquarium trade.

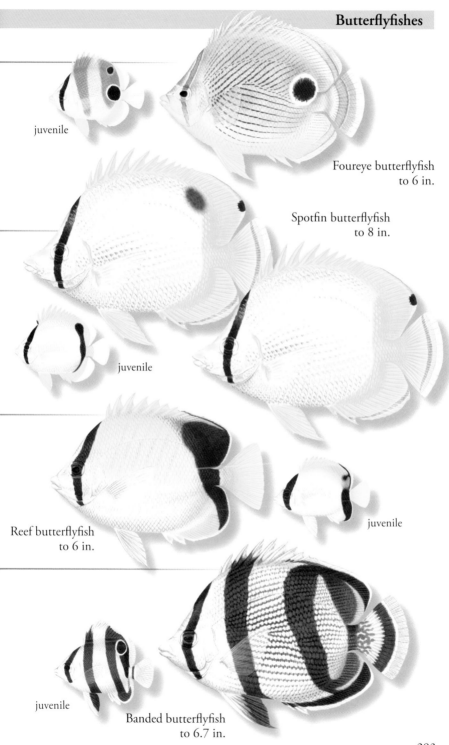

juvenile

Foureye butterflyfish
to 6 in.

Spotfin butterflyfish
to 8 in.

juvenile

Reef butterflyfish
to 6 in.

juvenile

juvenile

Banded butterflyfish
to 6.7 in.

Chaetodontidae - Butterflyfishes, *cont.*

Longsnout butterflyfish - *Prognathodes aculeatus* (Poey, 1860)

FEATURES: Dark yellow dorsally fading to silvery below. A yellow band extends from eyes to dorsal-fin origin. Spiny dorsal fin, posterior portion of back, and part of soft dorsal fin dark brownish. A narrow yellow band bisects soft dorsal fin. Part of caudal peduncle, inner anal fin, and pelvic fins yellow. Remainder of fins colorless. Snout long, pointed. Dorsal-fin spines tall. HABITAT: S FL, Gulf of Mexico, Bahamas, Caribbean Sea to northern South America. Also Bermuda. Recently recorded from NC. Over moderate to deep reefs and rocky areas usually from about 50 to 180 ft. BIOLOGY: Longsnout butterflyfish use slender mouth to pick invertebrates from small crevices and pick flesh of urchins from between spines. May form schools.

Bank butterflyfish - *Prognathodes aya* (Jordan, 1886)

FEATURES: Body silvery white to silvery tan. Dark brown stripe along snout profile. Dark blackish brown band from dorsal-fin origin, through eyes, to corner of mouth. Broad, dark blackish brown band from about fourth dorsal-fin spine to anal-fin base. Yellowish band from about the fifth dorsal-fin spine through caudal peduncle. Pelvic fins yellow. Snout long, pointed. Dorsal-fin spines very tall. HABITAT: NC (rare) to FL and Gulf of Mexico to Campeche Banks. Found around rocky slopes between about 65 and 650 ft. BIOLOGY: Feed on small, bottom-dwelling invertebrates.

Pomacanthidae - Angelfishes

Cherubfish - *Centropyge argi* Woods & Kanazawa, 1951

FEATURES: Lower head and chest yellow. Top of head and rest of body dark indigo blue. Eyes with a blue ring. Dark spots on scales. Dorsal and anal fins with two blackish inner bands and pale blue margins. Three spines on cheek posterior to jaws. Preopercle with long spine at corner. Body somewhat deep, compressed. HABITAT: NE and S FL, northern and southern Gulf of Mexico, Bahamas, Caribbean Sea to northern South America. Also Bermuda. Recently recorded from NC. Around coastal rocky and reef areas from near shore to about 98 ft. BIOLOGY: Territorial and somewhat secretive. Feed on algae and invertebrates. Popular in aquarium trade.

Blue angelfish - *Holacanthus bermudensis* Goode, 1876

FEATURES: Tannish green to tannish blue on most of body and fins. Nape and chest pale blue. Pectoral fins with inner yellow band. Dorsal and anal fins with inner yellow margin, outer blue margin. Caudal fin with yellow margin. Preopercular margin spiny with prominent spine at corner. Juveniles dark greenish blue with pale blue bars on sides that are nearly horizontal; yellow snout and chest; dark band through eyes, bordered by pale blue bands. HABITAT: NJ (rare) to FL, Bahamas, Gulf of Mexico to Yucatán. Also Bermuda. Occur around coral reefs from near shore to about 196 ft. Also around oil platforms. BIOLOGY: Blue angelfish feed on bottom-dwelling invertebrates. Known to hybridize with Queen angelfish. The resulting fish is known as the Townsend angelfish and shares several visual attributes with each.

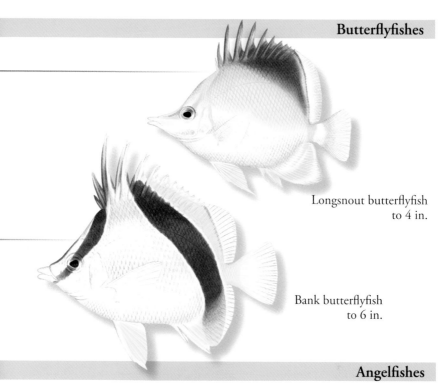

Longsnout butterflyfish
to 4 in.

Bank butterflyfish
to 6 in.

Angelfishes

Cherubfish
to 2 in.

Blue angelfish
to 15 in.

juvenile

Queen angelfish - *Holacanthus ciliaris* (Linnaeus, 1758)

FEATURES: Body greenish to bluish with yellow-margined scales. Nape with dark blue, ocellated blotch. Chest blue. Pectoral fins and caudal fin yellow. Dorsal and anal fins fade to yellowish orange near tips; margins blue. Preopercle spiny with prominent spine at corner. Juveniles dark greenish blue with pale blue bars on sides that are slightly curved; yellow snout and chest; dark band through eyes bordered by pale blue bands. HABITAT: NC (rare) to FL, Gulf of Mexico, Bahamas, and Caribbean Sea to Brazil. Also Bermuda. Found around shallow coral reefs and hard structures from near shore to about 196 ft. BIOLOGY: Queen angelfish feed on bottom-dwelling invertebrates. Juveniles pick ectoparasites from other fishes. Popular in aquarium trade. Hybridizes with Blue angelfish - see previous page.

Rock beauty - *Holacanthus tricolor* (Bloch, 1795)

FEATURES: Nape, head, and anterior portion of body yellow. Jaws bluish to blackish. Sides of body and most of dorsal and anal fins dark blackish blue. Posterior margins of dorsal and anal fins yellow. Other fins yellow. Preopercular margin spiny, with prominent spine at corner. Juveniles yellow with dark spot bordered by blue ring on posterior upper sides. Intermediate specimens yellow with dark area on rear upper portion of body. Both with yellow jaws. HABITAT: NC (rare) to FL, Gulf of Mexico, Bahamas, and Caribbean Sea to Brazil. Also Bermuda. Found over coral reefs, rocky bottoms, and hard structures from about 3 to 295 ft. Juveniles associated with stinging corals. BIOLOGY: Rock beauty feed on sponges.

Gray angelfish - *Pomacanthus arcuatus* (Linnaeus, 1758)

FEATURES: Body and fins gray to brownish gray. Jaws and chin whitish. Head pale gray, chest dark gray. Scales with dark centers, pale edges. Preopercle with small spines on margin and prominent spine at corner. Middle soft rays of the dorsal and anal fins are long and trailing. Juveniles blackish with yellow bars on head and body; yellow bar extending along forehead splits over jaws to form an upside-down cross shape; caudal fin with a yellow band at base and broken yellowish to whitish margin. HABITAT: NY (rare) to FL, Gulf of Mexico, Bahamas, and Caribbean Sea to Brazil. Found over shallow coral reefs and around hard structures. BIOLOGY: Gray angelfish feed on a variety of invertebrates and algae. Juveniles pick ectoparasites from other fishes.

French angelfish - *Pomacanthus paru* (Bloch, 1787)

FEATURES: Body and fins blackish, head dark bluish gray. Jaws and chin whitish. Eyes circled with yellow. Opercular margin, pectoral-fin base, and scale margins yellow. Middle soft rays of dorsal and anal fins long and trailing. Preopercle with small spines at margin, prominent yellow spine at corner. Juveniles blackish with yellow bars on head and body; yellow bar extending along forehead splits over jaws to form an upside-down Y-shape; caudal fin with a yellow border that forms a ring. HABITAT: NY (rare) to FL, Bahamas, Antilles, and northern South America to Brazil. Possibly Bermuda, and Ascension Island in eastern Atlantic. Over shallow reefs and hard structures. BIOLOGY: French angelfish feed on invertebrates and algae.

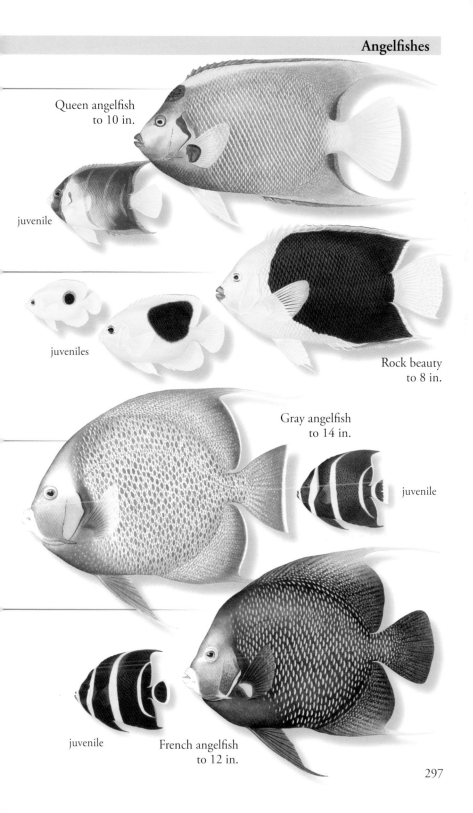

Queen angelfish
to 10 in.

juvenile

juveniles

Rock beauty
to 8 in.

Gray angelfish
to 14 in.

juvenile

juvenile

French angelfish
to 12 in.

Cirrhitidae - Hawkfishes

Redspotted hawkfish - *Amblycirrhitus pinos* (Mowbray, 1927)

FEATURES: Pale tannish to pale olivaceous. Five broad brownish to olivaceous bars on body extend into fins. Upper portion of fourth bar and bar on caudal peduncle are blackish. Bright orange to red spots scattered on head, forebody, and dorsal fin. Dorsal-fin spine tips with tufts of cirri. HABITAT: S FL, Gulf of Mexico, Bahamas, Caribbean Sea to northern South America. Also Bermuda. On reefs and hard substrates to about 147 ft. Also in areas of strong currents. BIOLOGY: Feed on small invertebrates. Use free pectoral-fin rays to move over substrate. Popular in aquarium trade.

Pomacentridae - Damselfishes

Sergeant major - *Abudefduf saxatilis* (Linnaeus, 1758)

FEATURES: Upper head and nape bluish green. Body usually bright yellow dorsally, bluish white below, with five black bars on sides. Juveniles similarly colored and patterned. Adult males turn dark blue during spawning and while guarding eggs. Body deep, laterally compressed. HABITAT: RI to FL, Gulf of Mexico, Bahamas, and Caribbean Sea to Brazil. Also Bermuda and eastern Atlantic. In shallow water around jetties and over coral and rocky reefs to about 45 ft. Juveniles associated with *Sargassum* seaweed, often far offshore. BIOLOGY: Feed on a variety of invertebrates.

Night sergeant - *Abudefduf taurus* (Müller & Troschel, 1848)

FEATURES: Yellowish tan above, paler below. Five wide brownish bars on sides extend into dorsal fins. May have sixth bar on caudal peduncle. Bars may be complete or incomplete. Small dark spot at upper pectoral-fin base. Juveniles similarly colored and patterned. HABITAT: S FL, Gulf of Mexico, Bahamas, and Caribbean Sea to Uruguay. Occur in very shallow and turbulent rocky areas. Occasional in low-salinity waters. BIOLOGY: Feed on algae, seagrasses, and hydroids.

Blue chromis - *Chromis cyanea* (Poey, 1860)

FEATURES: Head and body bright blue. Dorsal portion of head and body blackish. All fins blue. Dorsal and anal fins with blackish margins. Upper and lower caudal-fin margins blackish. Body moderately elongate, caudal fin deeply forked. HABITAT: FL, Gulf of Mexico, Bermuda, Bahamas, Caribbean Sea to Venezuela. Associated with shallow to deep coral reefs and slopes. BIOLOGY: Blue chromis are schooling with other species in the water column. Feed in aggregations on passing zooplankton.

Yellowtail reeffish - *Chromis enchrysura* Jordan & Gilbert, 1882

FEATURES: Dark blue dorsally from snout to posterior base of soft dorsal fin. Ventral area bluish white to grayish white. Outer portions of soft dorsal and anal fins yellow. Caudal fin yellow. Bright blue band runs from tip of snout to first dorsal-fin base. HABITAT: S FL, Gulf of Mexico, Bahamas, Caribbean Sea to Brazil. Also Bermuda. Occur over steep slopes of coral and patch reefs from about 65 to 330 ft. BIOLOGY: Feed on algae and zooplankton. Reported to form small groups.

Redspotted hawkfish
to 3.5 in.

Sergeant major
to 7 in.

Night sergeant
to 9.8 in.

Blue chromis
to 5 in.

Yellowtail reeffish
to 4 in.

299

Sunshinefish - *Chromis insolata* (Cuvier, 1830)

FEATURES: Brownish to olivaceous on upper two-thirds of body. Ventral area whitish to brownish. Blue line from tip of snout and above eyes may be present. Small dark spot at upper pectoral-fin base usually present. Outer portions of soft dorsal fin and caudal fin yellowish to pale. Juveniles yellow dorsally, bluish purple along midbody, whitish below. HABITAT: S FL, Gulf of Mexico, Bahamas, and Lesser Antilles to Brazil. Also Bermuda and St. Helena. Found over deep coral reefs. BIOLOGY: Sunshinefish feed on small invertebrates. Form clusters close to bottom.

Brown chromis - *Chromis multilineata* (Guichenot, 1853)

FEATURES: Olive brown to grayish brown dorsally, fading to whitish below. Dark spot at upper pectoral-fin base. Usually with white spot under rear dorsal-fin base. Outer margins of dorsal fin and tips of caudal fin yellow. Submargins of caudal-fin may be black or may be identical to body color. HABITAT: FL, Gulf of Mexico, Bahamas, Caribbean Sea to Brazil. Occur over coral reefs, slopes, and rubble from near shore to about 130 ft. BIOLOGY: Brown chromis form aggregations while feeding on zooplankton. Other name: Yellowedge chromis.

Purple reeffish - *Chromis scotti* Emery, 1968

FEATURES: Adults bluish green to grayish blue with few to many bright blue spots on scales. Blue spots may blend to form V-shape on snout. Eyes always with blue to bright blue crescent. Smaller specimens dusky blue dorsally, pale below, with bright blue spots. Pectoral fins always transparent. HABITAT: NC to FL, Gulf of Mexico, Bahamas, and Caribbean Sea to Brazil. Also Bermuda. Found over steep, outer reef slopes and patch reefs to about 164 ft. BIOLOGY: Feed on invertebrates.

Yellowtail damselfish - *Microspathodon chrysurus* (Cuvier, 1830)

FEATURES: Body and all fins, except caudal fin, bluish brown to yellowish brown with dark-edged scales. Caudal fin always yellow or pale yellow. Small, bright blue spots scattered on head, upper body, and dorsal fin. Juveniles dark blue to brownish blue with bright blue spots on head, body, dorsal and anal fins; caudal fin transparent, becoming yellow with age. HABITAT: S FL, Gulf of Mexico, Bahamas, Caribbean Sea to Venezuela. Also Bermuda. Occur over shallow coral reefs. BIOLOGY: Territorial. Feed on algae, coral polyps, and a wide variety of invertebrates.

Dusky damselfish - *Stegastes adustus* (Troschel, 1865)

FEATURES: Body and fins dark bluish brown to purplish brown. Scale edges dark. May have faint blue spots on head, chest, and abdomen. Small, dark spot at upper pectoral-fin base. Soft dorsal- and anal-fin tips reach to or slightly beyond caudal-fin base. Juveniles bright orange dorsally, purplish blue on sides and below with bright blue spots and a large, black, ocellated spot on base of dorsal fin; small dark spot on peduncle. HABITAT: S FL, Gulf of Mexico, Bahamas, Caribbean Sea to Venezuela. Possibly Bermuda. Occur around shallow, turbulent, coral and rocky shores.

Sunshinefish
to 4 in.

juvenile

Brown chromis
to 6.5 in.

Purple reeffish
to 4 in.

Yellowtail damselfish
to 8 in.

juvenile

Dusky damselfish
to 6 in.

juvenile

Longfin damselfish - *Stegastes diencaeus* (Jordan & Rutter, 1897)

FEATURES: Body and fins dark bluish brown, gray brown, to blackish. May have greenish cast on head. Scale edges dark. Small dark spot on upper pectoral-fin base. Dorsal and anal fins extend well beyond caudal-fin base. Juveniles yellow with blue lines on head that merge with blue spots on upper body; blue lines on snout form a V-shape; black ocellated spot present on dorsal fin. HABITAT: S FL, Gulf of Mexico, Bahamas, Caribbean Sea to Venezuela. Around quiet, protected coral and rocky reefs from about 6 to 147 ft. BIOLOGY: Territorial. Feed on algae, seaweeds.

Beaugregory - *Stegastes leucostictus* (Müller & Troschel, 1848)

FEATURES: Dark olive brown to dusky gray. May be pale below. Caudal fin yellowish to pale. May have blue spots on head and upper sides. Juveniles with blue spots over dark dusky blue on upper head and body, yellow below; a black ocellated spot present on dorsal fin. Spot on caudal peduncle absent in all. Body comparatively shallow. HABITAT: ME (summer only) to FL, Gulf of Mexico, Bahamas, Caribbean Sea to Brazil. Also Bermuda. Found around shallow, quiet seagrass beds, coral and rocky reefs, and sandy areas. BIOLOGY: Territorial. Feed on algae and invertebrates.

Bicolor damselfish - *Stegastes partitus* (Poey, 1868)

FEATURES: Several color phases exist: dark brown to blackish on anterior half to two-thirds of body with pearly posterior portion; dark brown to blackish on upper anterior portion of body, yellow below, with posterior portion pearly; overall pearly to grayish. Brownish black and yellow areas variable in size. Caudal fin may be pale to grayish. Pectoral fins always yellowish. HABITAT: NC to FL, Gulf of Mexico, Bermuda, Bahamas, Caribbean Sea to Venezuela. Occur around shallow coral reefs and patch reefs in deeper water. BIOLOGY: Feed on algae and copepods.

Threespot damselfish - *Stegastes planifrons* (Cuvier, 1830)

FEATURES: Purplish brown to brownish gray dorsally, yellowish below. Always with large black spot at upper pectoral-fin base and on upper caudal peduncle. Upper rim of eye always yellowish. Snout may have a blue wash. Juveniles yellow with a small black spot on upper pectoral-fin base, a large black spot on rear upper side, and a black spot on caudal peduncle. Body comparatively deep. HABITAT: NC to FL, Gulf of Mexico, Bahamas, Caribbean Sea to Venezuela. Also Bermuda. Coral reef associated.

Cocoa damselfish - *Stegastes variabilis* (Castelnau, 1855)

FEATURES: Variable. Dark bluish brown to greenish brown dorsally, yellow below. May have blue spots on head and upper body. Pectoral fins yellow. Rear portion of dorsal fin yellowish. Small spot at upper pectoral-fin base. May have a dark spot on upper caudal peduncle. Juveniles with blue spots over dark dusky blue on upper head and body, yellow below; black spot present on rear dorsal fin; may also have a black spot on upper peduncle. Body comparatively deep. HABITAT: NC to FL, Gulf of Mexico, Bahamas, Caribbean Sea to Brazil. Also Bermuda. Coral reef associated to about 95 ft.

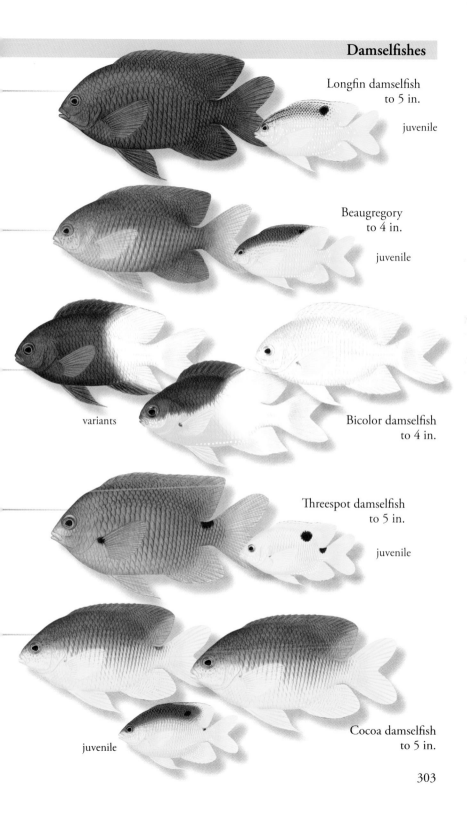

Longfin damselfish
to 5 in.

juvenile

Beaugregory
to 4 in.

juvenile

variants

Bicolor damselfish
to 4 in.

Threespot damselfish
to 5 in.

juvenile

juvenile

Cocoa damselfish
to 5 in.

303

Spotfin hogfish - *Bodianus pulchellus* (Poey, 1860)

FEATURES: Red on anterior two-thirds of body with white area on lower head. White area on head may blend with a faint to distinct white streak on lower body. Posterior dorsal fin, upper caudal peduncle, and caudal fin yellow. Two dark red to blackish streaks behind eyes. First two dorsal-fin spines and tips of upper pectoral-fin rays black. Intermediate specimens blackish anteriorly, yellow posteriorly, with white streak extending from lower head. Juveniles yellow with black on anterior dorsal-fin spines. HABITAT: SC south to Brazil. Also Gulf of Mexico and Bermuda. Over coral reefs and slopes between about 32 and 393 ft. BIOLOGY: Feed on mollusks and crustaceans. Juveniles sometimes pick parasites off other fishes.

Spanish hogfish - *Bodianus rufus* (Linnaeus, 1758)

FEATURES: Color and pattern varies with age. Purplish blue to reddish area on upper anterior portion of body; margins of area may be distinct or diffuse and may extend partially or entirely into dorsal fin. Lower body and fins yellow with areas of blue. Blue areas may be small or may extend over ventral area and fins. Adults become more mottled and blue with age. Some specimens almost entirely bluish. HABITAT: NC to FL, Gulf of Mexico, Bahamas, and Caribbean Sea to Brazil. Also Bermuda. Found over coral reefs and offshore banks to about 130 ft. BIOLOGY: Spanish hogfish feed on a variety of invertebrates. Juveniles pick parasites off other fishes.

Creole wrasse - *Clepticus parrae* (Bloch & Schneider, 1801)

FEATURES: Initial phase purple dorsally, lavender on sides, paler below. Head dark above eyes. May have irregular pale areas along back. Terminal phase purple to lavender anteriorly, yellowish posteriorly. Body variably mottled with paler and darker blotches. Head dark above eyes. Juveniles purple to lavender with regularly spaced pale areas along back. HABITAT: FL, Gulf of Mexico, Bahamas, and Caribbean Sea to Brazil. Also Bermuda. Over patch reefs and outer reef edges. Also around oil pylons. BIOLOGY: Use highly protrusible jaw to feed on plankton in the water column.

Red hogfish - *Decodon puellaris* (Poey, 1860)

FEATURES: Reddish dorsally, whitish below. Lips yellow. Yellow lines radiate from eyes. Yellow spots follow scales on sides. Fins pale reddish with yellow spots and lines. Upper pectoral-fin rays somewhat elongate. Upper and lower caudal-fin rays elongate. HABITAT: S FL, Gulf of Mexico, Bahamas, Caribbean Sea to Brazil. Also Bermuda. Reef associated from about 60 to 900 ft.

Dwarf wrasse - *Doratonotus megalepis* Günther, 1862

FEATURES: Shades of green with irregular reddish bands, lines, and spots on head and body. Oblique white band below eyes. Fins mottled. Snout pointed. Anterior dorsal-fin spines elongate. Caudal fin rounded. HABITAT: FL Keys, eastern Gulf of Mexico, Bahamas, and Caribbean Sea to Venezuela. Also Bermuda. Occur over shallow seagrass beds from shore to about 50 ft.

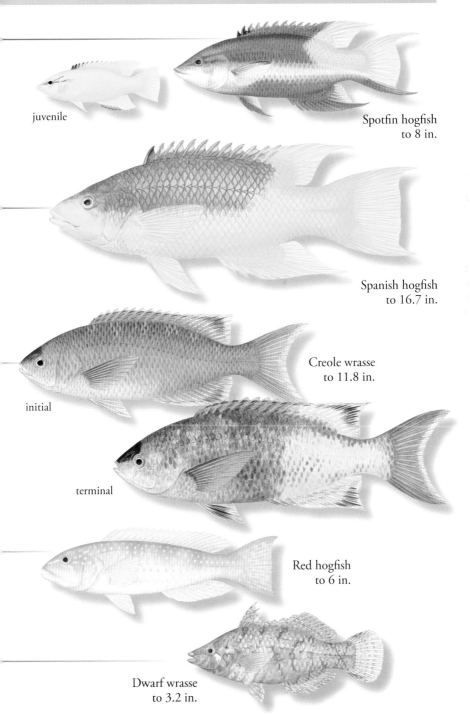

juvenile

Spotfin hogfish
to 8 in.

Spanish hogfish
to 16.7 in.

Creole wrasse
to 11.8 in.

initial

terminal

Red hogfish
to 6 in.

Dwarf wrasse
to 3.2 in.

Labridae - Wrasses, *cont.*

Greenband wrasse - *Halichoeres bathyphilus* (Beebe & Tee-Van, 1932)

FEATURES: Initial phase reddish dorsally, fading to pinkish below. Olive green band from snout to eye. Distinct broken yellow stripe on sides and small dark spot on caudal peduncle. May have small dark spot above pectoral fin. Terminal phase pinkish green dorsally, becoming pink at midline. Olive green band from snout to eye and from eye to opercular margin. Distinct broken yellow stripe on sides. Dark oblong spot above pectoral fin. HABITAT: NC to FL. Scattered in Gulf of Mexico from FL to Yucatán. Also Bahamas and Bermuda. Reef associated from about 164 to 393 ft.

Slippery dick - *Halichoeres bivittatus* (Bloch, 1791)

FEATURES: All phases variable. Initial phase with black stripe from snout to caudal peduncle. Stripe becomes single row of spots on sides. Bicolored spot on opercular margin. Small black spot on dorsal fin. Terminal phase greenish with pinkish wavy stripes on head and body. A purplish stripe on sides follows scale rows. Bicolored spot on opercular margin. Fins with pinkish stripes and bands. Intermediate phases share attributes of initial and terminal phases. HABITAT: NC south to Brazil. Also Gulf of Mexico and Bermuda. Over shallow reefs, rocky bottoms, and seagrass beds.

Mardi Gras wrasse - *Halichoeres burekae* Weaver & Rocha, 2007

FEATURES: Initial phase salmon pink dorsally with white stripe on lower sides. Snout yellow. Large black blotch on caudal peduncle. Terminal phase with purplish head and anterior upper sides. Bright blue lines and spots on head, body, and fins. Bright blue ocellated spot above pectoral fin. Large bright yellow area on sides. Dorsal fin with blackish band anteriorly. Snout short. HABITAT: In the Gulf of Mexico known from Stetson Bank, Flower Garden Banks, and Veracruz. Reef associated to about 160 ft. BIOLOGY: Form small mixed schools. Planktivorous.

Painted wrasse - *Halichoeres caudalis* (Poey, 1860)

FEATURES: Initial phase pale salmon pink dorsally, with diffuse pinkish and yellowish stripes on lower sides. Head with pale bluish lines. Pale blue spots follow scale rows on sides. Small black spot on lower dorsal fin. Terminal phase greenish dorsally, fading to pale reddish on sides. Head with bright blue lines and a dark spot behind eyes. Blue spots follow scale rows on sides. May have small black spot on lower dorsal fin. HABITAT: NC to FL, Gulf of Mexico, and Antilles to Venezuela. Reef associated from about 85 to 240 ft.

Yellowcheek wrasse - *Halichoeres cyanocephalus* (Bloch, 1791)

FEATURES: All phases with yellow on upper head and along back. Initial phase with a broad blue stripe from opercle to caudal fin. Terminal phase with blue along head profile and a broad dark blue to blackish stripe on sides. Dusky yellow band behind eye surrounded by blue, wavy lines. Lower portion of dorsal fin blue to blackish. HABITAT: NC to FL Keys, and Antilles to Brazil. Not recorded from Gulf of Mexico. Reef associated from about 90 to 300 ft.

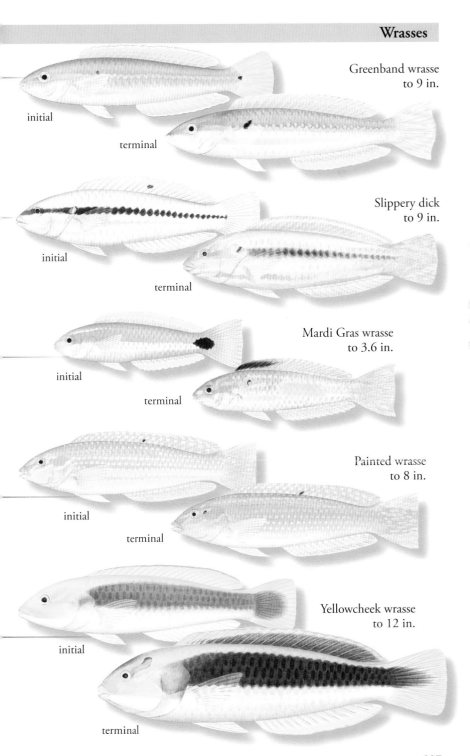

Greenband wrasse
to 9 in.

initial

terminal

Slippery dick
to 9 in.

initial

terminal

Mardi Gras wrasse
to 3.6 in.

initial

terminal

Painted wrasse
to 8 in.

initial

terminal

Yellowcheek wrasse
to 12 in.

initial

terminal

Labridae - Wrasses, *cont.*

Yellowhead wrasse - *Halichoeres garnoti* (Valenciennes, 1839)

FEATURES: Initial phase brownish on upper head and upper forebody, yellow on sides, whitish ventrally. Terminal phase variable; yellow to tannish head and forebody; posterior portion of body green to bluish; always with a dark mid-body bar that joins a dark stripe along posterior portion of back. Both phases with narrow black lines and spots radiating from eyes and banded fins. Juveniles yellow with bright blue stripe along midline. HABITAT: E FL, Gulf of Mexico, Bahamas, and Caribbean Sea to Brazil. Also Bermuda. Around shallow rocky and coral reefs to about 164 ft.

Clown wrasse - *Halichoeres maculipinna* (Müller & Troschel, 1848)

FEATURES: Initial phase greenish, with pinkish to orangish spots and wavy lines on sides. Terminal phase greenish to greenish blue on anterior upper sides, becoming greenish yellow posteriorly. Large, black blotch on lower sides above anal-fin insertion. Both with small blue spot at upper pectoral-fin base. Juveniles greenish yellow dorsally with broad blackish stripe on sides. All with orange to reddish stripes and bands on head and black blotch on anterior dorsal fin. HABITAT: NC south to Brazil. Also E and S Gulf of Mexico and Bermuda. Over shallow coral and rocky reefs.

Rainbow wrasse - *Halichoeres pictus* (Poey, 1860)

FEATURES: Initial phase tannish yellow dorsally, with brownish stripe from snout to opercle. Stripe becomes double row of spots on sides. Whitish ventrally. Terminal phase lavender green, bluish green, to yellowish green dorsally. Nape rosy. Whitish below. Upper and lower portions separated by blue lines and spots. Always with a large, black, ocellated blotch on caudal peduncle. HABITAT: S FL, Bahamas, Antilles to northern South America. Possibly southern Gulf of Mexico. Reef associated.

Blackear wrasse - *Halichoeres poeyi* (Steindachner, 1867)

FEATURES: Initial phase yellowish to greenish yellow. May be almost uniformly colored or with a pale stripe on lower sides. May have black spots on dorsal-fin membrane and at caudal-fin base. Terminal phase greenish yellow, with pinkish and bluish wavy lines and spots on head and body. Caudal fin with pink V-shaped mark. Both phases with an orange to purplish mark behind eyes and small black spot at rear dorsal-fin base. HABITAT: S FL, southern Gulf of Mexico, Bahamas, Caribbean Sea to Brazil. Usually over shallow seagrass beds. Occasionally over coral reefs.

Puddingwife - *Halichoeres radiatus* (Linnaeus, 1758)

FEATURES: Initial phase greenish orange with whitish saddles on upper sides. Terminal phase olive to yellowish green with pale bluish bar near midflank. Both phases with bright blue lines and spots on head and fins, dark spot at upper pectoral-fin base, blue lines on scales, and yellowish caudal-fin margin. Juveniles (not shown) orangish with blue bars and stripes and with two large black ocellated spots. All are robust, deep-bodied. HABITAT: NC to FL, Gulf of Mexico, Bahamas, and Caribbean Sea to Brazil. Also Bermuda. Reef associated from near shore to about 164 ft.

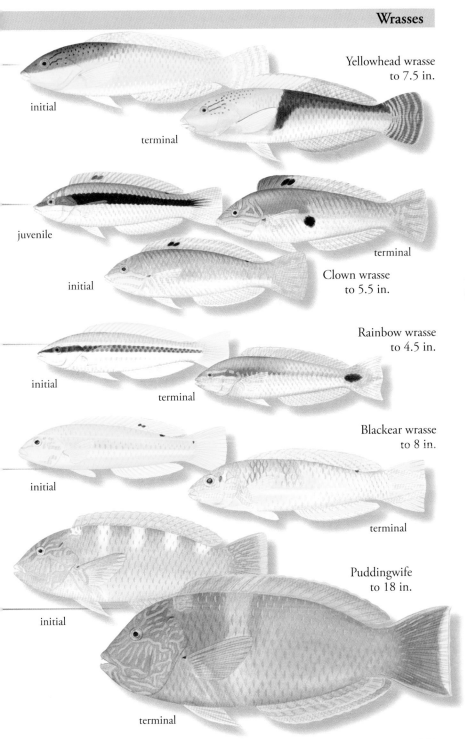

Yellowhead wrasse
to 7.5 in.

initial

terminal

juvenile

terminal

initial

Clown wrasse
to 5.5 in.

Rainbow wrasse
to 4.5 in.

initial

terminal

Blackear wrasse
to 8 in.

initial

terminal

Puddingwife
to 18 in.

initial

terminal

309

Labridae - Wrasses, *cont.*

Hogfish - *Lachnolaimus maximus* (Walbaum, 1792)

FEATURES: Adults shades of orange to pearly white on lower head and body. Upper head and nape reddish to blackish with paler lines and spots. Caudal fin dark at base and upper and lower margins. Juveniles pale or mottled and barred pearly white and reddish orange. All with dark spot at rear second dorsal-fin base. Anterior dorsal-fin spines free, elongate. All may pale or darken. HABITAT: NC to FL, Gulf of Mexico, Bermuda, Bahamas, Caribbean Sea to Brazil. Over soft bottoms and around deep reefs. BIOLOGY: Feed on variety of invertebrates. Popular game and aquarium fish. Vulnerable.

Tautog - *Tautoga onitis* (Linnaeus, 1758)

FEATURES: Color varies with habitat. Shades of brownish green, grayish green to blackish brown. Females and smaller specimens mottled and blotched with irregular bars. Large males somewhat uniformly colored dorsally and on sides, pale below, with a pale botch on each side. HABITAT: Nova Scotia to SC. Commonly between Cape Cod and Chesapeake Bay. Occur near bottom around inshore and coastal rocks, pilings, artificial reefs, and wrecks. BIOLOGY: Migratory and long-lived gamefish. Vulnerable.

Cunner - *Tautogolabrus adspersus* (Walbaum, 1792)

FEATURES: Color varies with habitat. Usually dusky orange to dusky reddish, but may be greenish. All with iridescent reflections. May be uniformly colored or may display faint bars or mottling on sides. May have a dusky spot on lower soft dorsal-fin base. Outer teeth canine-like. HABITAT: Labrador to VA. Near shore to about 420 ft. Around structures, rocky areas, and ledges. BIOLOGY: Cunner form loose congregations. Feed on crustaceans and mollusks.

Bluehead - *Thalassoma bifasciatum* (Bloch, 1791)

FEATURES: Initial phase with purplish band through eyes and purplish blotch on opercle. Diffuse, bicolored bars on sides. Black blotch on anterior dorsal-fin membranes. Dark blotch above pectoral-fin base. Terminal phase with bluish head and greenish body. Always with a black-and-white V-shaped wedge on forebody. Juveniles yellow dorsally, white ventrally, with or without a dark lateral stripe. Markings on head and dorsal fin similar to initial phase. HABITAT: E FL, Gulf of Mexico, Bahamas, and Caribbean Sea to northern South America. Also Bermuda and eastern Atlantic. Reef associated.

Rosy razorfish - *Xyrichtys martinicensis* Valenciennes, 1840

FEATURES: Females salmon on sides. Dark blotch on opercle. Scales on chest white with dark centers. Large white blotch and vertical red lines on abdomen. May display faint bars on sides. Males pearly yellow anteriorly, pinkish posteriorly. Head with pale blue bars. Scales with pale blue centers. Base of pectoral fins with dark, multicolored band. HABITAT: S FL, southern Gulf of Mexico, Bahamas, Caribbean Sea to northern South America. Also Bermuda. Over coral rubble bottoms and shallow seagrass beds. BIOLOGY: Burrow into bottom sediment when threatened.

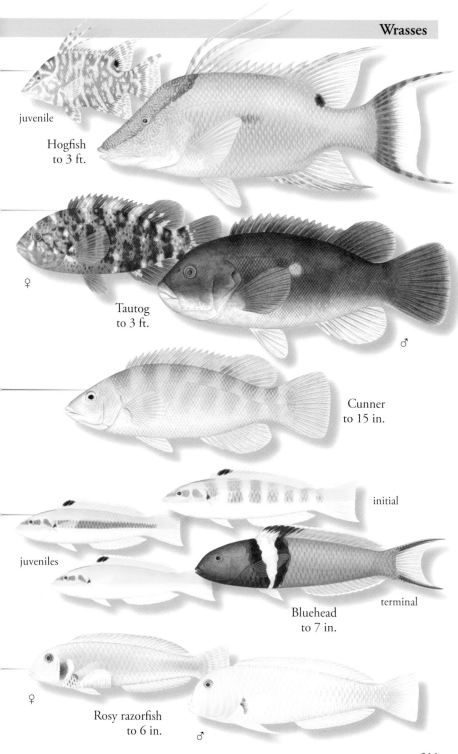

juvenile

Hogfish
to 3 ft.

♀

Tautog
to 3 ft.

♂

Cunner
to 15 in.

initial

juveniles

terminal

Bluehead
to 7 in.

♀

Rosy razorfish
to 6 in.

♂

Labridae - Wrasses, *cont.*

Pearly razorfish - *Xyrichtys novacula* (Linnaeus, 1758)

FEATURES: Females pearly pinkish, orangish, to whitish with bright blue lines on head. Distinct pearly patch on abdomen. First dorsal-fin spine taller than second. Males pearly greenish yellow with bright blue lines on head. Diffuse dusky purplish to reddish blotch on sides. Both with blue lines on scales rows, steep snout profile, and laterally compressed body. HABITAT: NC to FL, Gulf of Mexico, Bahamas, Caribbean Sea to Brazil. Also Bermuda and eastern Atlantic. Over sandy and coral bottoms from shore to about 260 ft. BIOLOGY: Burrow into bottom to avoid predators.

Green razorfish - *Xyrichtys splendens* Castelnau, 1855

FEATURES: Females greenish to pearly or reddish. Sides of head with pale green and orange bars. First two dorsal-fin spines tall. Males iridescent greenish to orange. Sides of head with greenish and orangish bars. One to several blackish, ocellated spots on sides. Caudal fin with pinkish outer margin. Pelvic fins very long. Both with reddish irises and bluish spot between first and second dorsal-fin spines. Both may display pale bars on sides. HABITAT: S FL, Dry Tortugas, Bahamas, Antilles to northern South America. Also Bermuda. Over coral rubble, sandy bottoms, and seagrass beds.

Scaridae - Parrotfishes

Bluelip parrotfish - *Cryptotomus roseus* Cope, 1871

FEATURES: Initial phase may be overall pale, iridescent salmon red, or dull brown dorsally, pearly white ventrally. May have broken whitish stripes and pale flecks and spots on upper body. Terminal phase dull to iridescent greenish dorsally, with broken salmon stripe from opercle to caudal fin. Blue and pinkish bands run from mouth to eyes. Dark spot at upper pectoral-fin base. HABITAT: FL, Bahamas, Antilles to Brazil. Also Bermuda. Found over weedy or sandy bottoms to about 190 ft.

Midnight parrotfish - *Scarus coelestinus* Valenciennes, 1840

FEATURES: All phases and ages dark indigo blue. Pale blue blotches on head. Scale centers variably pale blue. All fins dark blue. Snout gently sloping in both phases. Upper and lower caudal-fin tips somewhat to greatly elongate. HABITAT: S FL, southern Gulf of Mexico, Bahamas, and Antilles to Brazil. Also Bermuda. Reef associated to about 65 ft. BIOLOGY: Midnight parrotfish use fused teeth to scrape algae from rocks and coral polyps from coral heads.

Blue parrotfish - *Scarus coeruleus* (Edwards, 1771)

FEATURES: Head, body, and fins entirely bluish in initial and terminal phases. Lips may be blackish. Initial phase with evenly rounded snout. Terminal phase with distinct hump on snout. Juveniles pearly blue with yellow over upper head and anterior dorsal area. HABITAT: MD to FL, Gulf of Mexico, Bahamas, Caribbean Sea to Brazil. Also Bermuda. Over shallow coral reefs to about 65 ft. BIOLOGY: Blue parrotfish use fused teeth to scrape algae from hard substrates and coral polyps from coral heads.

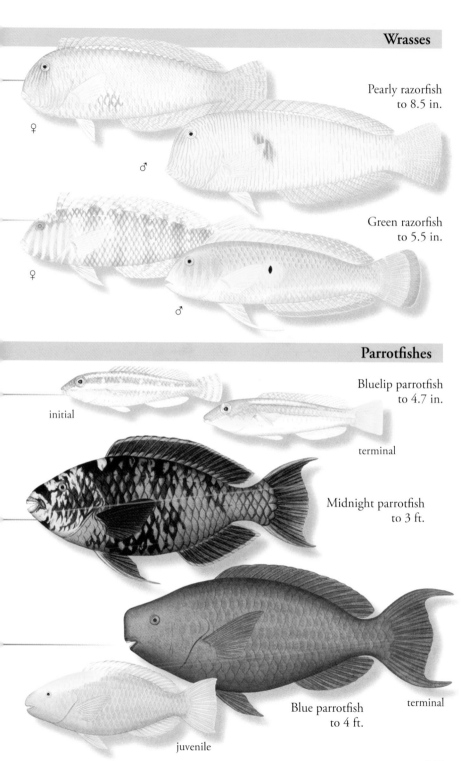

♀

♂

Pearly razorfish
to 8.5 in.

Green razorfish
to 5.5 in.

♀

♂

initial

Bluelip parrotfish
to 4.7 in.

terminal

Midnight parrotfish
to 3 ft.

terminal

Blue parrotfish
to 4 ft.

juvenile

Scaridae - Parrotfishes, *cont.*

Rainbow parrotfish - *Scarus guacamaia* Cuvier, 1829

FEATURES: Juveniles and initial phase rusty orange with green lines and spots radiating from eyes and green scale centers on body. Lower head and chest rusty orange. Upper and lower profiles similar in shape. Terminal phase rusty orange anteriorly, with purplish wash. Posterior portion of body green. Demarcation between anterior and posterior portions variable. Head profile concave above eyes. Body robust, comparatively deep. HABITAT: S FL, Bahamas, and Caribbean Sea to Brazil. Also Bermuda. Associated with shallow coral reefs. Juveniles occur around mangroves. BIOLOGY: Rainbow parrotfish retreat into a 'home' cave at night. Use fused teeth to scrape algae and coral polyps from reef surfaces.

Striped parrotfish - *Scarus iseri* (Bloch, 1789)

FEATURES: Initial phase with three broad, dark brown stripes separated by white stripes running from head to caudal fin. Abdomen pale with white streaks. Snout always yellow. Dorsal, caudal, and anal fins yellowish. Terminal phase somewhat variable. Always with two wavy, blue green bands above and below eyes. Pinkish on sides of head. Upper operuclar margin dark. May have a broad yellow to pink or grayish stripe on sides at midline. Posterior portions of body with blue green scale centers. Caudal fin with blue upper and lower margins, and pinkish to orange inner stripes. HABITAT: S FL, eastern Gulf of Mexico, Bahamas, and Caribbean Sea to Brazil. Also Bermuda. Strays to MA. Coral reef associated. BIOLOGY: Striped parrotfish scrape algae from coral surfaces with fused teeth. Feed in groups. May school with Princess parrotfish.

Princess parrotfish - *Scarus taeniopterus* Desmarest, 1831

FEATURES: Initial phase with three brownish stripes separated by pale yellow running from the head to caudal fin. Abdomen pale with white streaks. Snout brownish. Dorsal, caudal and anal fins yellowish. Upper and lower margins of caudal fin always brownish. Terminal phase with two blue, wavy bands above and below eyes. Sides of head orange. Lower sides with a broad, yellow patch. Dorsal and anal fins with broad, orange inner band. Caudal fin blue with orange upper and lower inner margins. Both phases become mottled while on the bottom or at night. HABITAT: FL, eastern Gulf of Mexico, Bahamas, and Caribbean Sea to Brazil. Also Bermuda. Strays to MA. Coral reef associated. BIOLOGY: Princess parrotfish form feeding aggregations. Take bites of algae from reef surfaces. Form a mucus cocoon at night.

Queen parrotfish - *Scarus vetula* Bloch & Schneider, 1801

FEATURES: Initial phase reddish to purplish brown on upper body. Broad whitish stripe on lower sides. Ventral area whitish with reddish brown scale centers. Head whitish with pale green hues. Terminal phase green to blue green with dusky orange to rosy scale margins on body. Green to blue green bands above and below jaws. Pectoral fins with orange to rosy inner band. Caudal fin with orange on inner upper and lower lobes. HABITAT: S FL, Bahamas, and Caribbean Sea to Brazil. Also Bermuda. Associated with shallow coral reefs. BIOLOGY: Use fused teeth to scrape algae from reef surfaces. Create mucus cocoon at night.

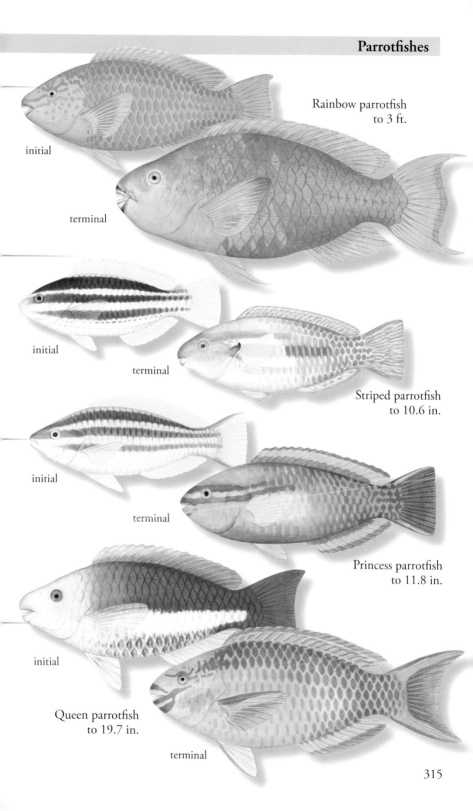

initial

Rainbow parrotfish
to 3 ft.

terminal

initial

terminal

Striped parrotfish
to 10.6 in.

initial

terminal

Princess parrotfish
to 11.8 in.

initial

Queen parrotfish
to 19.7 in.

terminal

Greenblotch parrotfish - *Sparisoma atomarium* (Poey, 1861)

FEATURES: Initial phase entirely reddish to yellowish red or with broken whitish stripes ventrally. Eyes and pectoral and anal fins yellowish. Terminal phase greenish yellow to greenish dorsally. Sides may be reddish to greenish. Abdomen with bright blue along scales. Narrow yellow band from corner of mouth to eye. Greenish blotch above pectoral fin. Small dark blotch between first and second dorsal-fin spines. HABITAT: FL Keys, Dry Tortugas, northern Gulf of Mexico, Bahamas, Caribbean Sea to northern South America. Around coral reefs and over seagrass beds from about 55 to 252 ft.

Redband parrotfish - *Sparisoma aurofrenatum* (Valenciennes, 1840)

FEATURES: Initial phase variable. May be dark olivaceous overall; greenish dorsally, reddish below; or greenish with two broken whitish stripes on sides. May be slightly to highly mottled. Always with distinct white blotch on upper caudal peduncle. Terminal phase greenish with pinkish area on lower sides. Upper head and forebody may be darker green. Always with reddish band from corner of mouth to opercle and small black spots above larger yellow blotch on upper side above pectoral fins. Anal fin reddish. Tips of caudal fin blackish. HABITAT: S FL, Gulf of Mexico, Bahamas, and Caribbean Sea to Brazil. Also Bermuda. Occur around coral reefs and over seagrass beds. BIOLOGY: Redband parrotfish feed on benthic algae and sea grasses.

Redtail parrotfish - *Sparisoma chrysopterum* (Bloch & Schneider, 1801)

FEATURES: Initial phase faintly or intensely mottled dusky reddish to dusky olivaceous. Always with distinct blackish blotch on upper pectoral-fin base. Colors may quickly fade or intensify. Terminal phase greenish with blue area on lower side. Diffuse blue line below eyes. Chin blue. Dark spot at upper pectoral-fin base. Pectoral fins with yellowish rays. Inner portion of caudal fin yellowish, bordered in reddish. May become mottled while on the bottom or at night. HABITAT: S FL, Gulf of Mexico, Bahamas, Caribbean Sea to Brazil. Also Bermuda. Occur over shallow coral reefs and seagrass beds. BIOLOGY: Feed on benthic algae and seagrasses by taking single large bites. Swim by flapping pectoral fins.

Bucktooth parrotfish - *Sparisoma radians* (Valenciennes, 1840)

FEATURES: Initial phase olivaceous to yellowish brown. Dorsal area densely mottled and speckled; may show irregular bars and diffuse stripes. Ventral area pale with dense speckles that follow scale rows. Opercular margin and pectoral-fin base with patch of blue green. May also have blue green band from mouth to behind eyes. Terminal phase highly variable. Uniform to mottled olivaceous to greenish brown. Dorsal area densely speckled. Always with a blue green and pinkish band from mouth to behind eyes. Pectoral-fin base always with blackish blotch. Caudal-fin margin always blackish. All with distinct tentacle on each nostril. HABITAT: S FL, eastern Gulf of Mexico, Bahamas, and Caribbean Sea to Brazil. Also Bermuda. Found primarily over seagrass beds. BIOLOGY: Bucktooth parrotfish swim rapidly away when disturbed, then come to rest on the bottom, where they assume a mottled color pattern.

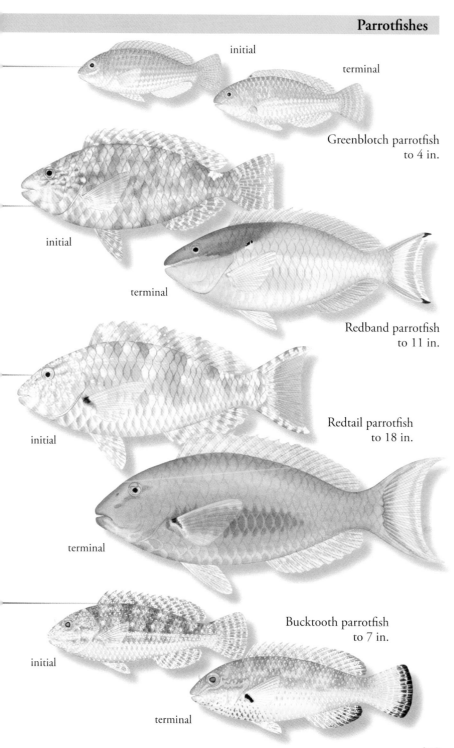

initial

terminal

Greenblotch parrotfish
to 4 in.

initial

terminal

Redband parrotfish
to 11 in.

initial

Redtail parrotfish
to 18 in.

terminal

Bucktooth parrotfish
to 7 in.

initial

terminal

Scaridae - Parrotfishes, *cont.*

Yellowtail parrotfish - *Sparisoma rubripinne* (Valenciennes, 1840)

FEATURES: Initial phase pale brownish gray with dark-rimmed scales and irregular bars on body. Dark bars on chin. Caudal peduncle and caudal fin yellow. Pelvic and anal fins pinkish. Terminal phase overall greenish blue on head and body. Dark blotch at upper pectoral-fin base. Inner portion of caudal fin yellow. Assume a mottled pattern while resting on the bottom or at night. HABITAT: MA to FL, E Gulf of Mexico, Bahamas, Caribbean Sea to Brazil. Also Bermuda and tropical West Africa. Over shallow seagrass beds and coral reefs. BIOLOGY: Feed on benthic vegetation. When pursued may swim into surf or lie on bottom. Other name: Redfin parrotfish.

Stoplight parrotfish - *Sparisoma viride* (Bonnaterre, 1788)

FEATURES: Initial phase mottled blackish brown and white on head and upper body. Abdomen and fins reddish. Inner portion of caudal fin whitish. Terminal phase blue to greenish. Upper portion of head and scale margins dusky pink to purplish. Dusky pinkish to purplish stripe runs from mouth to opercle. Always with bright yellow spot at upper opercular margin, yellow blotch at caudal-fin base, and yellow crescent on caudal-fin rays. Juveniles (not shown) reddish brown with evenly-spaced white spots on sides and white blotch at caudal-fin base. HABITAT: S FL, Gulf of Mexico, Bahamas, Caribbean Sea to Brazil. Also Bermuda. Over shallow coral reefs and seagrass beds. BIOLOGY: Feed on benthic vegetation.

ALSO IN THE AREA: Emerald parrotfish, *Nicholsina usta*, p. 431.

Zoarcidae - Eelpouts

Wolf eelpout - *Lycenchelys verrillii* (Goode & Bean, 1877)

FEATURES: Grayish brown above, pale below. Seven to ten dark, round patches on sides that are bisected by lateral line. Snout long. Mouth large. Upper jaw slightly projects over lower jaw. Pelvic fins very small. HABITAT: Newfoundland to NC. Demersal on muddy to sandy bottoms from about 150 to 3,600 ft.

Atlantic soft pout - *Melanostigma atlanticum* Koefoed, 1952

FEATURES: Head and abdomen silvery blue. Remainder of body tannish to creamy and translucent. Area around mouth blackish. Snout rounded. Anal fin taller than dorsal fin. Pelvic fins absent. Body soft, scaleless. HABITAT: Gulf of St. Lawrence to Cape Hatteras. Over edges of continental shelves to about 1,800 ft. Also NE Atlantic.

Ocean pout - *Zoarces americanus* (Bloch & Schneider, 1801)

FEATURES: Varying shades of yellowish brown to olivaceous. Dark bars radiate from eyes. Upper sides with dark, chain-like pattern. Pectoral fins and margins of dorsal and anal fins may be yellowish. Upper jaw projects over lower jaw. Posterior portion of dorsal fin is much lower than rest of fin; continuous with caudal fin. HABITAT: Labrador to NJ. Over variety of bottoms from shore to about 1,100 ft.

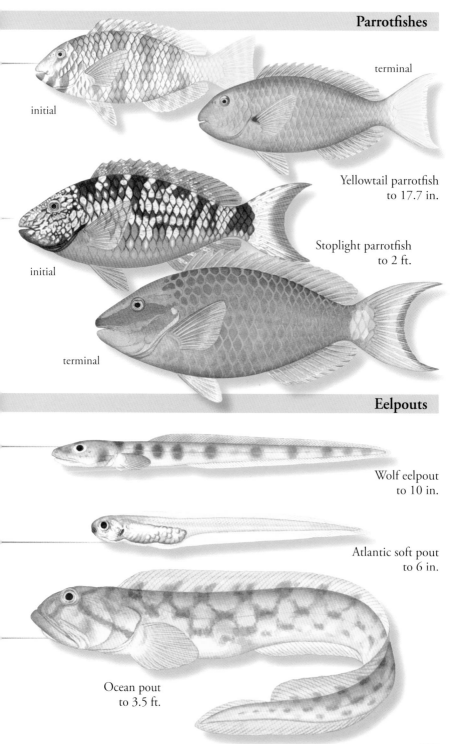

Parrotfishes

initial

terminal

initial

terminal

Yellowtail parrotfish
to 17.7 in.

Stoplight parrotfish
to 2 ft.

Eelpouts

Wolf eelpout
to 10 in.

Atlantic soft pout
to 6 in.

Ocean pout
to 3.5 ft.

Stichaeidae - Pricklebacks

Daubed shanny - *Leptoclinus maculatus* (Fries, 1838)

FEATURES: Tannish with darker spots and blotches. Abdomen pale. Dorsal fin banded. Caudal fin barred. Anterior dorsal-fin spines very short, gradually becoming taller at mid-body. Lower pectoral-fin rays elongate, free at tips. Pelvic fins long, very slender. HABITAT: From Arctic waters to Cape Cod. Also in northeastern Atlantic. Demersal over muddy bottoms from about 124 to 1,500 ft.

Snakeblenny - *Lumpenus lampretaeformis* (Walbaum, 1792)

FEATURES: Tannish with darker spots and blotches on upper sides. Abdomen pale. Dorsal fin banded. Caudal fin long, barred. First few dorsal fin spines somewhat shorter than tallest spines. Pelvic fins long, very slender. Anal fin base comparatively long. Body comparatively elongate, slender. HABITAT: Arctic and northern Atlantic Ocean to Cape Cod. Also in Atlantic. Found on or near soft to hard bottoms from about 6 to 300 ft. BIOLOGY: Form burrows in soft bottoms.

Radiated shanny - *Ulvaria subbifurcata* (Storer, 1839)

FEATURES: Brownish with alternating, paler blotches dorsally and on sides. Dark bands radiate from eyes. Dark blotch on dorsal fin between fifth or sixth to eighth or tenth spine. Anal fin margin dark. Body comparatively stout. Dorsal and anal fins continuous with caudal fin. Lateral line broken. HABITAT: Newfoundland to MA. Demersal over rocky shores and hard bottoms from low tide line to about 180 ft.

ALSO IN THE AREA: Arctic shanny, *Stichaeus punctatus*, p. 431.

Cryptacanthodidae - Wrymouths

Wrymouth - *Cryptacanthodes maculatus* Storer, 1839

FEATURES: Varying shades of reddish brown to brown. Two or three rows of irregular, darker spots on upper body from head to caudal fin. Top of head and fins spotted. Top of head flattened. Snout short. Mouth large, oblique. Pectoral fins well-developed. Dorsal and anal fins continuous with caudal fin. Pelvic fins absent. HABITAT: Labrador to NJ. Demersal over soft, muddy bottoms from intertidal zone to about 1,900 ft. BIOLOGY: Construct a system of branching burrows.

Pholidae - Gunnels

Rock gunnel - *Pholis gunnellus* (Linnaeus, 1758)

FEATURES: Color varies with habitat. May be greenish, yellowish, reddish, or brownish. All have a single row of 10–15 dark, occellated spots along dorsal-fin base that blend into dorsal profile. Two dark bars radiate from eyes. Sides variably mottled. Dorsal and anal fins continuous with caudal fin. Pectoral fins broad. Pelvic fins very small and short. HABITAT: Labrador to DE. More commonly to MA. Also in eastern Atlantic. Demersal from intertidal zone to about 240 ft.

Daubed shanny
to 7 in.

Snakeblenny
to 7 in.

Radiated shanny
to 7 in.

Wrymouth
to 3 ft.

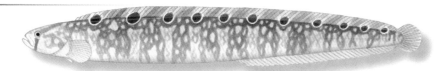

Rock gunnell
to 12 in.

321

Anarhichadidae - Wolffishes

Atlantic wolffish - *Anarhichas lupus* Linnaeus, 1758

FEATURES: Body and fins grayish, brownish, olivaceous to bluish. Larger specimens with dark blotches which form five to ten irregular darker bars on sides. Bars fade posteriorly. Smaller specimens appear more blotched. Head large, rounded. Mouth with large, protruding, canine-like teeth. Pelvic fins absent. HABITAT: Greenland to Cape Cod. Strays to NJ. Also in northeastern Atlantic. Found over hard bottoms from about 50 to 500 ft. BIOLOGY: A voracious and strong-jawed predator. Feed on wide variety of prey. Usually solitary; form pairs during breeding season.

Spotted wolffish - *Anarhichas minor* Olafsen, 1772

FEATURES: Olive brown to dark brown. Darker, somewhat occellated spots cover upper head and body and may form clusters. Dorsal fin spotted. Smaller specimens appear barred. Head large, rounded. Mouth with large, canine-like teeth. Pelvic fins absent. HABITAT: Primarily north of the Arctic circle. Greenland to Gulf of Maine; straying to MA and possibly NJ. Also in northeastern Atlantic. Demersal from about 300 to 1,500 ft. BIOLOGY: Feed on a variety of invertebrates and fishes.

Percophidae - Duckbills

Duckbill flathead - *Bembrops anatirostris* Ginsburg, 1955

FEATURES: Tannish dorsally, yellowish tan to whitish ventrally. Scales darkly outlined. Irregular, brownish blotches on sides. First dorsal and anal fins with dark margins. Head and anterior body flattened. Upper jaw with tentacle at rear corner. Eyes comparatively small. Second dorsal-fin spine elongate. HABITAT: Gulf of Mexico and Caribbean Sea. Demersal from about 270 to 1,700 ft.

Goby flathead - *Bembrops gobioides* (Goode, 1880)

FEATURES: Tannish dorsally, yellowish tan to whitish ventrally. Scales darkly outlined. First dorsal fin with dark anterior margin. Second dorsal fin dark along anterior margin and base. Caudal fin with dark spot on upper base and dark outer margin. Head and anterior body flattened. Upper jaw with short tentacle at rear corner. Eyes comparatively large. HABITAT: NY to FL, Gulf of Mexico, Bahamas, and Antilles. On outer continental shelves and slopes from about 270 to 1,800 ft.

Ammodytidae - Sand lances

American sand lance - *Ammodytes americanus* DeKay, 1842

FEATURES: Olive, bluish green, to brownish dorsally. Silvery on sides, white below. Head long, with pointed snout. Lower jaw protrudes. Dorsal fin single, long-based. Anal fin about half as long as dorsal fin. Caudal fin forked. HABITAT: Labrador to VA. Strays to NC. Found over soft bottoms of shallow coastal and estuarine waters. BIOLOGY: Able to quickly bury in soft substrates. May remain buried through low tide cycle. SIMILAR SPECIES: Northern sand lance, *Ammodytes dubius*, p. 431.

Wolffishes

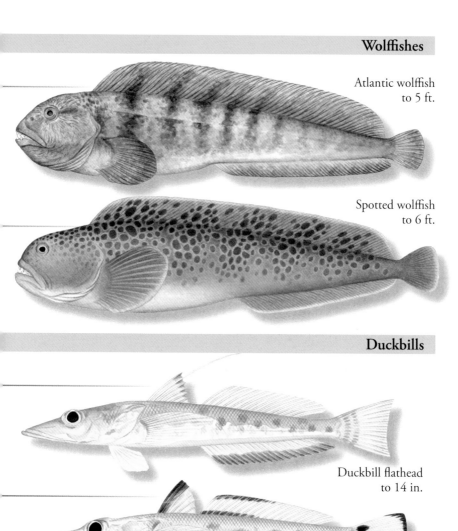

Atlantic wolffish
to 5 ft.

Spotted wolffish
to 6 ft.

Duckbills

Duckbill flathead
to 14 in.

Goby flathead
to 12 in.

Sand lances

American sand lance
to 8 in.

Uranoscopidae - Stargazers

Northern stargazer - *Astroscopus guttatus* Abbott, 1860

FEATURES: Brownish gray to olive brown dorsally to about midline. Whitish below. Dark area on caudal peduncle forms a stripe that merges with middle caudal-fin stripe. Dorsal area with close-set pale spots from lower jaw to caudal peduncle; spots become fewer and larger posteriorly. First dorsal fin blackish. Soft dorsal fin with two to four dark bands. Pectoral fins dark, pale at margin. Mouth broad, nearly vertical. Head flattened dorsally. Eyes small, set on top of head. Small venomous spine above each pectoral fin. Body robust. HABITAT: NY to NC. Found coastally over soft bottoms from near shore to about 120 ft. BIOLOGY: Lie buried in bottom sediment with only top of head exposed. Use electric organ to stun prey.

Southern stargazer - *Astroscopus y-graecum* (Cuvier, 1829)

FEATURES: Brownish dorsally to about midline. Whitish below. Dark portion does not form distinct stripe at caudal peduncle. Small, dark-ringed white spots cover dorsal portion of body from lower jaw to caudal peduncle. Spots stay relatively uniform in size posteriorly. First dorsal fin blackish. Soft dorsal fin with one to three dark bands. Pectoral fins dark, pale at margin. Mouth broad, nearly vertical. Head flattened dorsally. Eyes small, set on top of head. Small venomous spine above each pectoral fin. Body robust. HABITAT: NC to FL, Gulf of Mexico to Brazil. Absent from Antilles. Found coastally over soft bottoms from near shore to about 230 ft. BIOLOGY: Southern stargazer lie buried in bottom sediment with only top of head exposed. Use electric organ to stun prey. Feed on fishes.

Freckled stargazer - *Gnathagnus egregius* (Jordan & Thompson, 1905)

FEATURES: Brownish to grayish dorsally with darker, olivaceous lines, spots, and vermiculations. Paler ventrally. Dorsal, caudal, anal, and pectoral fins dark at base. Mouth broad, almost vertical. Eyes large, set on top of head. Head with knobby ridges, bumps, and spines. Preopercular corner protrudes laterally away from head. Broad, flattened spine above each pectoral fin. Spiny dorsal fin absent. Body robust. HABITAT: GA to FL, northwestern and southern Gulf of Mexico. Possibly Bermuda. Found on or near bottom from about 65 to 1,700 ft. BIOLOGY: Adults are found in deeper waters. Juveniles are found in shallow waters.

Lancer stargazer - *Kathetostoma albigutta* Bean, 1892

FEATURES: Brownish dorsally to about midline. Whitish below. Small, white spots on upper head. Spots become much larger posteriorly, forming blotches. Soft dorsal fin with several blackish blotches. Caudal fin with blackish blotches. Pectoral fins blackish, white along lower margin. Mouth broad, nearly vertical. Head flattened dorsally. Eyes small, set on top of head. Lower preopercular corner with three spines. Long, pointed, venomous spine above each pectoral fin. Spiny dorsal fin absent. Body robust. HABITAT: NC to FL and Gulf of Mexico to Yucatán. Demersal in offshore waters from about 88 to 1,400 ft. BIOLOGY: Adults are found in deeper waters. Juveniles are found in shallow waters.

Northern stargazer
to 22 in.

Southern stargazer
to 17.5 in.

Freckled stargazer
to 13 in.

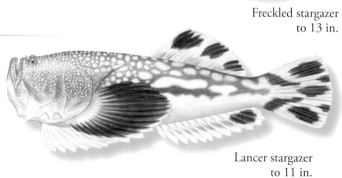

Lancer stargazer
to 11 in.

Tripterygiidae - Triplefin blennies

Lofty triplefin - *Enneanectes altivelis* Rosenblatt, 1960

FEATURES: Translucent with four oblique, reddish to brownish bars on sides. Bar on caudal peduncle not usually darker than others. Diffuse reddish bands radiate from eyes. Anal fin distinctly barred. Upper snout profile gently sloping. First dorsal fin considerably taller than anterior portion of second dorsal fin. Pectoral-fin base and abdomen scaled. Anal fin with two spines and 14–16 (usually 15) rays. HABITAT: S FL, FL Keys, Bahamas, Greater and Lesser Antilles to Nicaragua. Also Campeche Bay. Over clear, shallow coral reefs and rocky shores to about 32 ft.

Roughhead triplefin - *Enneanectes boehlkei* Rosenblatt, 1960

FEATURES: Translucent with four oblique, reddish to brownish oblique bars on sides. Bar on caudal peduncle usually darker than others. Diffuse reddish bands radiate from eyes. Anal fin distinctly barred. Upper snout profile sloping. First dorsal-fin spine shorter than anterior portion of second dorsal fin. Pectoral-fin base and abdomen scaleless. Anal fin with two spines and 16–17 rays. HABITAT: S FL, Dry Tortugas, Bahamas, Greater and Lesser Antilles to Venezuela. Also Veracruz, Mexico. Reef associated. Observed perching on sponges and corals.

Redeye triplefin - *Enneanectes pectoralis* (Fowler, 1941)

FEATURES: Translucent with four oblique, reddish to brownish bars on body. Bar on caudal peduncle blackish. Caudal-fin base reddish. Diffuse bands radiating from eyes. Anal fin mostly evenly pigmented. Upper snout profile sloping. First dorsal-fin spine considerably shorter than anterior portion of second dorsal fin. Pectoral-fin base and abdomen scaled. Anal fin with two spines and 14–16 (usually 15) rays. HABITAT: S FL, Bahamas, Antilles to Venezuela. Reef associated to about 36 ft.

Dactyloscopidae - Sand stargazers

Bigeye stargazer - *Dactyloscopus crossotus* Starks, 1913

FEATURES: Pale tannish. Eyes usually blackish. Body may be unmarked or may have eight to twelve indistinct brownish bars along dorsal profile. Cheek and opercle may be iridescent in some specimens. Eyes large, unstalked. First several dorsal-fin spines free. Body comparatively shallow. HABITAT: SE FL, Bahamas, Cuba, Barbados, and coast of Brazil. Demersal in shallow water along sandy beaches. BIOLOGY: Bigeye stargazer burrow into sandy bottoms to lie in wait for prey.

Speckled stargazer - *Dactyloscopus moorei* (Fowler, 1906)

FEATURES: Pale. Brownish, rather evenly scattered speckles on upper sides, often forming faint stripes along scales. Some specimens described as blotched or with irregular barring dorsally. Eyes on short stalks. Dorsal fin usually continuous, with first membranes incised but not separate. HABITAT: NC to FL Keys, and in Gulf of Mexico from FL to Texas. Demersal from about 10 to 115 ft. BIOLOGY: Speckled stargazer burrow into soft bottoms with only eyes, nostrils, and mouth exposed.

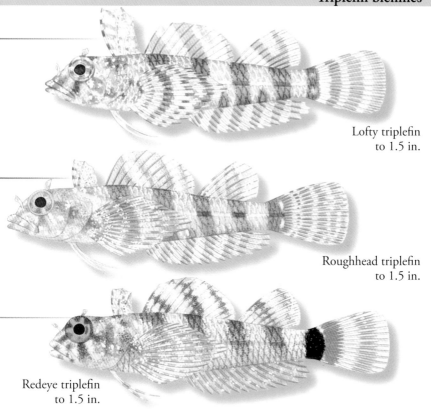

Lofty triplefin
to 1.5 in.

Roughhead triplefin
to 1.5 in.

Redeye triplefin
to 1.5 in.

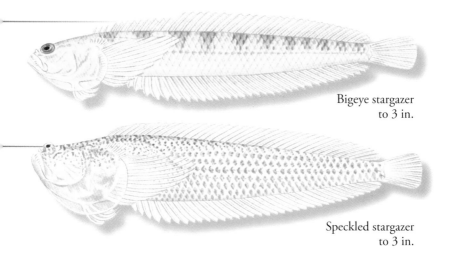

Bigeye stargazer
to 3 in.

Speckled stargazer
to 3 in.

Sand stargazer - *Dactyloscopus tridigitatus* Gill, 1859

FEATURES: Pale whitish with irregular brownish and tannish mottling on head. Upper body with brownish flecks. Some specimens display 11 to 14 short brownish bars along dorsal profile. Eyes very small, on tall stalks. First three to four dorsal-fin spines separate. Arched portion of lateral line comparatively short. HABITAT: SE FL, Bahamas, and Antilles to Venezuela. Also reported from Bermuda. On sandy bottoms from shore to about 95 ft. BIOLOGY: Burrow into soft bottoms to await prey.

Arrow stargazer - *Gillellus greyae* Kanazawa, 1952

FEATURES: Tannish dorsally, translucent below. Upper potion of body mottled in white and tan with small dark flecks and five narrow brownish bars. A sixth bar is present on caudal peduncle. Dark band across eyes. First dorsal fin reddish brown. First three dorsal-fin spines form a separate finlet. Remainder of dorsal fin notched near midlength. HABITAT: SE FL, FL Keys, Bahamas and Antilles to Brazil. Possibly Bermuda. Found over shallow sandy areas around coral reefs, seagrass beds, boulders and pilings from shore to about 88 ft. BIOLOGY: Burrow into soft bottoms.

Masked stargazer - *Gillellus healae* Dawson, 1982

FEATURES: Pale with brownish bar below and between eyes. Seven brownish bars on upper sides. Bars generally broader at midline and paler near centers; may have a dark blotch between each bar. Eyes with circle of dermal flaps; elevated but not stalked. First three dorsal-fin spines form a separate finlet. HABITAT: SC to FL Keys, and in Gulf of Mexico from Dry Tortugas to AL. Also Puerto Rico, Aruba, and Lesser Antilles to Belize. Demersal from about 69 to 240 ft.

Warteye stargazer - *Gillellus uranidea* Böhlke, 1968

FEATURES: Pale with four to five narrow brownish saddles on upper sides, or variably plain with traces of brown dorsally and on sides. Head with dark spots and blotches. Eyes elevated but not stalked and with a circle of dermal flaps, and a tentacle posteriorly. Fimbrae absent from upper lip. First three dorsal-fin spines form separate finlet. Remainder of dorsal fin notched at about mid-length. HABITAT: SE FL and Bahamas, Antilles to Panama. Demersal over sandy bottoms around rocks and patch reefs. BIOLOGY: Burrow into soft bottoms to await prey.

Saddle stargazer - *Platygillellus rubrocinctus* (Longley, 1934)

FEATURES: Pale with four broad brownish bars from first dorsal fin to caudal peduncle. Bars generally reach lower profile. Brownish mottling on head. Pectoral fin may have brownish flecks. Eyes with incomplete circle of dermal flaps. First three dorsal-fin spines form semi-isolated finlet. Remainder of dorsal fin notched near mid-length. HABITAT: SE FL, FL Keys, Dry Tortugas, Bahamas, Antilles to Venezuela. Also western Caribbean Sea. Demersal from shore to about 98 ft. BIOLOGY: Saddle stargazer burrow into soft bottoms to await prey.

ALSO IN THE AREA: Reticulate stargazer, *Dactyloscopus foraminosus*, p. 431.

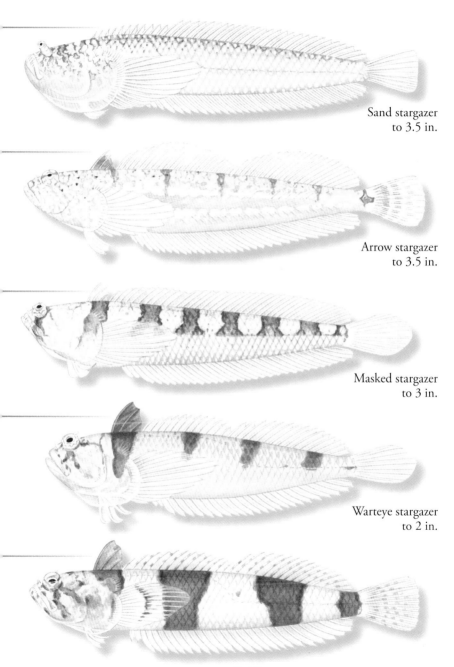

Sand stargazer
to 3.5 in.

Arrow stargazer
to 3.5 in.

Masked stargazer
to 3 in.

Warteye stargazer
to 2 in.

Saddle stargazer
to 2.5 in.

Blenniidae - Combtooth blennies

Striped blenny - *Chasmodes bosquianus* (Lacepède, 1800)

FEATURES: Color varies. Females shades of dusky greenish blue to dusky brownish with irregular, pale lines and spots on sides. Males shades of olive to brownish with irregular lines and spots on sides. Bright blue blotch between first and second dorsal-fin spines. An orange streak extends from the first to about the tenth dorsal-fin spine. Jaws extend to about middle of eyes in all. HABITAT: NY (rare) to N FL. Found in shallow nearshore and estuarine waters over soft bottoms, seagrass beds, and oyster reefs. BIOLOGY: Spawn in summer. Empty shells are preferred spawning sites and males aggressively defend them. Migrate to deeper flats and channels in winter.

Stretchjaw blenny - *Chasmodes logimaxilla* Williams, 1983

FEATURES: Color varies. Females shades of dark brown with irregular, pale lines and spots on sides. Males shades of olive to brownish with irregular, pale lines and spots on sides. Bright blue blotch between first and second dorsal-fin spines. An orange streak extends from first to about tenth dorsal fin spine. Jaws extend well past rear margin of eyes in all. Snout comparatively long, pointed. HABITAT: Gulf of Mexico from Pensacola, FL, to Veracruz, Mexico. Typically found over nearshore oyster reefs, but may be found over seagrass beds. BIOLOGY: Males aggressively defend territories and clutches of eggs during spawning. Females are less territorial.

Florida blenny - *Chasmodes saburrae* Jordan & Gilbert, 1882

FEATURES: Olivaceous to brownish with irregular pale lines and mottling on sides. Males with bright blue blotch on anterior dorsal fin and an orange streak from the first to about the tenth dorsal-fin spine. All with short jaws not reaching rear margin of eyes. Head profile evenly rounded. Snout comparatively short, blunt. HABITAT: E FL to Chandeleur Islands, LA. Primarily over nearshore seagrass beds, sometimes on oyster beds. BIOLOGY: Feed on variety of invertebrates and fish larvae.

Pearl blenny - *Entomacrodus nigricans* Gill, 1859

FEATURES: Color and pattern variable. Tannish with dark brownish to reddish brown bars on sides. Bars often subdivided. Bluish to white spots pepper pale areas and edges of bars. Snout and upper lip with narrow dark lines. Diffuse dark spot on head behind eyes. Head blunt. Snout steeply sloping. Tuft of cirri over nostrils and eyes. Single cirrus on nape. Body elongate. Dorsal fin distinctly notched. HABITAT: S FL, S Gulf of Mexico, Bahamas, Caribbean Sea to Venezuela. Also Bermuda. Found in tide pools and over shallow rocky and hard bottoms from near shore to about 20 ft.

Barred blenny - *Hypleurochilus bermudensis* Bebee & Tee-Van, 1933

FEATURES: Head grayish orange with darker and paler smudges. Body pale grayish to pale tannish with wide, brownish bars on sides. Bars merge along midline. Pale areas with dark smudges. Cirri on eyes may have one to six branches. HABITAT: FL Keys, Dry Tortugas, Flower Garden Banks, and off Veracruz, Mexico. Also Bahamas and Bermuda. Occur over rocky bottoms and coral reefs to about 88 ft.

Striped blenny
to 4 in.

♀

♂

Stretchjaw blenny
to 4 in.

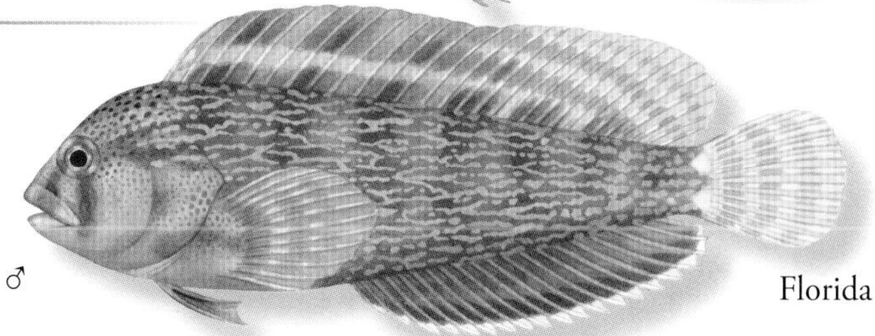

♀

♂

Florida blenny
to 4 in.

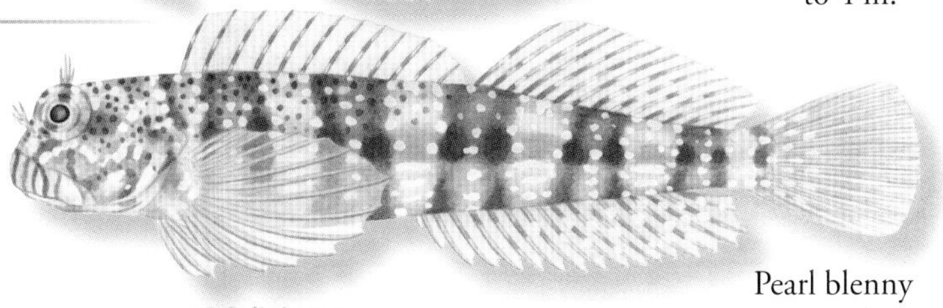

♂

Pearl blenny
to 4 in.

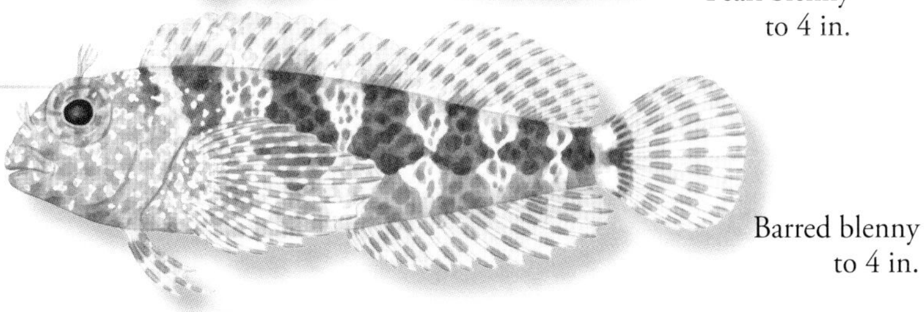

Barred blenny
to 4 in.

Zebratail blenny - *Hypleurochilus caudovittatus* Bath, 1994

FEATURES: Dark brown to dark grayish brown on head and body. Often with dark bluish spot between first and second dorsal-fin spines. Caudal fin pale, translucent, and with three to five dark bands. Rear portions of soft dorsal and anal fins may have translucent blotches. Cirri on eyes may have one to seven branches. HABITAT: In Gulf of Mexico from Sarasota, FL, to St. Andrews Bay, FL. Associated with sandy bottoms. Also reported over rocky and rubble bottoms and artificial reefs.

Crested blenny - *Hypleurochilus geminatus* (Wood, 1825)

FEATURES: Grayish to brownish gray with orange brown spots and smudges on head and body. Spots may cluster on sides, forming irregular, oblique bars. Dark cluster of spots on preopercle behind eyes. Males darker than females; distinct crest on nape. Females with rather evenly scattered spots and smudges. Both with a dark spot on membrane between first and second dorsal-fin spines. Cirri over eyes broad-based with many smaller branches at base. Dorsal fin lacks distinct notch. HABITAT: NJ to E FL. Found over hard bottoms, on jetties, and around oil rigs.

Featherduster blenny - *Hypleurochilus multifilis* (Girard, 1858)

FEATURES: Color variable. Observed specimen reddish brown, others described as grayish brown. Females have dark bars on the jaws and spotting on the head; the body is spotted and banded, mostly anteriorly. Males are generally darker overall, with faint spots and bands; may appear almost uniformly dark. Both with or without a dark spot on membrane between first and second dorsal-fin spines. Caudal fin may be banded or uniformly colored. Cirri on eyes multi-branched, with a central stalk. Males with very tall cirri. HABITAT: Gulf of Mexico from Panama City, FL, to Rockport, TX. Reported from around pilings and over oyster beds in shallow water.

Oyster blenny - *Hypleurochilus pseudoaequipinnis* Bath, 1994

FEATURES: Females with about six pairs of diffuse, dark brown blotches that form loose bars on sides. Males are described as greenish gray with small orange spots on upper body. Chin with dark bands. Both with spots on head. A darker area below a pale stripe present behind eyes. Pale spots may be present on midline below dark bars. Dorsal fin with a dark spot on membrane between first and second spines. Cirri on eyes with a tall central tentacle and several small branches at base. HABITAT: E and S FL, Yucatán, Bahamas, Antilles to Brazil. Around mangrove roots, rocky outcroppings, and pilings. BIOLOGY: Previously known as *Hypleurochilus aequipinnis*.

Orangespotted blenny - *Hypleurochilus springeri* Randall, 1966

FEATURES: Pale bluish to creamy, with orange spots on head and forebody. Spots cluster and form bars on sides and become darker and larger posteriorly in young specimens. Dorsal, caudal, and anal fins evenly spotted. Distinct spot between first and second dorsal-fin spines absent. Cirri above eyes short, with few branches. HABITAT: S FL and Bahamas to northern South America. Found over rocky and reef areas along quiet, shallow shorelines. BIOLOGY: Feed on worms, crustaceans, and algae.

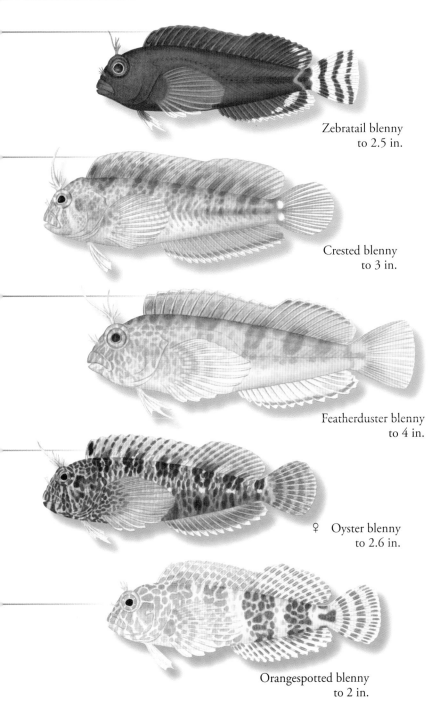

Zebratail blenny
to 2.5 in.

Crested blenny
to 3 in.

Featherduster blenny
to 4 in.

♀ Oyster blenny
to 2.6 in.

Orangespotted blenny
to 2 in.

333

Blenniidae - Combtooth blennies, *cont.*

Feather blenny - *Hypsoblennius hentz* (Lesueur, 1825)

FEATURES: Olive brown with darker spots and smudges that cluster to form oblique bars on head and body. Bluish spot ususally present on second dorsal-fin spine. Dark blotch on head behind eyes. Pelvic fins dark. Anal fin dark in males, banded in females. Cirri above eyes have many branches off main stalk. HABITAT: Nova Scotia to N FL. In Gulf of Mexico from W FL to Campeche Bay. Occur over soft, muddy bottoms, oyster reefs, and seagrass beds. BIOLOGY: Feather blenny spawn in summer over oyster reefs. Feed on a variety of invertebrates and plant material.

Tesselated blenny - *Hypsoblennius invemar* Smith-Vaniz & Acero P., 1980

FEATURES: Head and body covered with reddish to orange spots and polygons. Spots surrounded by blackish rings dorsally. Interspaces bluish to greenish blue. Black blotch on head behind eyes. Dorsal and anal fins spotted. Pectoral- and caudal-fin bases spotted. Fleshy cirri above each eye with up to four branches. HABITAT: Northern Gulf of Mexico, Lesser Antilles, Venezuela, and Colombia. Occupy large, empty barnacle shells that are attached to pilings, buoys, and rocks, in shallow, clear waters. BIOLOGY: Live in barnacles with head protruding. Leave barnacle to feed.

Redlip blenny - *Ophioblennius macclurei* (Silvester, 1915)

FEATURES: Overall reddish brown; or, reddish brown to gray anteriorly, grayish posteriorly. Lower lip reddish. Lower pectoral-fin rays, dorsal- and anal-fin margins reddish. Simple cirrus above each eye. Nostrils with branched cirri. Cirri on nape. Snout profile almost vertical. HABITAT: NY (rare) to FL, in Gulf of Mexico from W FL, Dry Tortugas, TX, Veracuz, and Yucatán. Also eastern Atlantic. Occur over coral reefs and rocky shores. BIOLOGY: Feed almost exclusively on filamentous algae.

Seaweed blenny - *Parablennius marmoreus* (Poey, 1876)

FEATURES: Color and pattern highly variable. Tannish with series of brownish spots; brownish with series of pale spots; heavily spotted along sides; or, overall yellowish. Always with bluish lines on snout and lower cheeks. Lines may be straight, wavy, or broken. Pale to bright blue spot may be present on anterior dorsal fin. Cirri above eyes with several branches at base. HABITAT: NY to FL, Gulf of Mexico, Bahamas, Caribbean Sea to northern South America. Also Bermuda. Occur over rocky reefs.

Molly miller - *Scartella cristata* (Linnaeus, 1758)

FEATURES: Tannish with series of diffuse, dark smudges along back and sides interlaced with pale wavy lines. May be overall brownish. Red spots usually present on head. Tuft of cirri over each eye. Row of cirri along nape profile. Cirri usually banded in red and white. Specimens from Eleuthera reported to have few to no cirri on nape. HABITAT: FL, Gulf of Mexico, Bahamas, Caribbean Sea to Brazil. Also Bermuda and eastern Atlantic. Occur along shallow rocky shores, in tide pools, and on jetties.

ALSO IN THE AREA: Freckled blenny, *Hypsoblennius ionthas*; Highfin blenny, *Lupinoblennius nicholsi*; Mangrove blenny, *Lupinoblennius vinctus*, p. 431.

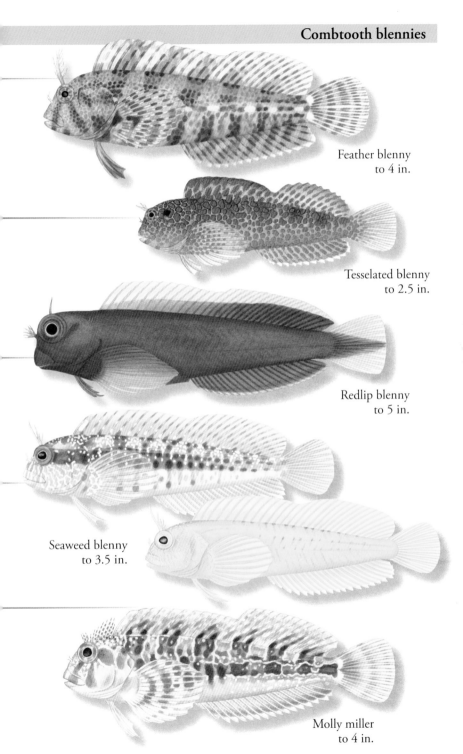

Feather blenny
to 4 in.

Tesselated blenny
to 2.5 in.

Redlip blenny
to 5 in.

Seaweed blenny
to 3.5 in.

Molly miller
to 4 in.

Puffcheek blenny - *Labrisomus bucciferus* Poey, 1868

FEATURES: Yellowish brown, brownish, to reddish brown with four to five dark bars on sides; anterior bars extend into dorsal fin. Bars are darker toward midline and are bordered by pale blotches. Two dark spots or small blotches behind eyes. Distinct bar below eyes. Females with spotted fins, males with reddish fins. Ocellus on opercle absent. Nostrils and eyes with tuft of cirri. Nape with transverse row of cirri. First dorsal-fin spine taller than fifth. HABITAT: S FL, Dry Tortugas, Bahamas, Caribbean Sea to Nicaragua. Possibly Bermuda. Occur in tide pools and over shallow mixed sand and rocky bottoms, seagrass beds, patch reefs, and other coral formations.

Palehead blenny - *Labrisomus gobio* (Valenciennes, 1836)

FEATURES: Pale with four or five blackish brown to brownish bars on sides. Bars are darker above midline and widest at midline and are bordered by pale areas. Sides of head spotted and mottled. Females with pale spots on fins. Males with unspotted fins. Ocellus on opercle absent. Tuft of cirri on nostrils and above eyes. Transverse row of cirri on nape. First dorsal-fin spine slightly shorter than fourth and fifth spines. HABITAT: SE FL, Bahamas, Caribbean Sea, Yucatán to Nicaragua. From near shore to about 50 ft. over rocky and sandy bottoms, coral formations, and seagrass beds.

Mimic blenny - *Labrisomus guppyi* (Norman, 1922)

FEATURES: Females greenish gray to buff with darker bars on sides. Males grayish green with grayish red bars; may be very dark and change color quickly. Bars extend into lower dorsal fins. Pale to dark ocellus always present on opercle. Head and fins variably spotted and banded. Tuft of cirri on nostrils and above eyes. Transverse row of cirri on nape. HABITAT: S FL, FL Keys, Dry Tortugas, Campeche Bay, Bahamas, Caribbean Sea to Brazil. In tide pools and over mixed sand and rocky bottoms, coral and patch reefs, and limestone slopes to about 25 ft.

Longfin blenny - *Labrisomus haitiensis* Beebe & Tee-Van, 1928

FEATURES: Tannish with six to seven brownish, broken bars on sides. Bars wider along midline. Anterior bars extend into dorsal fin. Fins banded or spotted in both sexes. Dark bar below eyes. Tuft of cirri on nostrils and above eyes. Transverse row of cirri on nape. First dorsal-fin spine taller than third and fourth spines. Pelvic fins long, with third ray half, or less than half, the length of the longest ray. HABITAT: S FL, Dry Tortugas, Campeche Bay, Bahamas, Antilles to Honduras. Usually occur on patch reefs and coral formations to about 50 ft.

Downy blenny - *Labrisomus kalisherae* (Jordan, 1904)

FEATURES: Head reddish to pinkish. Body with brown, reddish brown to greenish brown bars that extend into dorsal fin. Bars separated by diffuse, pale bars. Fins densely spotted. Lower opercle whitish. Juveniles with blotch on opercle. Tuft of cirri on nostrils and over eyes. Transverse row of cirri on nape. First dorsal-fin spine taller than third and fourth spines. HABITAT: SE FL, Dry Tortugas to Campeche, Greater and Lesser Antilles to Brazil. Over rocky and rubble bottoms, reefs, and seagrass beds.

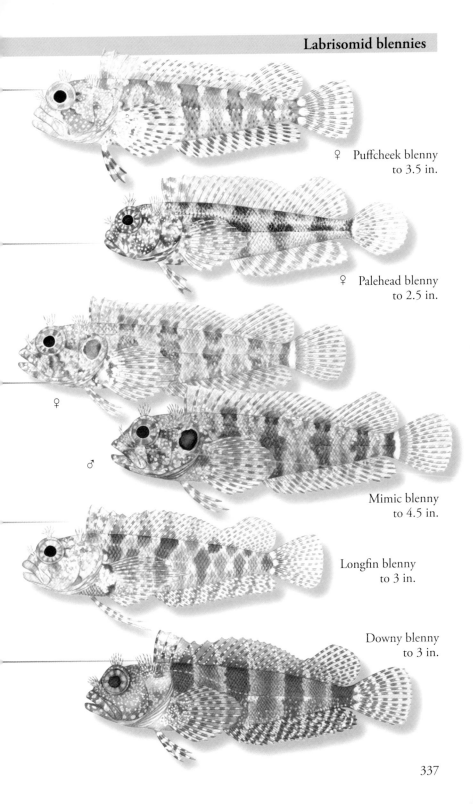

♀ Puffcheek blenny
to 3.5 in.

♀ Palehead blenny
to 2.5 in.

♀

♂

Mimic blenny
to 4.5 in.

Longfin blenny
to 3 in.

Downy blenny
to 3 in.

Spotcheek blenny - *Labrisomus nigricinctus* Howell Rivero, 1936

FEATURES: Females whitish to buff with blackish to brownish bars on sides. Males yellowish to buff with orange bars on sides. Bars extend into dorsal fins. Dark bands radiate from eyes. Dark, ocellated spot on opercle in all. Snout pointed. First dorsal-fin spine shorter than remaining spines. Tuft of cirri on nostrils and over eyes. Transverse row of cirri on nape. HABITAT: S FL, FL Keys, Dry Tortugas, Bahamas, Antilles to Venezuela. In clear, shallow waters over mixed bottoms and reefs and in tide pools.

Hairy blenny - *Labrisomus nuchipinnis* (Quoy & Gaimard, 1824)

FEATURES: Color highly variable. Females tannish with irregular brownish bars on sides. Bars extend into dorsal fins. Head, body, and fins heavily and irregularly spotted. Males reddish on lower head and on abdomen. Body diffusely mottled and blotched in shades of green. Outer margins of fins reddish. Both with ocellated spot on opercle. Smaller specimens with dark spot on anterior dorsal fin. Breeding males with bright reddish head and alternating blackish and grayish bars on sides. Tuft of cirri on nostrils and over eyes. Transverse row of cirri on nape. HABITAT: FL, southern Gulf of Mexico, Bahamas, and Antilles to Brazil. Also Bermuda. Occur in shallow coastal waters over sand and seagrass beds, rocky areas, and coral rubble.

Goldline blenny - *Malacoctenus aurolineatus* Smith, 1957

FEATURES: Tannish to whitish with dark brownish to blackish bars on sides. Anterior bars wide, defined, may merge. Orange to golden wavy stripes and spots overlay bars. Lower head banded. Anal fin golden. Males more distinctly marked. Females may have spotting on dorsal fin. Tuft of cirri over eyes and row of cirri on nape. HABITAT: S FL, Dry Tortugas, Campeche Bank, south to Venezuela. In shallow coastal waters. Associated with sea urchins.

Rosy blenny - *Malacoctenus macropus* (Poey, 1868)

FEATURES: Color, pattern highly variable. Females pale tannish; finely spotted and diffusely blotched. Pinkish ventrally. Males grayish to reddish or brownish, with diffuse to well-defined dark bars on upper sides. Bars may blend together. Spotted ventrally. Both with single cirrus on nostril and over each eye. Single, erect cirrus on both sides of nape. HABITAT: S FL, Dry Tortugas, Campeche Bank, Bahamas, Caribbean Sea to Venezuela. Also Bermuda. In shallow coastal waters over mixed sand, rubble, and seagrass beds and with sponges.

Saddled blenny - *Malacoctenus triangulatus* Springer, 1959

FEATURES: Color, pattern variable. Three to five saddles on upper sides. Saddles may be pale, dark, uniformly colored, reticulating, or absent. May have dark blotches on lower sides between saddles. Lower portion of head banded or blotched. May have dark blotch on anterior dorsal fin that merges with first bar. Females with spotted fins. Tuft of cirri over eyes and row of cirri on nape. HABITAT: S FL, Dry Tortugas, Campeche Bank, Bahamas, Caribbean Sea to Brazil. Over shallow rocky and reef areas from shore to about 45 ft.

♀

♂

Spotcheek blenny
to 3 in.

♀

Hairy blenny
to 9 in.

♂

Goldline blenny
to 2.3 in.

♀

Rosy blenny
to 2.2 in.

♂

Saddled blenny
to 2.5 in.

339

Barfin blenny - *Malacoctenus versicolor* (Poey, 1876)

FEATURES: Color variable. Females yellowish to pale grayish with five olivaceous to brownish bars on sides. Small, scattered, dark spots on areas between bars. Dorsal, caudal, and anal fins may have scattered spots. Males shades of brown with brownish bars. Dark spots between bars and on fins absent. Males darker than females, bars more defined. Bars extend to dorsal-fin margins in all. All with tuft of cirri above eyes and a transverse row of cirri on both sides of nape. HABITAT: S FL, Cuba, Bahamas, Antilles. Also reported from Belize and Honduras. Found over shallow coastal coral reefs and rocky, sandy, and rubble areas. BIOLOGY: Feed on invertebrates and algae.

Threadfin blenny - *Nemaclinus atelestos* Böhlke & Springer, 1975

FEATURES: Observed specimen reddish orange. Others described as brownish. Males with dark spot on anterior dorsal fin and dark anterior anal fin. Females without dark markings. Nostrils with short tube. Long, flattened cirrus above eyes. Simple cirrus on both sides of nape. Middle pectoral-fin rays elongate. Pelvic fins very long. HABITAT: In Gulf of Mexico from Flower Garden Banks and Yucatán Peninsula. Bahamas, Antilles to Nicaragua. Also Bermuda. Found from about 110 to 840 ft.

Coral blenny - *Paraclinus cingulatus* (Evermann & Marsh, 1899)

FEATURES: Mottled whitish with five broad brown bars on sides. Anterior three bars well-defined, posterior two diffuse. All extend into fins. Anterior head reddish. Dark bar below each eye. Pale, lappet-like, pronged cirrus over eyes and on both sides of nape. Pectoral-fin base scaleless. HABITAT: S FL, FL Keys, Dry Tortugas, Bahamas, and Greater Antilles to Honduras. Occur in tide pools and over shallow, coastal coral rubble areas from shore to about 20 ft.

Banded blenny - *Paraclinus fasciatus* (Steindachner, 1876)

FEATURES: Uniformly pale to dark or densely mottled in shades of buff to brown. Faint, diffuse, darker bars on sides. Fins spotted and banded. Zero to four (usually one to two) pale to dark ocelli on rear dorsal fin. Dark bar at caudal-fin base. One to four cirri above eyes. Cirrus on nape is dark, pale-edged, lappet-like, and pronged. HABITAT: S FL, eastern and southern Gulf of Mexico, Bahamas, and Antilles to Venezuela. Occur over shallow sandy, rocky, coral, and turtle-grass bottoms.

Marbled blenny - *Paraclinus marmoratus* (Steindachner, 1876)

FEATURES: Color and pattern variable. Tannish to brownish or olivaceous. May be uniformly to irregularly marbled. May have about six bars or rows of pale blotches on sides. Bars and blotches may be well-defined or obscure. Variable number of ocelli may be present on rear dorsal and anal fins. One to four cirri above eyes. Cirrus on nape is flat, erect, and fringed. First three dorsal-fin spines always very tall and almost entirely separated from remainder of fin. HABITAT: S FL, Bahamas, Cuba to Venezuela. Found over shallow seagrass beds and coral reefs, and with sponges. BIOLOGY: Males guard egg masses. Other name: Rhino blenny.

Barfin blenny
to 2.5 in.

♀

♂

Threadfin blenny
to 1.4 in.

Coral blenny
to 1 in.

Banded blenny
to 2.5 in.

Marbled blenny
to 4 in.

Labrisomidae - Labrisomid blennies, *cont.*

Blackfin blenny - *Paraclinus nigripinnis* (Steindachner, 1867)

FEATURES: Shades of greenish brown, reddish brown, to blackish. May be pale to very dark. Most show six or seven bars on sides that extend into dorsal fin. Bars may be obscure. Always with a dark bar at caudal-fin base. Usually with one ocellus on rear dorsal fin. Several cirri above each eye. Cirrus on either side of nape is pale and lappet-like, may have several prongs. HABITAT: S FL, Dry Tortugas, Bahamas, Antilles to Brazil. Possibly Bermuda. Occur in a variety of shallow-water coastal habitats.

Checkered Blenny - *Starksia ocellata* (Steindachner, 1876)

FEATURES: Varies from almost uniformly brownish to having three rows of evenly spaced blotches on sides. Always with dark-ringed, golden to orange spots on sides of head and at pectoral-fin base. May have ringed spots along dorsal-fin base. Simple cirrus on each nostril, over eyes, and on both sides of nape. HABITAT: NC to W FL, Veracruz, and Bahamas. Found over rocky and coral bottoms and with sponges.

ALSO IN THE AREA: Horned blenny, *Paraclinus grandicomis*; Bald blenny, *Paraclinus infrons*; Key blenny, *Starksia starcki*, p. 431.

Chaenopsidae - Tube blennies

Roughhead blenny - *Acanthemblemaria aspera* (Longley, 1927)

FEATURES: Females yellow to pale pinkish with pale spots anteriorly. Dark spots or broken lines on lower head. Males brownish to blackish with scattered pale spots anteriorly; some are pale overall. Dark, ocellated spot on anterior dorsal fin. Both with spines on top of head forming a V-shape from nape to eyes. Complex, highly branched cirrus over nostrils and eyes. HABITAT: S FL, FL Keys, Dry Tortugas, Yucatán, Bahamas, Antilles to Panama. Reef associated. Inhabit holes in coral heads.

Papillose blenny - *Acanthemblemaria chaplini* Böhlke, 1957

FEATURES: Yellowish to orangish anteriorly. Tannish posteriorly, with a series of diffuse pale and dark spots on sides. A few minute spines present below eyes. Few to numerous simple cirri on head. Cirri on nostrils and over eyes branched, but on a single plane. HABITAT: SE FL and Bahamas. Occur over shallow coastal slopes with coral heads. BIOLOGY: Papillose blenny inhabit holes in coral heads. Dart out to catch prey.

Spinyhead blenny - *Acanthemblemaria spinosa* Metzelaar, 1919

FEATURES: Females with dark head and numerous brownish blotches on sides. Males (not shown) almost uniformly dark on head and most of body. Both with varying number of pale spots on head; greenish to yellowish eyes; and a large number of small, short, close-set spines on head. Cirri over nostrils and eyes simple to weakly branched at tips. HABITAT: S FL (rare; reports may be erroneous), Bahamas, Antilles to Curaçao. Occur over relatively shallow, small patch reefs. BIOLOGY: Spinyhead blenny inhabit vacant holes in coral heads.

Blackfin blenny
to 2 in.

Checkered blenny
to 2 in.

Roughhead blenny
to 1.4 in.

♀

♂

Papillose blenny
to 1.8 in.

♀ Spinyhead blenny
to 1.1 in.

Yellowface pikeblenny - *Chaenopsis limbaughi* Robins & Randall, 1965

FEATURES: Shades of tan with pale speckles on head and body. Dark areas may form wide, diffuse bars. Midlateral stripe always present; may be faint and pale to dark. Males more variable in color; always with a black, ocellated spot below orange blotch between first and second dorsal-fin spines and dark gill membranes. Females with low dorsal fin, males with tall dorsal fin. Snout U-shaped from above. HABITAT: FL Keys, Bahamas, and Caribbean Sea. Reef associated over sand and rubble to about 69 ft. Inhabit holes and abandoned worm tubes. BIOLOGY: Form colonies.

Bluethroat pikeblenny - *Chaenopsis ocellata* Poey, 1865

FEATURES: Shades of tan with pale speckles on head and body. Dark areas may form diffuse bars. Midlateral stripe absent. Males more variable in color; always with a curved black mark partially enclosing an orange spot between first and second dorsal-fin spines and blue gill membranes. Females with low dorsal fin, males with tall dorsal fin. Snout V-shaped from above. HABITAT: S FL, Bahamas, Cuba, Virgin Islands. Occur over sandy areas with grasses and algae to about 10 ft. Inhabit holes and abandoned worm tubes. BIOLOGY: Usually solitary.

Banner blenny - *Emblemaria atlantica* Jordan & Evermann, 1898

FEATURES: Females brownish with six pale bars on sides that are outlined with dark brown. Males more uniformly colored, may have faint, pale blotches on sides. Both with sloping snout, a long, unbranched cirrus over each eye, and 14 pectoral-fin rays. Males with sail-like dorsal fin. HABITAT: GA and FL. In Gulf of Mexico from W FL and TX. Also Bermuda. Absent from S FL. Associated with rocky reefs from about 100 to 360 ft. BIOLOGY: Males use sail-like dorsal fin to attract females.

Sailfin blenny - *Emblemaria pandionis* Evermann & Marsh, 1900

FEATURES: Females and immature males pale brownish to grayish with dark spots and clusters of pale spots. Males dark brown to bluish black with a variably banded dorsal fin. Anterior dorsal fin ray may have blue to yellowish streaks. Both with steep snout, a long cirrus over each eye, and 13 pectoral-fin rays. Cirri over eyes usually branched at tip. Males with sail-like dorsal fin. HABITAT: S FL, Gulf of Mexico from FL to TX, Bahamas, and Antilles to Venezuela. Occur in clear, shallow waters over rocky shores, coral rubble, and sandy bottoms. Inhabit holes and abandoned worm tubes. BIOLOGY: Males use sail-like dorsal fin to attract females. Territorial.

Pirate blenny - *Emblemaria piratula* Ginsburg & Reid, 1942

FEATURES: Females (not shown) mottled brown anteriorly, becoming pale posteriorly. Males mottled blackish brown anteriorly, pale posteriorly; lower portion of anterior dorsal fin orange. Dorsal fin tall in males, low in females. Both with a long, fleshy, uniformly colored cirrus over each eye and 12–13 pectoral-fin rays. HABITAT: NC to FL, Dry Tortugas, W FL, and AL. Reported to inhabit holes in coral rubble over sandy areas near deep patch reefs.

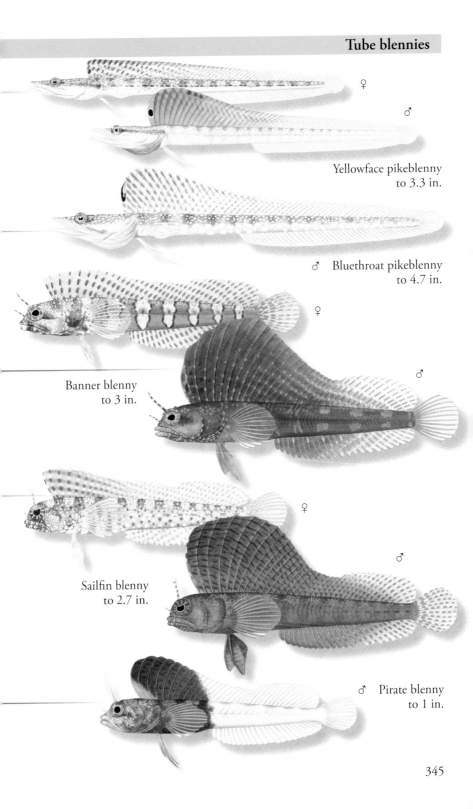

♀

♂

Yellowface pikeblenny
to 3.3 in.

♂ Bluethroat pikeblenny
to 4.7 in.

♀

♂

Banner blenny
to 3 in.

♀

♂

Sailfin blenny
to 2.7 in.

♂ Pirate blenny
to 1 in.

Blackhead blenny - *Emblemariopsis bahamensis* Stephens, 1961

FEATURES: Females translucent. Males dark brown to blackish on head and forebody; translucent posteriorly. Both with reddish patch on cheeks and pale spots on head. Organs, spinal column, and ribs pigmented. Lower jaw does not protrude. First dorsal-fin spine shorter than second and third. Cirri above eyes always absent. HABITAT: FL Keys, Dry Tortugas, Veracruz, Bahamas, Antilles to northern South America. Reef associated in shallow water.

Glass blenny - *Emblemariopsis diaphana* Longley, 1927

FEATURES: Translucent. Females sparsely pigmented on head and anterior dorsal fin. Males similar to females, but more densely pigmented on head and anterior dorsal fin; head becoming blackish while inhabiting burrows. Organs, spinal column, and ribs pigmented. First dorsal-fin spine taller than second and third. Together, first three spines form a spike. Cirri above eyes always absent. HABITAT: FL Keys and Dry Tortugas. Coral reef associated.

Wrasse blenny - *Hemiemblemaria simulus* Longley & Hildebrand, 1940

FEATURES: Three color phases: yellow dorsally, fading to white below; yellow to greenish yellow dorsally with dark midlateral stripe; pale with midlateral stripe broken into segments. All with purplish to blackish stripe through eyes and a black spot on anterior dorsal fin. Juveniles translucent with dark midlateral stripe. All stages with protruding lower jaw. HABITAT: S FL, FL Keys, Dry Tortugas, Cuba, Bahamas to Honduras. Coral reef associated. BIOLOGY: Wrasse blenny inhabit holes in coral made by boring mollusks. Mimic female and juvenile Bluehead, *Thassoma bifasciatum*, which it swims with while in search of prey.

Blackbelly blenny - *Stathomonotus hemphillii* Bean, 1885

FEATURES: Males highly variable: may be pale, reddish orange, blackish, or greenish. Some with black-margined white bands on head and/or a white ocellated spot behind eyes. Fins may have irregular white bands and a reddish margin. Females described as pale, with dark lines and spots radiating from eyes. Cirri on head absent. Pectoral fins greatly reduced. Dorsal fin entirely spiny. Body scaleless. HABITAT: S FL, FL Keys, Dry Tortugas, Bahamas, and Antilles to Nicaragua. Associated with rock and rubble near coral reefs and with seagrass beds from shore to about 20 ft.

Eelgrass blenny - *Stathomonotus stahli* (Evermann & Marsh, 1899)

FEATURES: Color variable. Usually greenish, but may be brownish to grayish. May have series of whitish, reddish to blackish spots or mottling on sides. A reddish blotch on cheeks present in most. Single, flap-like cirrus above eyes. Simple cirrus on nape. Pectoral fins comparatively large. Dorsal fin entirely spiny. Body scaled. HABITAT: S FL, FL Keys, Dry Tortugas, Bahamas, Caribbean Sea to Venezuela. Over seagrass beds and rock and rubble near coral reefs from shore to about 20 ft.

ALSO IN THE AREA: Freckled pikeblenny, *Chaenopsis roseola*, p. 431.

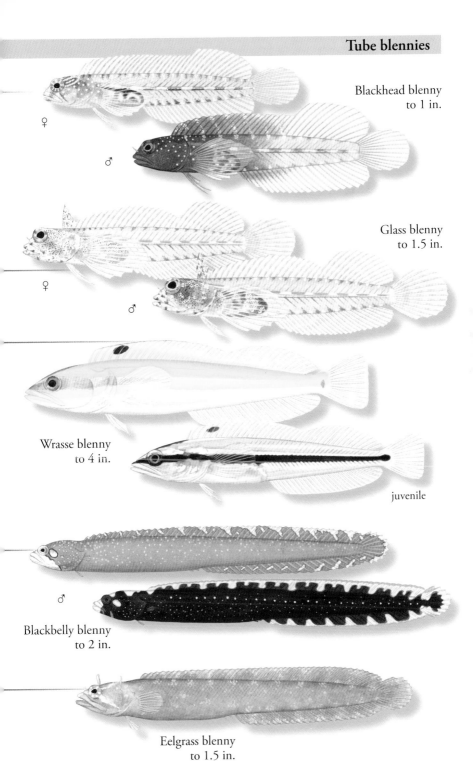

Blackhead blenny
to 1 in.

♀

♂

Glass blenny
to 1.5 in.

♀

♂

Wrasse blenny
to 4 in.

juvenile

♂

Blackbelly blenny
to 2 in.

Eelgrass blenny
to 1.5 in.

Gobiesocidae - Clingfishes

Emerald clingfish - *Acyrtops beryllinus* (Hildebrand & Ginsburg, 1927)

FEATURES: Usually emerald green; may also be brownish. May have diffuse, pale whitish spots and lines dorsally or may show reticulating pattern. A few golden to brassy lines may radiate from eyes. Snout broadly rounded. Head slightly broader than body. Dorsal fin with 5–7 rays. Pelvic fins form sucking disc. HABITAT: S FL, FL Keys, Bahamas, and Lesser Antilles to Belize. Associated with turtle-grass beds. BIOLOGY: Feed on invertebrates.

Stippled clingfish - *Gobiesox punctulatus* (Poey, 1876)

FEATURES: Tannish to grayish or olive. Dorsal surface stippled with small black spots. Some have diffuse bands radiating from eyes. May have a few diffuse bands across back. Upper lip broad. Head broadly rounded and wider than body. Dorsal fin with 11–12 (usually 11) rays. Pelvic fins form relatively large sucking disc. HABITAT: Northern Gulf of Mexico to northern South America. Rare to absent from coast of AL and FL. Associated with limestone rocks and ledges in clear, shallow waters.

Skilletfish - *Gobiesox strumosus* Cope, 1870

FEATURES: Brownish to olive brown with mottled, net-like pattern dorsally. About six faint lines radiate from each eye. Always with dark band at caudal-fin base. Upper lip broad. Head broadly rounded and wider than body. Dorsal fin with 10–13 (usually 12) rays. Pelvic fins form large sucking disk. HABITAT: NJ to FL, Gulf of Mexico, and possibly Bermuda. Occur in rocky tide pools and over grassy areas in shallow waters.

Callionymidae - Dragonets

Spotted dragonet - *Diplogrammus pauciradiatus* (Gill, 1865)

FEATURES: Females tannish to whitish and densely mottled and speckled. Males similarly patterned but usually darker; may have orange patches and blackish spots on head. First dorsal-fin spines long and filamentous in males. Both with laterally expanded head, three-pronged preopercular spine, seven second dorsal-fin rays, and lateral keel extending from abdomen to caudal fin. Keel always with row of spots. HABITAT: NC to FL, eastern and southern Gulf of Mexico to Colombia. Also Bahamas and Bermuda. Associated with seagrass beds in shallow water. BIOLOGY: Spotted dragonet feed during the day, lie on or bury themselves in bottom sediment at night.

Spotfin Dragonet - *Foetorepus agassizii* (Goode & Bean, 1888)

FEATURES: Reddish with greenish brown mottling dorsally. White ventrally. First dorsal fin with black, ocellated spot. Second dorsal fin and caudal fins with yellow bands. Anal fin with blackish submarginal band. First dorsal-fin spines never filamentous. Second dorsal-fin margin straight in females, slightly convex in males. Males with elongate lower caudal-fin rays. HABITAT: Canada to FL, Gulf of Mexico, Antilles to Brazil. Demersal. Occur at depths between about 300 and 2,100 ft.

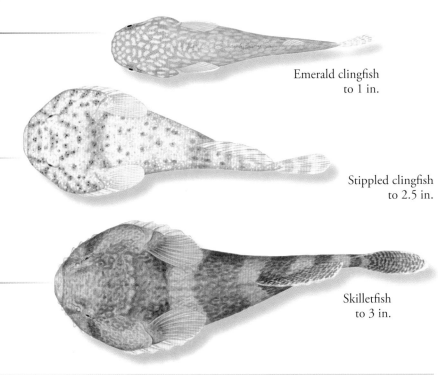

Emerald clingfish
to 1 in.

Stippled clingfish
to 2.5 in.

Skilletfish
to 3 in.

Dragonets

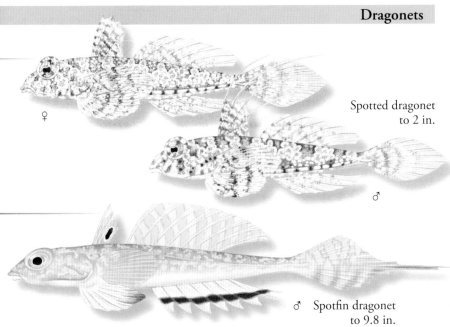

♀

Spotted dragonet
to 2 in.

♂

♂ Spotfin dragonet
to 9.8 in.

Callionymidae - Dragonets, *cont.*

Palefin dragonet - *Foetorepus goodenbeani* Nakabo & Hartel, 1999

FEATURES: Red with pale greenish brown mottling dorsally. Two yellowish, wavy stripes on sides. White ventrally. First dorsal fin dark yellowish with pinkish white marbling and a whitish spot on third spine. Second dorsal fin with yellowish spots. Anal fin with reddish submarginal band. Females and males with comparatively tall dorsal fin; first spine tall, filamentous. Second dorsal fin with slightly concave margin. Males with elongate lower caudal-fin rays. HABITAT: MA to FL and northern Gulf of Mexico to Texas. Demersal from about 154 to 1,300 ft.

Lancer dragonet - *Paradiplogrammus bairdi* (Jordan, 1888)

FEATURES: Variably mottled and spotted in shades of tannish to grayish. Lower head faintly to vividly barred. Females with very dark to pale first dorsal fin. First dorsal fin of males is tall, expanded, and variably spotted and banded in yellow to gray. Both with relatively expanded head, and second dorsal fin with 9 rays. Preopercle with one forward-projecting lower spine and 3–9 upward-facing prongs. Caudal keels always absent. HABITAT: Cape Hatteras to FL, Gulf of Mexico to northern South America. Also Bahamas and Bermuda. Over sandy bottoms around coral reefs to about 30 ft.

Eleotridae - Sleepers

Fat sleeper - *Dormitator maculatus* (Bloch, 1792)

FEATURES: Color variable. Tan, brown to bluish with dark reddish to blackish scales. Dark bands radiate from eyes. Dorsal and anal fins banded. Always with a bright to faint, bluish ocellated blotch above pectoral-fin base. Juveniles with obscure bars and blotches along midline. Mouth relatively small. Snout profile convex. Scales large. Body relatively deep. HABITAT: Chesapeake Bay to FL, Gulf of Mexico, Bahamas, Antilles to Brazil. Occur in shallow coastal brackish and fresh waters.

Largescaled spinycheek sleeper - *Eleotris amblyopsis* (Cope, 1871)

FEATURES: Head and body shades of tan to brown and densely speckled and blotched. Dorsal portion may be paler than sides and ventral area. Several dark bands radiate from eyes. Fins banded. Mouth large, oblique. Head broad, flattened dorsally. Preopercle with a sharp, concealed spine. Caudal fin rounded. HABITAT: NC to FL and Gulf of Mexico to Brazil. Also Bermuda and Greater Antilles. Possibly from Bahamas. Occur in shallow coastal marine and brackish waters. Also in freshwater ponds and ditches. BIOLOGY: Previously known as *Eleotris pisonis*.

Emerald sleeper - *Erotelis smaragdus* (Valenciennes, 1837)

FEATURES: Dark olivaceous to brown above. Tan below. Dark spot at upper pectoral-fin base present. Jaws large and oblique with thin lips. Head laterally expanded with dorsal profile almost straight. Caudal fin pointed, with anterior rays originating on body. HABITAT: SE FL, Gulf of Mexico, Bahamas, Antilles to Brazil. Also Bermuda. Occur in coastal waters over sandy to silty bottoms.

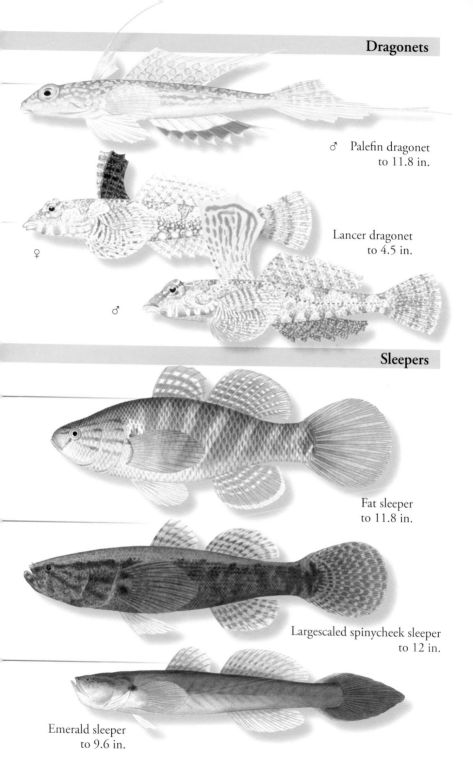

♂ Palefin dragonet
to 11.8 in.

Lancer dragonet
to 4.5 in.

♀

♂

Sleepers

Fat sleeper
to 11.8 in.

Largescaled spinycheek sleeper
to 12 in.

Emerald sleeper
to 9.6 in.

Bigmouth sleeper - *Gobiomorus dormitator* Lacepède,1800

FEATURES: Olivaceous to yellowish or olive brown dorsally. Paler below. Obscure bands radiate from eyes. An obscure stripe runs from pectoral-fin base to caudal fin. Dorsal, caudal, and pectoral fins banded. Mouth very large and moderately oblique, with sharp teeth arranged in bands. Head is large and flattened dorsally. Preopercular margin is fleshy. HABITAT: S FL, southern Gulf of Mexico, and Bahamas to eastern Brazil. Associated with coastal fresh and brackish waters over soft bottoms. Adults in fresh water, may be found inland.

Gobiidae - Gobies

Bearded goby - *Barbulifer ceuthoecus* (Jordan & Gilbert, 1884)

FEATURES: Color and pattern somewhat variable. Head and body with pale patches and dark melanophores forming speckles. Dark band below eye. Dark bar at pectoral-fin base. Barbels present on chin, snout, and area below eyes. Upper lip with a free fold. Nostrils tubular. Gill openings restricted. HABITAT: S FL, FL Keys, Bahamas to Venezuela. Found near shore in very shallow salt water over rocks and silty rubble.

Notchtongue goby - *Bathygobius curacao* (Metzelaar, 1919)

FEATURES: Tan to brownish with pale scale centers. Irregular bands radiate from eyes. Sides with somewhat regularly spaced, dark blotches. Second dorsal and anal fins with tannish inner margin. Tongue deeply notched at tip. Pectoral fins with 15−18 rays; upper 4−5 rays filamentous, free of membrane at tips. Scales in horizontal series on sides number 31−34. HABITAT: S FL, southern Gulf of Mexico, Bahamas, Antilles to northern South America. Also Bermuda. Found in tide pools, around mangroves, and over sheltered seagrass beds.

Island frillfin - *Bathygobius mystacium* Ginsburg, 1947

FEATURES: Tannish with pale to white scale centers. Irregular bands radiate from eyes. Upper sides with obscure saddles, lower sides with series of dark blotches. Tongue rounded at tip. Pectoral fins with 19−20 rays; upper 4−5 filamentous, free of membrane at tips. Scales in horizontal series on sides number 33−36. HABITAT: S FL, southern Gulf of Mexico, Bahamas, Antilles to Central America. Associated with unsheltered rocky and sandy bottoms that are exposed to wave action.

Frillfin goby - *Bathygobius soporator* (Valenciennes, 1837)

FEATURES: Varying shades of brown; may be very dark. Head blotched. Five dark saddles along back; saddle below first dorsal fin broadest. Dark blotches on sides. Tongue slightly notched at tip. Pectoral fins with 18−21 rays; upper 4−5 filamentous, free of membrane at tips. Scales in horizontal series on sides number 37−41. HABITAT: NC to FL, Gulf of Mexico, Bahamas, Antilles to southeastern Brazil. Also Bermuda. Found in a variety of shallow water habitats from muddy river bottoms to estuaries and rocky tide pools.

Bigmouth sleeper
to 2.4 ft.

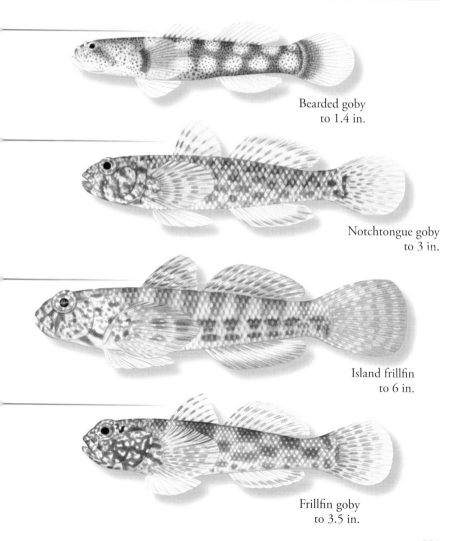

Bearded goby
to 1.4 in.

Notchtongue goby
to 3 in.

Island frillfin
to 6 in.

Frillfin goby
to 3.5 in.

White-eye goby - *Bollmannia boqueronensis* Evermann & Marsh, 1899

FEATURES: Pale with orange brown blotches along midline. Blotch at caudal-fin base darker than other blotches. Eyes with a bright white, reflective iris. Dorsal and upper caudal fins with orange brown stripes. First dorsal fin with a black, ocellated spot. Second dorsal fin with 1 spine and 12 rays. HABITAT: S FL, Dry Tortugas, Lesser Antilles to northern South America. Offshore from about 88 to 180 ft.

Barfin goby - *Coryphopterus alloides* Böhlke & Robins, 1960

FEATURES: Translucent. Pale anteriorly, yellow posteriorly; areas divided by a dark bar below first dorsal fin. Dark smudges on head and body. Cheeks reddish. Dark smudge at caudal-fin base. Females with dark smudge between second and third dorsal-fin spines. In males, dark smudge on dorsal fin merges with a dark band. HABITAT: S FL, Bahamas, and Belize. Over reef areas deeper than 40 ft.

Sand-canyon goby - *Coryphopterus bol* Victor, 2008

FEATURES: Color variable, corresponds with habitat. Translucent whitish to tannish; sparsely to densely marked. All with: spotted eyes; variable orangish spots or stripes on head; white 'bridle' from upper jaw to preopercle; pale to dark patch above opercle; few to many white and orangish spots on sides, dark spots may form Xs; dark bar on caudal-fin base. Most have distinct spot on lower pectoral-fin base. HABITAT: FL to southern Caribbean Sea. Deep reefs and canyons with exposed and strong currents.

Colon goby - *Coryphopterus dicrus* Böhlke & Robins, 1960

FEATURES: Translucent with brownish and whitish stripes behind eyes. Sides with rows of fairly regularly spaced brown and whitish spots. Usually two distinct brown spots at pectoral-fin base. Brownish bar at caudal-fin base. HABITAT: S FL, FL Keys, southern Gulf of Mexico, Puerto Rico, Antilles to Central America. Reef associated.

Pallid goby - *Coryphopterus eidolon* Böhlke & Robins, 1960

FEATURES: Translucent bluish gray. Orange yellow stripe behind eye bordered with dark pigment. Orange yellow blotches below eyes and in series along back and on sides. Black bar at caudal-fin base. Spinal column with dark smudges. Pelvic-fin spines joined. HABITAT: S FL, southern Gulf of Mexico, Bahamas, Haiti, and Antilles. Over reefs and on coralline sand around reefs.

Bridled goby - *Coryphopterus glaucofraenum* Gill, 1863

FEATURES: Color highly variable, corresponds to habitat. Translucent whitish to tannish; sparsely to densely marked. All with: spotted eyes; variable orangish spots or stripes on head; white 'bridle' from upper jaw to preopercle; a two-pronged spot or bar above opercle; few to many white spots and orangish spots on sides, dark spots may form Xs; a pair of elongate spots or a barbell-shaped mark at caudal-fin base. HABITAT: NC to FL Keys, AL, Campeche Bay, Bahamas, and Antilles to Brazil. Also Bermuda. Inshore in shallow sandy bays and around reefs and mangroves.

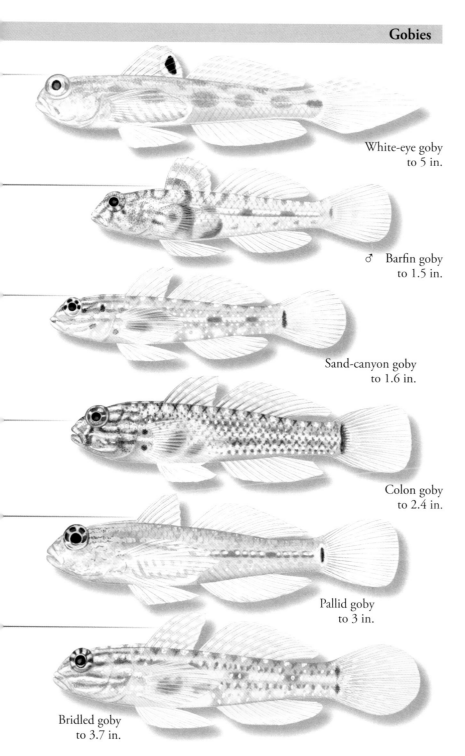

White-eye goby
to 5 in.

♂ Barfin goby
to 1.5 in.

Sand-canyon goby
to 1.6 in.

Colon goby
to 2.4 in.

Pallid goby
to 3 in.

Bridled goby
to 3.7 in.

Glass goby - *Coryphopterus hyalinus* Böhlke & Robins, 1962

FEATURES: Translucent. Upper snout usually blue; lower snout and upper lip dark brownish. Patch of melanophores form a variable dark wedge behind and below eyes. Large orange area from head to mid-body. Orange spots along spinal column. Anus near rear of large black blotch. HABITAT: FL, Bahamas, Antilles to Colombia. Also Bermuda. Reef associated. BIOLOGY: Swim in small schools above bottom.

Kuna goby - *Coryphopterus kuna* Victor, 2007

FEATURES: Translucent grayish white. Two rows of small spots on upper portion of eyes. White and brownish to reddish brown flecks and spots on head and body. Dark bar or cluster of spots forming a bar below eyes. Dark spots cluster below first dorsal fin; may form a broken arc. Fins may be spotted. Pectoral fins with 15 rays. HABITAT: S FL and in western Caribbean from Colombia, Honduras, Belize, Yucatán, and Panama. Over sandy bottoms in relatively deep water.

Peppermint goby - *Coryphopterus lipernes* Böhlke & Robins, 1962

FEATURES: Translucent yellowish to gold on head and body. Bright blue markings on snout and over eyes. Pale blue lines radiate behind eyes, may extend to forebody. A dark ring surrounds the anus. Several pale blotches along spinal column. Second dorsal-fin spine elongate. HABITAT: FL Keys, Bahamas, and Antilles to Central America. Coral reef associated from about 50 to 78 ft. BIOLOGY: Usually solitary.

Masked goby - *Coryphopterus personatus* (Jordan & Thompson, 1905)

FEATURES: Translucent. Upper snout usually blue, lower snout and upper lip orange brown. A patch of dark melanophores form a variable dark wedge behind and below eyes. Pale blue spot at pectoral-fin base. Large orange area from head to mid-body. Orange spots along spinal column. Anus near center of large black blotch. HABITAT: S FL, Bahamas, Haiti, and Virgin Islands to Yucatán. Also Bermuda. Coral reef associated from about 6 to 100 ft. BIOLOGY: Form large schools.

Spotted goby - *Coryphopterus punctipectophorus* Springer, 1960

FEATURES: Whitish to bluish white with a series of orange brown blotches on sides. Several dark-margined lines extend from snout through eyes to below first dorsal fin. Lower pectoral-fin base with a distinct dark spot. Dorsal fins banded. Caudal fin spotted. HABITAT: S FL, AL, and Veracruz. Found near rocky areas from about 60 to 120 ft. BIOLOGY: Reported to inhabit burrows.

Bartail goby - *Coryphopterus thrix* Böhlke & Robins, 1960

FEATURES: Translucent whitish. Eyes spotted. Body with fairly evenly spaced brownish and white spots. Intensity of spots may vary. Usually with a large, dark spot on upper pectoral-fin base. Spot may be faint to dark. Dark bar at caudal-fin base. Second dorsal-fin spine elongate. HABITAT: S FL and the Bahamas. Also captured off Texas. Occur in clear water around coral heads from about 30 to 60 ft.

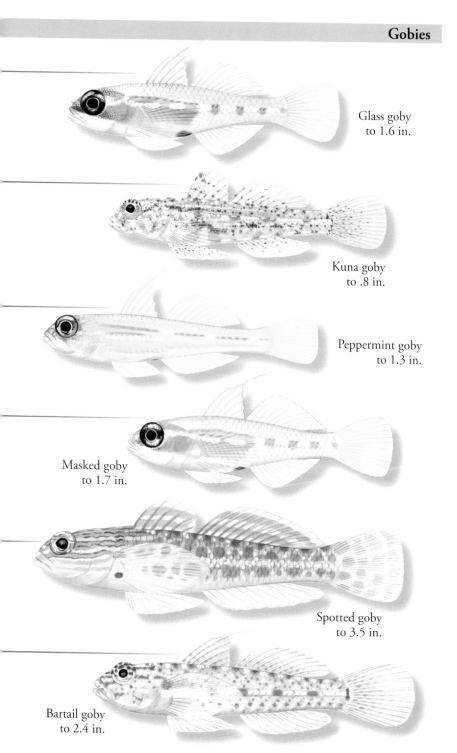

Glass goby
to 1.6 in.

Kuna goby
to .8 in.

Peppermint goby
to 1.3 in.

Masked goby
to 1.7 in.

Spotted goby
to 3.5 in.

Bartail goby
to 2.4 in.

Patch-reef goby - *Coryphopterus tortugae* (Jordan, 1904)

FEATURES: Color variable, corresponds to habitat. Translucent to tannish; sparsely to densely marked. All with: spotted eyes; a white 'bridle' from jaw to preopercle; a dark triangular spot or single-pronged bar over opercle; few to many white and orangish spots on sides, dark spots may form Xs; a variable dark bar at caudal-fin base. Usually with no spot on lower pectoral-fin base. HABITAT: S FL, Caribbean Sea. In shallow, sandy bays, around patch reefs, and in clear water and low-current areas.

Darter goby - *Ctenogobius boleosoma* (Jordan & Gilbert, 1882)

FEATURES: Tan to dusky. Dark blotches on head. Dark blotch on upper pectoral-fin base. Four to five dark blotches along midline that merge with V-shaped saddles under second dorsal fin. Dark spot at caudal-fin base. Dorsal and caudal fins banded. HABITAT: DE to FL, nothern and southern Gulf of Mexico, Bahamas to Brazil. Also Bermuda. Over muddy bottoms of bays and estuaries and in fresh water.

Slashcheek goby - *Ctenogobius pseudofasciatus* (Gilbert & Randall, 1971)

FEATURES: Tannish. Dark mark from corner of mouth to preopercular corner. Dark bands on head. Body with irregular cross-hatching dorsally and four dark blotches along midline. Distinct dark spot on caudal peduncle. Dorsal and caudal fins banded. HABITAT: FL, Bahamas, Antilles, Belize, Costa Rica, and Trinidad. Demersal in brackish to marine waters and in rivers.

Dash goby - *Ctenogobius saepepallens* (Gilbert & Randall, 1968)

FEATURES: Translucent whitish gray to bluish white with iridescent reflections. May have diffuse to distinct brownish to reddish brown spots dorsally. Always with a variable dark bar below eye; a dark, triangle-shaped mark on lower opercle; and five elongate 'dashes' along midline. Dashes may be large to very faint. Fins variably streaked. Third dorsal-fin spine elongate. HABITAT: S FL and Bahamas to Venezuela. Over sandy bottoms. BIOLOGY: Dash goby share burrows with shrimps.

Freshwater goby - *Ctenogobius shufeldti* (Jordan & Eigenmann, 1887)

FEATURES: Tannish yellow to light brown. Dark band from corner of mouth to upper opercle. Head and upper sides with dark flecks. Four squarish blotches along midline and a dark spot at caudal-fin base. Dorsal and caudal fins banded. HABITAT: NC to S FL, Gulf of Mexico. Occur in low-salinity bays, estuaries, and fresh water.

Emerald goby - *Ctenogobius smaragdus* (Valenciennes, 1837)

FEATURES: Tannish with irregular dark smudges dorsally and along midline. A dark blotch is present on shoulder above pectoral-fin base. Dark-ringed greenish spots are scattered on head and sides. Dorsal, caudal, and pectoral fins banded. Males with elongate third dorsal-fin spine and very long caudal fin. HABITAT: NC to FL, in Gulf of Mexico from SW FL. Also Cuba to Brazil. Occur around stagnant, weedy backwaters and mangrove habitats.

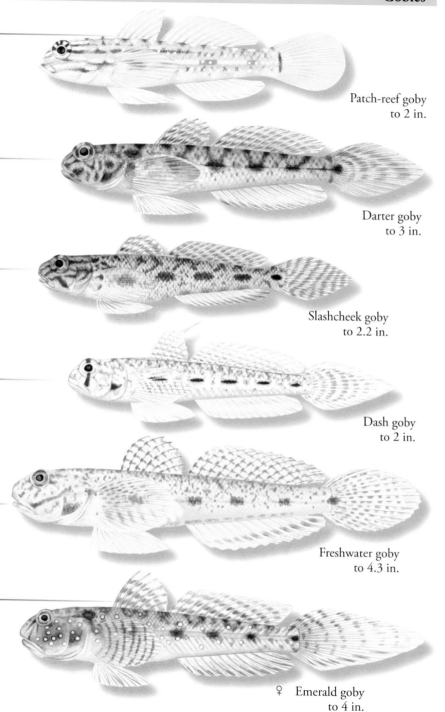

Patch-reef goby
to 2 in.

Darter goby
to 3 in.

Slashcheek goby
to 2.2 in.

Dash goby
to 2 in.

Freshwater goby
to 4.3 in.

♀ Emerald goby
to 4 in.

Marked goby - *Ctenogobius stigmaticus* (Poey, 1860)

FEATURES: Tannish. Three distinct dark reddish bars on cheeks below eyes. A large dark blotch present on shoulder above pectoral-fin base. Reddish smudges on upper head, along back, and along midline. Pale bars on sides. First dorsal fin with at least one elongate spine. HABITAT: SC to FL. FL Keys to Pensacola, FL, and off TX. Also south to Brazil. Reported from shallow, muddy, and sandy bottoms of estuaries.

Spottail goby - *Ctenogobius stigmaturus* (Goode & Bean, 1882)

FEATURES: Pale tannish gray with irregular, darker smudges and lines on head and body. Dark bar below eyes. Preopercular margin dark. Dark spot at upper pectoral-fin base. Dark blotches along midline resemble commas. Dark spot at caudal-fin base. Fins banded. HABITAT: SE FL to Panama. Also Cuba. Reported from Key West and Bermuda. Associated with shallow inshore and brackish-water habitats.

Yellowline goby - *Elacatinus horsti* (Metzelaar, 1922)

FEATURES: Dark brown dorsally, slightly paler below. Upper portion of eyes yellow. A distinct yellow to whitish stripe extends from eyes to caudal-fin base. Florida specimens have a yellow stripe, those from Caribbean have a white stripe. Fins transparent. Spots or lines absent from snout. HABITAT: S and SW FL, Bahamas, Haiti, Cayman Islands, and Belize to Curaçao. Associated with vase-like and cylindrical sponges. Reported to prefer medium-depth reefs.

Tiger goby - *Elacatinus macrodon* (Beebe & Tee-Van, 1928)

FEATURES: Translucent. Narrow, dark, blackish brown stripes on head and encircling body. Cheeks reddish. Spinal column with alternating dark and pale smudges. Males with elongate first dorsal-fin spine. Body scaleless except on caudal peduncle. HABITAT: S and SW FL. Cuba, Haiti, and Veracruz. Also Bermuda. Over rock faces, coral rubble, pilings, and large sponges. Also in tide pools.

Neon goby - *Elacatinus oceanops* Jordan, 1904

FEATURES: Upper body blackish; underside of head and ventral profile whitish. A neon blue stripe extends from snout through eyes to posterior portion of caudal fin. Blue stripes do not meet on snout. Cheeks may have reddish wash. Spots absent from snout. Snout bluntly pointed. Mouth inferior, 'shark-like.' HABITAT: SE FL, FL Keys. In Gulf of Mexico from the Flower Garden Banks, off TX, and Campeche Bay. Also south to Colombia. Reef associated from shore to about 120 ft.

Leopard goby - *Elacatinus saucrus* (Robins, 1960)

FEATURES: Pale, translucent whitish. Cheeks with reddish tint. Head and body with orange brown to reddish brown spots overlaid by darker melanophores; spots form rows posteriorly. Spots on abdomen larger than other spots, may be squarish. HABITAT: FL Keys, Bahamas, Jamaica, Antilles, and Belize. Reported to perch on coral heads at depths between about 15 and 100 ft.

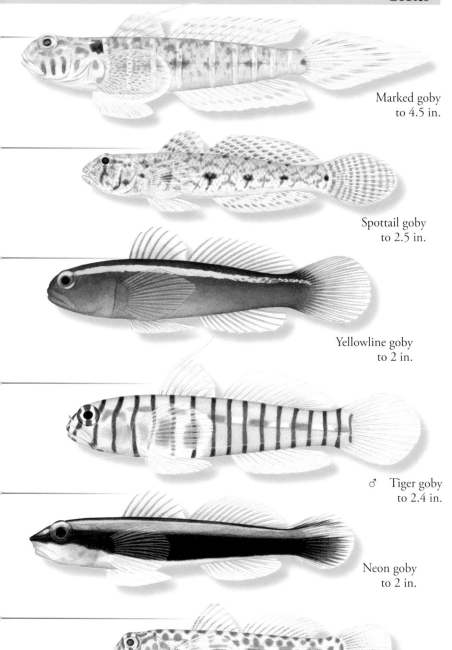

Marked goby
to 4.5 in.

Spottail goby
to 2.5 in.

Yellowline goby
to 2 in.

♂ Tiger goby
to 2.4 in.

Neon goby
to 2 in.

Leopard goby
to 1 in.

Gobiidae - Gobies, *cont.*

Yellowprow goby - *Elacatinus xanthiprora* (Böhlke & Robins, 1968)

FEATURES: Blackish from dorsal profile to around midline; lower body pale. A bright yellow to whitish stripe extends from eyes to caudal fin. Upper portion of eyes yellow. Distinct, yellow bar on snout. Cheeks may have reddish wash. Snout blunt. Lower head lacks bluish stripe. HABITAT: FL Keys, Dry Tortugas, Florida Middlegrounds to Jamaica. Also reported from Central America. Associated with coral reefs; may live in sponges.

Roughtail goby - *Evermannichthys metzelaari* Hubbs, 1923

FEATURES: Translucent with reddish brown on head and forebody. Irregular dark reddish brown bars and smudges along sides. First dorsal fin with 4–5 spines. Scales present above base of anal fin. HABITAT: NC, northeast Gulf of Mexico, Bahamas, Curaçao, and Colombia. Demersal in marine waters.

Sponge goby - *Evermannichthys spongicola* (Radcliff, 1917)

FEATURES: Translucent with irregular blackish gray bands, bars, and smudges on head and body. Markings become more broken posteriorly. First dorsal fin with 6–7 spines. Scales absent above anal-fin base. HABITAT: NC to Fort Pierce, FL. Also Panama City, FL, to Dry Tortugas and Campeche Bay. Sponge associated.

Lyre goby - *Evorthodus lyricus* (Girard, 1858)

FEATURE Tannish to grayish. Five or six dark smudges along dorsal profile that merge with smudges along midline. Smudge under first dorsal fin darkest, merges into lower dorsal-fin base. Two distinct, squarish blotches at caudal-fin base. Snout short, rounded. Males with elongate dorsal-fin spines; long caudal fin; and pinkish stripe on upper and lower caudal-fin lobes. HABITAT: VA to FL, Gulf of Mexico, Bahamas, Antilles to northern South America. In bays and estuaries over muddy and grassy bottoms.

Goldspot goby - *Gnatholepis thompsoni* Jordan, 1904

FEATURES: Variably marked. Translucent with brownish spots on upper head and sides. Spots may overlie dark smudges on back and lower sides. Narrow, blackish brown to pale brown bar between and through eyes to lower head. Typically with a bright golden spot almost encircled by a blackish brown to pale brown ring over pectoral-fin base. Snout short, steeply sloping. HABITAT: FL, Caribbean Sea to northern South America. In Gulf of Mexico from SW FL, Flower Garden Banks, off Veracruz, and Campeche Bank. Also Bermuda. Over sand, rock, and rubble bottoms.

Violet goby - *Gobioides broussonnetii* Lacepède, 1800

FEATURES: Bluish gray to purplish brown dorsally. Variously creamy to whitish on sides and below. Dark chevron-like marks along midline. Mouth large, gaping. Eyes small. Dorsal and anal fins continuous with pointed caudal fin. Body elongate. HABITAT: SC to FL, Gulf of Mexico to Brazil. In muddy bays, estuaries, and river mouths to almost fresh water. Also offshore on muddy bottoms.

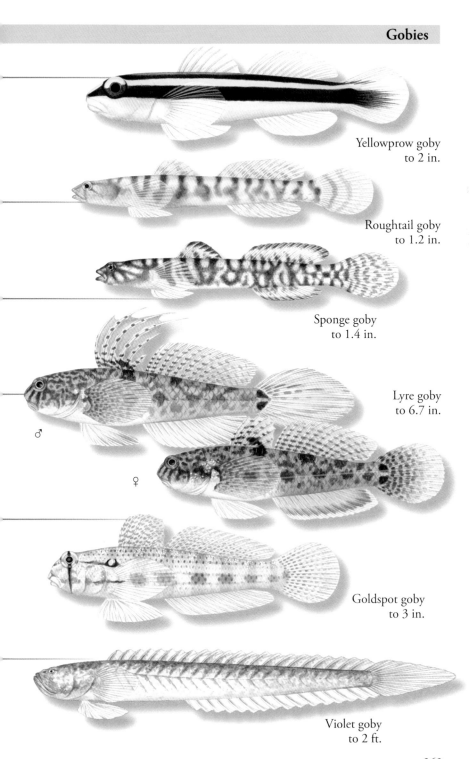

Yellowprow goby
to 2 in.

Roughtail goby
to 1.2 in.

Sponge goby
to 1.4 in.

Lyre goby
to 6.7 in.

♂

♀

Goldspot goby
to 3 in.

Violet goby
to 2 ft.

Highfin goby - *Gobionellus oceanicus* (Pallas, 1770)

FEATURES: Pale brown on upper head and posterior body. Dark bar below eyes. Dark blotch at upper pectoral-fin base. Large, blackish, ocellated spot under first dorsal fin. Smaller, paler spots anterior and posterior to large spot. Row of faint smudges along midline. Dark spot at caudal-fin base. First dorsal fin spines elongate. HABITAT: S FL, Gulf of Mexico to Brazil. In estuaries with submerged vegetation. BIOLOGY: Feed on filamentous algae, marine invertebrates, and insect larvae.

Naked goby - *Gobiosoma bosc* (Lacepède, 1800)

FEATURES: Shades of tan to gray with 9-11, broad, dark bars on body. May be very dark. Width between bars narrow to very narrow. Bars extend into dorsal fins. Head mottled, depressed, and somewhat laterally expanded. Body entirely scaleless. HABITAT: MA to N FL. In Gulf of Mexico from SW FL to Campeche. Occur in estuaries and over protected coastal, vegetated, and rubble bottoms and oyster beds.

Seaboard goby - *Gobiosoma ginsburgi* Hildebrand & Schroeder, 1928

FEATURES: Tannish to grayish with eight darker bars on sides from nape to caudal peduncle. Bars with pale inner blotches. Spaces between bars narrow. Dark dashes or spots bisect bars along midline. Head moderately depressed, appears swollen from above. Two scales on caudal-fin base. HABITAT: MA to Jacksonville, FL. Also SW FL. From near shore to about 180 ft. Also in estuaries.

Rockcut goby - *Gobiosoma grosvenori* (Robins, 1964)

FEATURES: Tannish with nine irregular, densely speckled bars on sides from nape to caudal peduncle. Bars may be very obscure. Dark dashes bisect bars along midline. Variable dark bar radiating from eye to mouth. Head moderately depressed. Second dorsal fin with one spine and nine rays. Scales on posterior body. HABITAT: SE FL and Jamaica to Venezuela. Reported from shallow marine waters over rocky areas.

Code goby - *Gobiosoma robustum* Ginsburg, 1933

FEATURES: Pale to greenish tan or tannish. Ten to twelve irregular, speckled bars on sides. Bars become broken on lower sides, appear chain-like. Series of dark dots and dashes along midline. Variable dark bar from eye to mouth. Second dorsal fin with 1 spine and 10-12 rays. Body entirely scaleless. HABITAT: E FL, FL Keys, and Gulf of Mexico. In shallow, protected areas over algal mats and seagrass beds; prefer saline waters.

Paleback goby - *Gobulus myersi* Ginsburg, 1939

FEATURES: Pale tannish dorsally, brownish below. Upper sides with scattered, dark flecks and two poorly defined saddles. May have obscure, dark blotch on upper pectoral-fin base. Mouth oblique. Top of head depressed. Body scaleless. HABITAT: S FL, northeast Gulf of Mexico, Bahamas, and Antilles to Venezuela. Occur around coral reefs and shell rubble from about 16 to 155 ft.

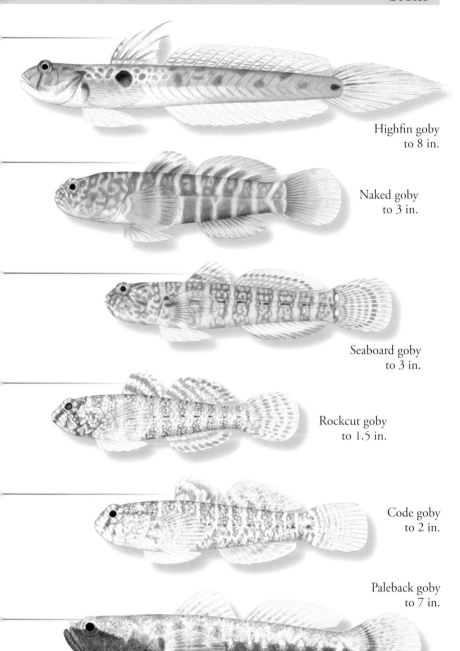

Highfin goby
to 8 in.

Naked goby
to 3 in.

Seaboard goby
to 3 in.

Rockcut goby
to 1.5 in.

Code goby
to 2 in.

Paleback goby
to 7 in.

Crested goby - *Lophogobius cyprinoides* (Pallas, 1770)

FEATURES: Females tannish with brownish bands on head that extend to below dorsal fins. Four dark blotches on lower sides. First dorsal fin with a blackish smudge and a distinct orange blotch. Males more uniform and darker in color; become blackish while breeding. Dorsal fin dark, typically lacking orange blotch. All with prominent crest on head. HABITAT: S FL, western and southern Gulf of Mexico, Bahamas, Caribbean Sea to northern South America. Also Bermuda. In shallow, protected, estuarine, and coastal waters, including bays and mangrove areas.

Island goby - *Lythrypnus nesiotes* Böhlke & Robins, 1960

FEATURES: Body with alternating dark brown and pale bars. A dark, whole or broken line bisects pale bars. Dark brown spots or broken bands on sides of head. One (possibly two) dark blotches at pectoral-fin base. Tip of tongue pointed. HABITAT: S FL, Bahamas, Antilles, western Caribbean, and northern South America. Scattered in Gulf of Mexico. Over shallow rocky bottoms and coral reefs to about 30 ft.

Convict goby - *Lythrypnus phorellus* Böhlke & Robins, 1960

FEATURES: Body with alternating rusty brown to dark brown and pale bars. A dark, broken or whole line bisects pale bars. Dark spots or broken bands radiate from eyes. Two faint to dark spots on pectoral-fin base. Tip of tongue is bluntly pointed. HABITAT: NC to FL Keys, Flower Garden Banks, and off coast of Central America. Associated with coral reefs from about 6 to 110 ft.

Bluegold goby - *Lythrypnus spilus* Böhlke & Robins, 1960

FEATURES: Head and body with alternating orangish to gold and blue gray bars. Blue gray bars bisected by a dark line. A large, dark blotch covers most of pectoral-fin base. First two dorsal-fin spines elongate. HABITAT: S FL, FL Keys, Bahamas to Haiti. In the Gulf of Mexico from the Flower Garden Banks. Also Bermuda. Associated with coral reefs from about 10 to 85 ft.

Seminole goby - *Microgobius carri* Fowler, 1945

FEATURES: Pinkish white with iridescent highlights. Bright blue stripe behind eyes. Dusky green stripe along dorsal profile. Broad yellow orange stripe on midline merges with yellow on caudal fin. Second, fourth, and fifth dorsal-fin spines elongate in males. HABITAT: NC to S FL. In Gulf of Mexico from FL to AL. Also Lesser Antilles. Over sandy bottoms from about 20 to 70 ft. BIOLOGY: Construct burrows.

Clown goby - *Microgobius gulosus* (Girard, 1858)

FEATURES: Pale grayish tan. Obscure dark blotches on upper sides. Pearly white band from mouth to opercle. White bar below first dorsal fin. Males with elongate first dorsal-fin spines. Caudal fin lacks spots. Females without elongate spines; caudal fin spotted. Both with large, gaping mouth; broad, slightly compressed head. HABITAT: Chesapeake Bay to TX. In quiet, muddy, shallow, marine, estuarine, and fresh waters.

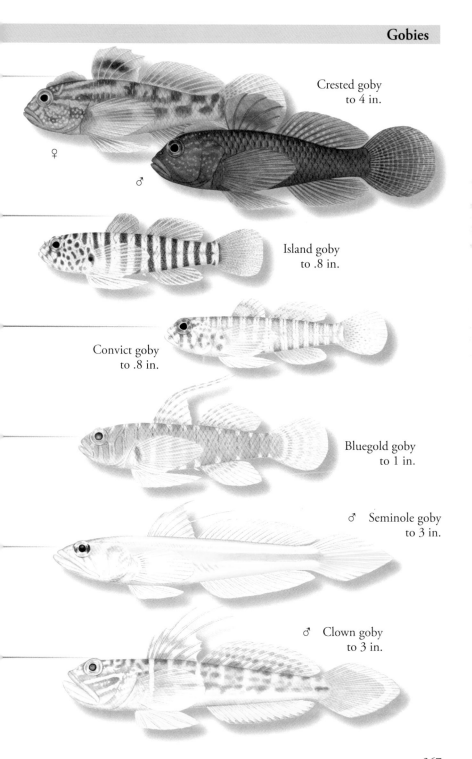

Crested goby
to 4 in.

♀

♂

Island goby
to .8 in.

Convict goby
to .8 in.

Bluegold goby
to 1 in.

♂ Seminole goby
to 3 in.

♂ Clown goby
to 3 in.

Banner goby - *Microgobius microlepis* Longley & Hildebrand, 1940

FEATURES: Males pale grayish with a greenish, bluish, or lavender cast dorsally; alternating blue and orangish streaks on cheeks; dorsal fins rosy with a clear to yellow submarginal band. Females (not shown) similarly colored to males but with a white area above anal fin that is bound by an oblique dusky bar on abdomen and a dusky stripe from about mid-body to caudal fin. Both with oblique mouth, nuchal crest, pointed caudal fin. HABITAT: S FL, Bahamas to Belize. Over shallow, sandy bottoms.

Green goby - *Microgobius thalassinus* (Jordan & Gilbert, 1883)

FEATURES: Pale greenish with some iridescent luster. Two iridescent bands on lower head. Females with a red spot on anterior first dorsal fin; anal fin without spots. Males with several iridescent blue bars on sides below first dorsal fin and black spots along anal-fin margin. Mouth strongly oblique. HABITAT: Chesapeake Bay to FL; Gulf of Mexico to Galveston Bay. Absent from S FL. Over shallow, protected soft bottoms.

Orangespotted goby - *Nes longus* (Nichols, 1914)

FEATURES: Color, pattern variable. Pale with five to seven pairs of brown blotches along lower sides. May have few to many yellowish blotches between dark blotches; all may be small to large, and sparse to densely clustered. Dark melanophores peppered over blotches. Fins spotted to banded. First dorsal-fin spine elongate. HABITAT: S FL, Dry Tortugas, Yucatán, Bahamas, Puerto Rico to Venezuela. Over sandy bottoms from shore to about 30 ft. BIOLOGY: Share burrows with alpheid shrimps.

Spotfin goby - *Oxyurichthys stigmalophius* (Mead & Böhlke, 1958)

FEATURES: Pearly with oblong orange blotches dorsally. Five large, dark orange blotches along midline. First dorsal fin with large black blotch at rear. HABITAT: FL, Bahamas, and southern Gulf of Mexico to Surinam. Over sandy and muddy bottoms from about 6 to 200 ft. BIOLOGY: Share burrows with alpheid shrimps.

Rusty goby - *Priolepis hipoliti* (Metzelaar, 1922)

FEATURES: Dark reddish orange with pale bluish bars on head and body. Body densely peppered with minute, dark melanophores. Dorsal and caudal fins with spots forming bands. Mouth oblique; lower jaw protrudes. Second dorsal-fin spine elongate. HABITAT: S FL, FL Keys, Bahamas to northern South America. Also Flower Garden Banks and Veracruz. Over reefs and rocky areas from about 100 to 330 ft.

Tusked goby - *Risor ruber* (Rosén, 1911)

FEATURES: Translucent pale grayish to brownish with head and body covered in closely-set brown to dark gray speckles. May be very dark. Cheeks reddish. May have obscure, broken bars on sides. Mouth very small, located under very blunt snout. Posterior portion of body scaled. HABITAT: S FL, Bahamas to Surinam. In Gulf of Mexico from W FL, AL, and Flower Garden Banks. Associated with large sponges.

NOTE: There are eight other Gobies in the area, p. 431.

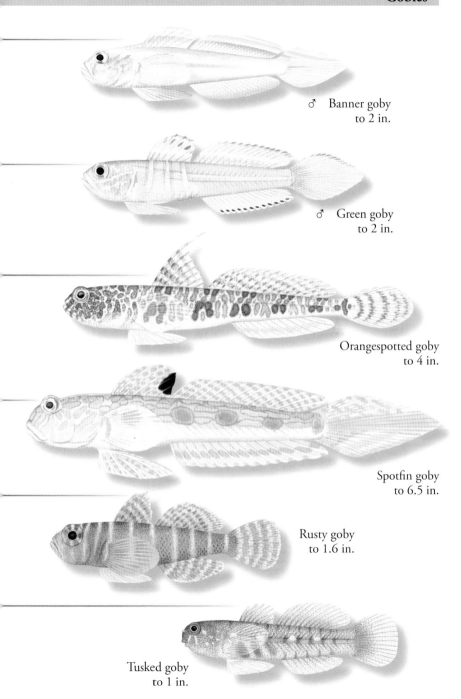

♂ Banner goby
to 2 in.

♂ Green goby
to 2 in.

Orangespotted goby
to 4 in.

Spotfin goby
to 6.5 in.

Rusty goby
to 1.6 in.

Tusked goby
to 1 in.

Microdesmidae - Wormfishes

Pugjaw wormfish - *Cerdale floridana* Longley, 1934

FEATURES: Pale whitish, somewhat translucent. Upper head and body sparsely peppered with minute, dark flecks. Eyes white. Snout short. Mouth oblique with protruding lower jaw. Body elongate and somewhat laterally compressed. Dorsal and anal fins continuous with caudal fin. HABITAT: S FL, Dry Tortugas, Bahamas, Greater and Lesser Antilles to Colombia. Coral reef associated to about 100 ft.

Pink wormfish - *Microdesmus longipinnis* (Weymouth, 1910)

FEATURES: Orangish to tannish with pinkish cast and iridescent sheen. Lower jaw barred. Snout short. Mouth oblique, lower jaw protrudes. Body eel-like. Dorsal and anal fins continuous with rounded caudal fin. HABITAT: NC to FL. Cayman Islands, Gulf of Mexico from FL to TX and Veracruz. Also Bermuda. Demersal.

ALSO IN THE AREA: Lancetail wormfish, *Microdesmus lanceolatus*, p. 431.

Ptereleotridae - Dartfishes

Blue dartfish - *Ptereleotris calliura* (Jordan & Gilbert, 1882)

FEATURES: Pale bluish gray to lavender. Blue line from snout to dorsal-fin origin and behind eyes to opercle. Outer margin of dorsal fins blackish; inner margin blue. Upper and lower margins of caudal fin may be black. Caudal fin long and pointed. HABITAT: S FL and Gulf of Mexico. Possibly north to NC. Over shelly rubble and sand around coral reefs. BIOLOGY: Hover over bottom. Inhabit burrows.

Hovering dartfish - *Ptereleotris helenae* (Randall, 1968)

FEATURES: Pale bluish gray with lavender highlights. Blue lines from snout to dorsal-fin origin and on side of head. Fins yellowish. Dorsal, caudal, and anal fins with blue outer margin and red inner margin. Caudal fin oblong. HABITAT: Reported from S FL, Bahamas, Antilles, and western Caribbean Sea. Over coral rubble and sandy and silty bottoms around reefs. BIOLOGY: Hover over bottom. Inhabit burrows.

Ephippidae - Spadefishes

Atlantic spadefish - *Chaetodipterus faber* (Broussonet, 1782)

FEATURES: Silvery with broad, brownish gray, or brownish to blackish bars on body. Bars may be absent, especially in large adults. Juveniles dark brown or blackish with some white. Snout very short, mouth small. Preopercular margin finely serrated, lacks long spine at corner. Opercle with blunt spine at posterior margin. Spiny dorsal fin tall in juveniles, shortens with age. Second dorsal and anal fins with elongate anterior rays. Body disc-shaped in profile. HABITAT: MA to FL, Gulf of Mexico, Bahamas, Caribbean Sea to Brazil. Occur coastally over wrecks, reefs, pilings. Also around buoys and mangroves, in harbors, and under bridges. BIOLOGY: Feed on invertebrates. Juveniles mimic dead leaves or floating debris. Sought as foodfish.

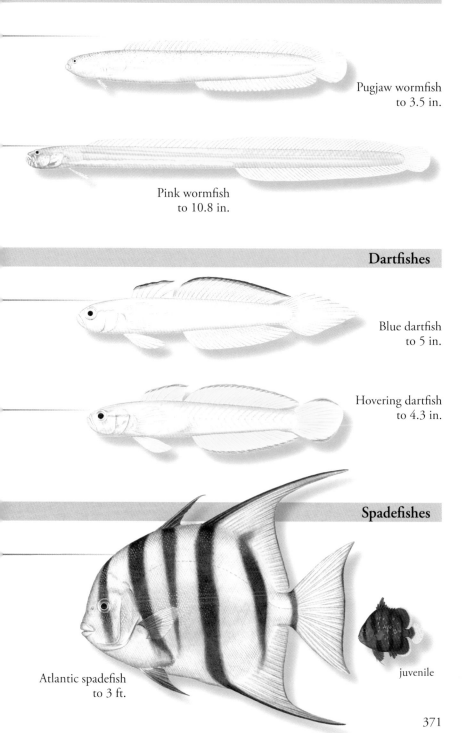

Wormfishes

Pugjaw wormfish
to 3.5 in.

Pink wormfish
to 10.8 in.

Dartfishes

Blue dartfish
to 5 in.

Hovering dartfish
to 4.3 in.

Spadefishes

Atlantic spadefish
to 3 ft.

juvenile

Luvaridae - Louvar

Louvar - *Luvarus imperialis* Rafinesque, 1810

FEATURES: Silvery bluish to pinkish dorsally. Silvery white below. All fins except pelvic fins reddish. Snout very short. A deep groove is present from snout to above eyes. The head is elevated, the forehead is compressed and keel-like. Dorsal fin originates behind eyes in juveniles. In adults, dorsal fin originates posterior to midline. Caudal peduncle with slender keel. Pelvic fins reduced to a small flap. HABITAT: Worldwide in tropical to temperate seas. Records scattered from CT to FL. Also in Gulf of Mexico. Pelagic and oceanic in deep water off continental shelves to about 1,900 ft. Sometimes in shallower water near coast. BIOLOGY: Female Louvar may carry up to 47 million eggs. Solitary. Feed on plankton.

Acanthuridae - Surgeonfishes

Ocean surgeon - *Acanthurus bahianus* Castelnau, 1855

FEATURES: Yellowish, bluish gray, grayish brown to brown. May change color, pale, darken, or display a mottled pattern. Narrow blue lines radiate from eyes. Margin of opercle is blackish. Dorsal, caudal, and anal fins with narrow blue margins. Base of caudal fin often white but may be same color as body. Pelvic-fin membranes blackish. Caudal fin margin moderately to deeply concave. HABITAT: MA to FL, Bahamas, Caribbean Sea to Brazil. Also Bermuda, and reported from Flower Garden Banks. Found over coral reefs and rocky bottoms with sandy areas. BIOLOGY: Ocean surgeon feed on algae and seagrass. Usually in small groups and with Blue tang and Doctorfish. Swim by flapping the pectoral fins.

Doctorfish - *Acanthurus chirurgus* (Bloch, 1787)

FEATURES: Pale to dark brown, or grayish brown to grayish blue. May change color, pale, darken, or display dark blotches. Narrow blue lines radiate from eyes. Margin of opercle blackish. Usually with 8–12 dark bars on sides. Dorsal, caudal, and anal fins with narrow blue margins. Base of caudal fin may be pale or the same color as body. Pelvic fins with bluish rays. Caudal-fin margin somewhat to moderately concave. HABITAT: MA to FL, Bahamas, Caribbean Sea to Brazil. Also Bermuda and reported from Flower Garden Banks. Over coral reefs and rocky bottoms with sandy areas. BIOLOGY: Feed on algae and seagrass; also observed feeding on soft coral. School with Blue tang and Ocean surgeon. Swim by flapping pectoral fins.

Blue tang - *Acanthurus coeruleus* Bloch & Schneider, 1801

FEATURES: Pale to dark blue, or purplish. May appear uniformly colored or with narrow, wavy lines on sides. May also display pale bars on sides. Can pale or darken. Yellow spine on caudal peduncle. Juveniles yellow with blue on eyes and margins of dorsal and anal fins. Body very deep, laterally compressed. HABITAT: NY (rare) to FL, Gulf of Mexico, Bahamas, Caribbean Sea to Brazil. Also Bermuda. Found over shallow coral reefs and rocky bottoms. BIOLOGY: Blue tang feed on algae and seagrass. Occur singly or in large schools. Some juveniles may be as large as adults.

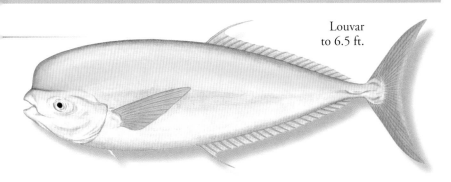

Louvar
to 6.5 ft.

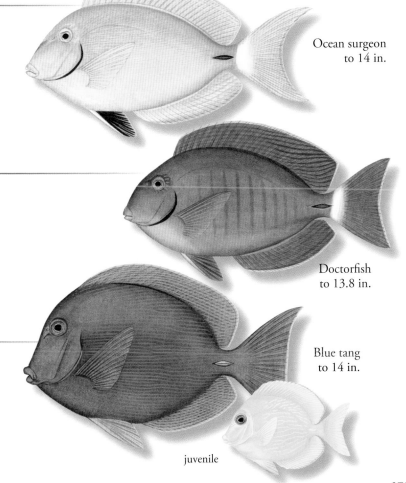

Ocean surgeon
to 14 in.

Doctorfish
to 13.8 in.

Blue tang
to 14 in.

juvenile

Scombrolabracidae - Longfin escolar

Longfin escolar - *Scombrolabrax heterolepis* Roule, 1921

FEATURES: Overall blackish brown. Snout length about equal to eye diameter. Jaws large, somewhat oblique, protrusible, and with many sharp teeth. Eyes very large. Dorsal fin long-based and notched. Second dorsal and anal fins similar in shape and size. Pectoral fins long, almost reaching anal-fin origin. Lateral line runs close to dorsal profile. HABITAT: In tropical to subtropical waters of the Atlantic, Pacific, and Indian oceans over continental shelves and slopes from about 330 to 2,900 ft.

Sphyraenidae - Barracudas

Great barracuda - *Sphyraena barracuda* (Edwards, 1771)

FEATURES: Bluish gray to gray dorsally. Upper sides with oblique bars. Silvery below. Lower sides with few to many variably-sized, gray to black blotches. Fins may be tipped white. Head large. Lower jaw protrudes. Mouth large with large canine-like teeth. Caudal fin slightly forked. HABITAT: In tropical to warm temperate waters of the Atlantic and Indo-West Pacific. Found from inshore to offshore waters over structures and reefs. MA to FL, Gulf of Mexico, Bahamas, Caribbean Sea to Brazil. BIOLOGY: Adults usually occur singly. Juveniles schooling. Voracious predator.

Northern sennet - *Sphyraena borealis* DeKay, 1842

FEATURES: Olivaceous dorsally, silvery on sides and below. Fins often yellowish. Head large. Mouth large and toothy. Lower jaw protrudes. Pectoral fins do not reach pelvic-fin origin. Pelvic-fin base under or slightly posterior to dorsal-fin origin. Caudal fin forked. HABITAT: Nova Scotia to FL, Gulf of Mexico to Panama. Also Bermuda. In coastal waters to about 215 ft. Found over a variety of bottoms.

Guaguanche - *Sphyraena guachancho* Cuvier, 1829

FEATURES: Olivaceous dorsally, silvery on sides and below. A yellow stripe is present on sides. Head large. Mouth large and toothy. Lower jaw protrudes. Margins of caudal, anal, and pelvic fins blackish. Pectoral fins reach past pelvic-fin origin. Pelvic-fin base under or slightly anterior to first dorsal-fin origin. Caudal fin forked. HABITAT: MA (rare) to FL, Gulf of Mexico, Bahamas, Caribbean Sea to Brazil. Occur in turbid coastal waters over muddy bottoms. Also in bays and estuaries.

Gempylidae - Snake mackerels

Snake mackerel - *Gempylus serpens* Cuvier, 1829

FEATURES: Steely gray to grayish brown with silvery luster. Fins with dark margins. Jaws large, toothy. Lower jaw with pointed tip. First dorsal fin long-based, with 26–32 spines. Second dorsal fin followed by 5–6 finlets. Anal fin followed by 6–7 finlets. Caudal fin forked. Body very elongate and laterally compressed. HABITAT: Worldwide in tropical to temperate seas from surface to about 650 ft. NY to FL, Gulf of Mexico, Bahamas, Caribbean Sea to northern South America. Offshore, oceanic.

Longfin escolar

Longfin escolar
to 11.8 in.

Barracudas

Great barracuda
to 10 ft.

Northern sennet
to 18 in.

Guaguanche
to 2 ft.

Snake mackerels

Snake mackerel
to 3.5 ft.

Escolar - *Lepidocybium flavobrunneum* (Smith, 1843)

FEATURES: Almost uniformly dark brown, becoming black with age. First dorsal fin low, with eight to nine spines. Second dorsal fin followed by four to six finlets. Lateral line sinuous. Caudal peduncle with one strong keel and two smaller keels. HABITAT: Worldwide in tropical to warm temperate seas. Georges Bank to FL, Gulf of Mexico, and Bahamas to Surinam. Over continental slopes to about 650 ft. BIOLOGY: Migrate to shallow water at night. Feed on squids and a variety of fishes.

Black snake mackerel - *Nealotus tripes* Johnson, 1865

FEATURES: Dark brownish gray to blackish brown. Dorsal and anal fins pale brownish gray to brownish. Mouth and head large. First dorsal fin with 20–21 spines. Second dorsal and anal fins similarly shaped; followed by two finlets. Body elongate, compressed. HABITAT: Worldwide in tropical to warm temperate seas from about 330 to 2,000 ft. Nova Scotia to FL, Gulf of Mexico, and Caribbean Sea to Brazil.

American sackfish - *Neoepinnula americana* (Grey, 1953)

FEATURES: Dark brown dorsally, silvery on sides and below. Fins dark. Two flat spines at opercular margin. First dorsal fin originates over opercle. Lateral line is double; both branches originate above upper gill opening. Body moderately deep, compressed. HABITAT: N and W Gulf of Mexico. Yucatán, Greater Antilles, Haiti, Surinam, Venezuela, and Bermuda. Occur near bottom from about 545 to 1,500 ft.

Black gemfish - *Nesiarchus nasutus* Johnson, 1862

FEATURES: Dark brown with violet sheen. Fins blackish. Mouth and head large. Jaws with conical, fleshy tips. Dorsal fin with 19–21 spines. Second dorsal and anal fins similarly shaped; second dorsal-fin base longer than anal-fin base; both followed by two finlets. Body very long. HABITAT: Probably worldwide in tropical to warm temperate seas. NC to FL, Gulf of Mexico, Caribbean Sea to Brazil. Near bottom over continental slopes from about 655 to 4,000 ft. BIOLOGY: Migrate upward at night.

Roudi escolar - *Promethichthys prometheus* (Cuvier, 1832)

FEATURES: Silvery gray to coppery brown. Fins blackish in large specimens; yellowish with dark tips in small specimens. First dorsal fin with 17–19 spines. Second dorsal fin and anal fins similar in shape; followed by two finlets. Lateral line single; arched anteriorly. HABITAT: Worldwide in tropical to warm temperate seas. Eastern US. Gulf of Mexico to Brazil. Near bottom over continental slopes to about 2,500 ft.

Oilfish - *Ruvettus pretiosus* Cocco, 1833

FEATURES: Brown to dark brown. Tips of pectoral and pelvic fins blackish. First dorsal fin low, with 13–15 spines. Second dorsal fin followed by two finlets. Lateral line straight posteriorly. Abdomen keeled with bony scales. HABITAT: Worldwide in tropical to warm temperate seas. Newfoundland to FL, Gulf of Mexico, Caribbean Sea, and Bermuda. Near bottom over continental slopes and rises to about 2,300 ft.

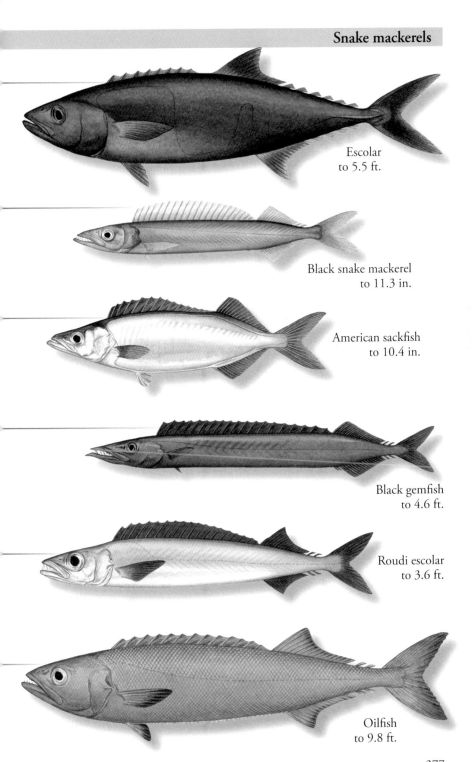

Escolar
to 5.5 ft.

Black snake mackerel
to 11.3 in.

American sackfish
to 10.4 in.

Black gemfish
to 4.6 ft.

Roudi escolar
to 3.6 ft.

Oilfish
to 9.8 ft.

Trichiuridae - Cutlassfishes

Atlantic cutlassfish - *Trichiurus lepturus* Linnaeus, 1758

FEATURES: Iridescent silvery blue to steely blue. Dorsal fin pale dusky yellow. Mouth large, lower jaw protrudes. Dorsal fin originates anterior to opercular opening and tapers posteriorly. Body tapers to a pointed tail. Caudal fin absent. HABITAT: Worldwide in tropical to warm temperate seas. Cape Cod to FL. Gulf of Mexico, Caribbean Sea to Argentina. Near bottoms on continental shelves to about 330 ft.

Scombridae - Mackerels and Tunas

Wahoo - *Acanthocybium solandri* (Cuvier, 1832)

FEATURES: Iridescent bluish green dorsally, silvery below. Numerous irregular bars on sides. Snout long, pointed. First dorsal fin with 23–27 spines; originates over pectoral-fin base. Second dorsal fin followed by nine finlets. Body elongate, slightly compressed. HABITAT: Worldwide in tropical to subtropical seas. NJ to FL, Gulf of Mexico, Caribbean Sea to northern South America. Offshore in open ocean from surface to mid-water depths. BIOLOGY: Feed on squids and a variety of fishes. Females may carry up to 6 million eggs. Sought commercially and as sportfish.

Bullet mackerel - *Auxis rochei* (Risso, 1810)

FEATURES: Dark blue dorsally, deep purple to blackish over head. Upper scaleless area with about 15 irregular, moderately broad, oblique bars; some may merge. Dorsal fins widely separated. Pectoral fins short, not extending past corselet. Corselet well developed posteriorly. Body comparatively shallow. HABITAT: Worldwide in tropical to warm temperate seas. Nova Scotia to FL, Gulf of Mexico, Caribbean Sea to Brazil. Occur in water column over continental shelves and slopes. BIOLOGY: Form large schools. Sought commercially. Other name: Bullet tuna.

Frigate mackerel - *Auxis thazard* (Lacepède, 1800)

FEATURES: Dark blue dorsally, deep purple to blackish over head. Upper scaleless area with 15 or more irregular, oblique, wavy bars. Bars may be whole, broken, or converging. Dorsal fins widely separated. Pectoral fins short, reaching slightly beyond corselet. Corselet narrow posteriorly. Body comparatively deep. HABITAT: Worldwide in tropical to subtropical seas. NC to FL, scattered south to Venezuela. Exact distribution uncertain. Occur in water column over continental shelves and slopes. BIOLOGY: Schooling. Other name: Frigate tuna.

Little tunny - *Euthynnus alletteratus* (Rafinesque, 1810)

FEATURES: Greenish to dark blue dorsally. Silvery below. Upper scaleless area with highly variable cluster of narrow, oblique, wavy dark lines. Several faint to dark spots on abdomen below pectoral fins. Dorsal fins connected at base. HABITAT: Occur in the water column and at surface mainly over continental shelves of the Atlantic. Also found coastally and inshore. Nova Scotia to FL, Gulf of Mexico to Brazil. Also Bermuda. BIOLOGY: Schooling. Other names: False albacore, Albie.

Atlantic cutlassfish
to 4 ft.

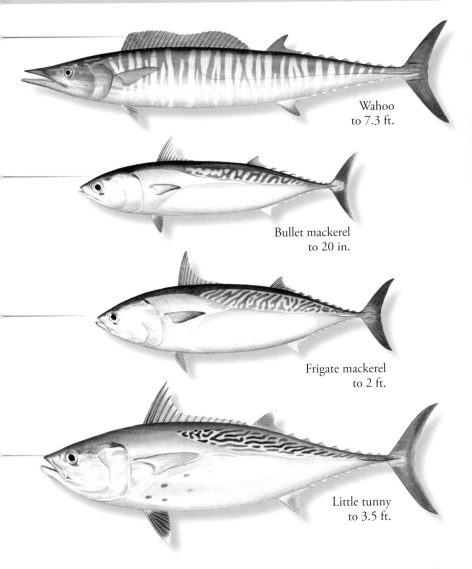

Wahoo
to 7.3 ft.

Bullet mackerel
to 20 in.

Frigate mackerel
to 2 ft.

Little tunny
to 3.5 ft.

Scombridae - Mackerels and Tunas, *cont.*

Skipjack tuna - *Katsuwonus pelamis* (Linnaeus, 1758)

FEATURES: Bluish black dorsally, silvery below. Upper scaleless area may have several oblique, iridescent bands. Lower sides with four to six grayish stripes that may appear as discontinuous lines of dark blotches. Dorsal fins separated by a small interspace. HABITAT: Worldwide in tropical to warm temperate seas. Nova Scotia to FL, Gulf of Mexico, Caribbean Sea to Brazil. Occur in the water column over continental shelves and slopes, generally above the thermocline. BIOLOGY: Form large schools. Often mix with Blackfin tuna. Sought commercially and as sportfish.

Atlantic bonito - *Sarda sarda* (Bloch, 1793)

FEATURES: Bluish to greenish blue dorsally, fading to silvery below. Upper sides with numerous oblique, dark stripes. May also have broad, diffuse, faint bars underlying dark stripes. First dorsal fin with straight to slightly concave margin; separated from second dorsal fin by small interspace. Body entirely covered in minute scales; scales on corselet well developed. HABITAT: Along tropical to temperate coasts of Atlantic. Nova Scotia to FL, Gulf of Mexico, Colombia, and Venezuela. Pelagic and at surface of coastal and inshore waters. BIOLOGY: Feed mostly on fishes. Form schools near the surface. Migratory. Sought commercially and as sportfish.

Atlantic chub mackerel - *Scomber colias* Gmelin, 1789

FEATURES: Blue dorsally, fading to silvery below. Back and upper sides with many oblique, faint to dark wavy lines. Lines break into dusky spots around midline. Eyes with adipose lids; comparatively large. Dorsal fins separated by distance about equal to first dorsal-fin base. Two small keels at caudal-fin base. HABITAT: Nova Scotia to FL, Bahamas to Venezuela. Uncommon in Gulf of Mexico and Caribbean Sea. Also in eastern Atlantic. Pelagic in coastal waters. BIOLOGY: Schooling, migratory.

Atlantic mackerel - *Scomber scombrus* Linnaeus, 1758

FEATURES: Iridescent dark blue dorsally, becoming greenish to about midline. Silvery below. Upper sides with many oblique wavy lines. A broken dark stripe usually present along midline. Eyes with adipose lids; comparatively small. Dorsal fins separated by a distance greater than the length of the first dorsal-fin base. Two small keels on caudal peduncle. HABITAT: Labrador to NC. Also in eastern Atlantic. Occur in estuaries to over outer continental shelves from surface to about 600 ft. BIOLOGY: Form dense schools. Migratory. Sought commercially and as bait.

King mackerel - *Scomberomorus cavalla* (Cuvier, 1829)

FEATURES: Iridescent bluish with greenish reflections dorsally. Silvery below. Small specimens with gray to yellow spots on sides. First dorsal fin uniformly colored. Dorsal fins scarcely separated. Pectoral fins scaled at base. Lateral line curves downward under second dorsal fin. HABITAT: MA (rare) to FL, Gulf of Mexico, throughout Antilles to northern South America. Occur in the water column over continental shelves and slopes. Also over outer reefs, wrecks, and hard structures. BIOLOGY: Occur singly or in small groups. Migratory. Sought commercially and as sportfish.

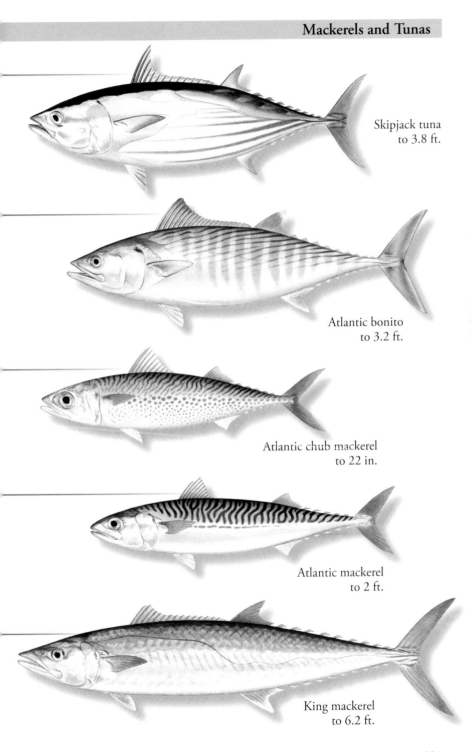

Skipjack tuna
to 3.8 ft.

Atlantic bonito
to 3.2 ft.

Atlantic chub mackerel
to 22 in.

Atlantic mackerel
to 2 ft.

King mackerel
to 6.2 ft.

Scombridae - Mackerels and Tunas, *cont.*

Spanish mackerel - *Scomberomorus maculatus* (Mitchill, 1815)

FEATURES: Iridescent greenish to blue green dorsally. Silvery below. Sides with several rows of fairly evenly-spaced rusty orange to yellowish oval spots. Anterior portion of first dorsal fin blackish, white at base. Second dorsal fin and pectoral fins with dusky tips. Pectoral fin scaled at base. Lateral line gently sloping. HABITAT: Nova Scotia to FL and Gulf of Mexico. Occur in water column over continental shelves and slopes. Also at surface and near shore. Enter estuaries. BIOLOGY: Form large and small schools. Migratory. Sought commercially and as sportfish.

Cero - *Scomberomorus regalis* (Bloch, 1793)

FEATURES: Iridescent greenish to blue green dorsally. Silvery below. Sides with rusty orange to yellowish streak along midline; similarly colored oblong spots present above and below streak. Anterior portion of dorsal fin blackish, white at base. Inner portion of second dorsal fin dark. Pectoral fins scaled. Lateral line gradually sloping. HABITAT: MA to FL, throughout Antilles to Brazil. Also recorded from Gulf of Mexico. Occur in clear water around reefs. Also reported from over seagrass beds. BIOLOGY: Usually solitary or in small groups. Sought as sportfish.

Albacore - *Thunnus alalunga* (Bonnaterre, 1788)

FEATURES: Metallic blackish blue dorsally, silvery on sides and below. Second dorsal fin yellowish. Caudal-fin margin white. Pectoral fins extremely long, reaching well beyond second dorsal-fin origin, usually to or beyond second dorsal finlet. HABITAT: Worldwide in tropical to temperate seas. Nova Scotia to FL, Caribbean Sea to Argentina. Not in Gulf of Mexico. Oceanic; usually below thermocline. BIOLOGY: Young often in large schools. Sought commercially and as sportfish.

Yellowfin tuna - *Thunnus albacares* (Bonnaterre, 1788)

FEATURES: Metallic blackish blue dorsally, becoming yellow on upper sides to silvery below. Lower sides often with about 20 pale spots and nearly vertical lines. Second dorsal and anal fins yellow. Finlets yellow with black margin. Pectoral fins moderately long. Second dorsal and anal fins tall, become taller and longer with age. HABITAT: Worldwide in tropical to subtropical seas. Nova Scotia to FL, Gulf of Mexico, Caribbean Sea to Brazil. Oceanic, above and below thermocline, in temperatures from about 64-87°F. BIOLOGY: School near surface, sometimes with other tuna species of similar size. Sought commercially and as sportfish.

Blackfin tuna - *Thunnus atlanticus* (Lesson, 1831)

FEATURES: Metallic blackish blue dorsally. May have yellow luster on upper sides. Lower sides silvery; may have almost vertical pale streaks. Second dorsal and anal fins dusky with silvery luster. Finlets dusky with a trace of yellow. Pectoral fins moderately long. HABITAT: In western Atlantic from MA to FL, Gulf of Mexico, Caribbean Sea to Brazil. Oceanic, in warm water; not usually below 68°F. BIOLOGY: Prefer warm waters. Migrate northward in summer months. Form mixed schools with Skipjack tuna of the same size. Sought commercially and as sportfish.

Spanish mackerel
to 2.8 ft.

Cero
to 3 ft.

Albacore
to 4.4 ft.

Yellowfin tuna
to 7.2 ft.

Blackfin tuna
to 3.5 ft.

Scombridae - Mackerels and Tunas, *cont.*

Bigeye tuna - *Thunnus obesus* (Lowe, 1839)

FEATURES: Metallic blackish blue dorsally. Silvery below. Pale spots and lines may be present on lower sides. Second dorsal and anal fins silvery yellow and comparatively short. Finlets yellow with black margin. Pectoral fins very long in small specimens, moderately long in large specimens. HABITAT: Worldwide in tropical to subtropical seas. Nova Scotia to FL, Gulf of Mexico, Caribbean Sea to Brazil. Oceanic and pelagic from surface to about 820 ft. BIOLOGY: Juveniles and small adults school at surface; may school with Yellowfin and Skipjack tunas. Listed as Vulnerable.

Atlantic bluefin tuna - *Thunnus thynnus* (Linnaeus, 1758)

FEATURES: Metallic dark blue to black dorsally. Silvery to iridescent coppery below. Oblique pale lines and spots on lower sides. Second dorsal fin brownish. Anal fin with dark margin. Finlets dusky yellow with dark margin. Lateral keel blackish in adults. Pectoral fins short. HABITAT: In Atlantic Ocean from Nova Scotia to FL, Gulf of Mexico, Caribbean Sea to Brazil. Pelagic. BIOLOGY: Very fast, highly migratory. Juveniles in warm water. Adults enter cold water. Listed as Endangered.

Xiphiidae - Swordfish

Swordfish - *Xiphias gladius* Linnaeus, 1758

FEATURES: Back and upper sides metallic bluish black to brownish black. Coppery below. Bars absent. Bill flattened in cross-section. Eyes large. First dorsal fin short-based. Second dorsal and anal fins small. Single, broad lateral keel present. Body robust. HABITAT: Worldwide in tropical to warm temperate seas. Newfoundland to FL, Gulf of Mexico to Brazil. Oceanic. From surface to about 2,000 ft.

Istiophoridae - Billfishes

Sailfish - *Istiophorus platypterus* (Latreille, 1804)

FEATURES: Dark blue dorsally, coppery on sides, silvery below. May display iridescent blue bars and spots on upper sides. Dorsal fin bluish black to dark brownish blue with dark spots; tall and sail-like. Pelvic fins very long. HABITAT: Worldwide. NY to FL, Gulf of Mexico, Caribbean Sea to S Brazil. Coastal and oceanic; usually above the thermocline. BIOLOGY: Very swift and migratory. Recorded to hunt in coordinated groups. Sought commercially and for sport. Previously known as *Istiophorus albicans*.

Blue marlin - *Makaira nigricans* Lacepède, 1802

FEATURES: Dark blackish blue to blackish brown dorsally. Silvery white ventrally. Several bars on upper body. First dorsal fin dark blue to black; anterior portion not taller than depth of body. Pelvic fins shorter than pectoral fins. Lateral line reticulating. HABITAT: Worldwide. Gulf of Maine to FL, Gulf of Mexico, Caribbean Sea to S Brazil. Oceanic; usually in or above the thermocline from about 71–88°F. BIOLOGY: Highly migratory. Enter colder water to feed. Sought commercially and for sport.

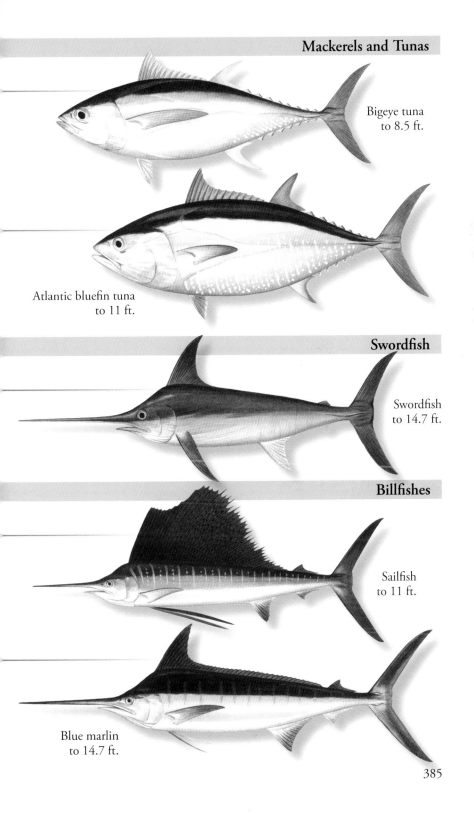

Mackerels and Tunas

Bigeye tuna
to 8.5 ft.

Atlantic bluefin tuna
to 11 ft.

Swordfish

Swordfish
to 14.7 ft.

Billfishes

Sailfish
to 11 ft.

Blue marlin
to 14.7 ft.

Istiophoridae - Billfishes, *cont.*

White marlin - *Kajikia albida* (Poey, 1860)

FEATURES: Dark blue to brown dorsally. Coppery on sides. Silvery white below. May display iridescent bars on sides. Dorsal fin dark blue with blackish spots. Dorsal and first anal fins rounded at tips. Pectoral fins usually rounded at tips. Anus located close to first anal-fin origin. Scales on sides pointed anteriorly, stiff and one- to two-pronged posteriorly. HABITAT: Atlantic Ocean. Nova Scotia to FL, Gulf of Mexico, Caribbean Sea to S Brazil. Oceanic. Usually found above the thermocline, in water above 72°F. BIOLOGY: Highly migratory. Found at higher latitudes during warm months. Sought commercially and for sport.

Roundscale spearfish - *Tetrapturus georgii* Lowe, 1841

FEATURES: Dark blue dorsally. Coppery on sides. Silvery white below. May display iridescent bars on sides. Dorsal fin dark blue; spots absent. Dorsal and first anal fins usually blunt at tips. Pectoral fins usually pointed at tips. Anus located at a distance equal to or exceeding the height of the first anal fin. Scales on sides rounded anteriorly, soft and two- to three-pronged posteriorly. HABITAT: Atlantic Ocean and Mediterranean Sea. In western Atlantic known from VA to FL and east of Antilles. Exact distribution uncertain. BIOLOGY: May be misidentified as White marlin.

Longbill spearfish - *Tetrapturus pfluegeri* Robins & de Sylva, 1963

FEATURES: Dark blue dorsally. Coppery on sides. Silvery white below. Bars and spots absent from sides. First dorsal fin blackish blue; spots absent. Anterior portion of dorsal fin taller than depth of body. Bill comparatively short. Body comparatively shallow. HABITAT: Atlantic Ocean. NJ to FL, Gulf of Mexico, Caribbean Sea to Venezuela. Oceanic in tropical to subtropical waters; usually above the thermocline. BIOLOGY: Feed on fishes, crustaceans, and cephalopods. Caught as bycatch.

Centrolophidae - Medusafishes

Black ruff - *Centrolophus niger* (Gmelin, 1789)

FEATURES: Almost uniformly dark bluish gray to blackish with abdomen slightly paler. Other specimens described as brownish to brownish pink. Females described as paler than males. Head small. Snout rounded, with nostrils near tip. Dorsal fin is continuous, originates over pectoral-fin base, and has four to five weak spines. HABITAT: Nova Scotia to NJ. Also in eastern North Atlantic and Mediterranean Sea. Oceanic. BIOLOGY: Young found near surface, adults near bottom.

Black driftfish - *Hyperoglyphe bythites* (Ginsburg, 1954)

FEATURES: Shades of gray to brown dorsally, pale below. Snout rounded, with nostrils near tip. Preopercle striated. Dorsal fin is continuous, originates over or slightly behind pectoral-fin base, and has seven to eight low spines. Body moderately deep and compressed. HABITAT: S FL, Gulf of Mexico, possibly to Brazil. BIOLOGY: Young found near surface, associated with flotsam. Adults schooling, near bottom.

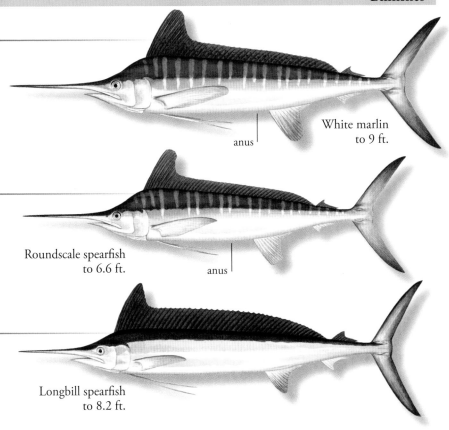

anus

White marlin
to 9 ft.

Roundscale spearfish
to 6.6 ft.

anus

Longbill spearfish
to 8.2 ft.

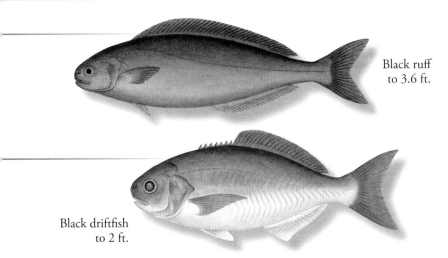

Black ruff
to 3.6 ft.

Black driftfish
to 2 ft.

387

Centrolophidae - Medusafishes, *cont.*

Barrelfish - *Hyperoglyphe perciformis* (Mitchill, 1818)

FEATURES: Shades of gray to brown dorsally, pale below. May be mottled. Snout rounded, with nostrils near tip. Preopercle striated. Dorsal fin is continuous, originates over or slightly behind pectoral-fin base, and has eight to nine low spines. Body moderately deep and compressed. HABITAT: Nova Scotia to northern Gulf of Mexico. Also Bermuda. BIOLOGY: Juveniles associated with drifting seaweed. Adults occur near bottom from about 196 to 397 ft.

Brown ruff - *Schedophilus medusophagus* (Cocco, 1839)

FEATURES: Uniformly dark brown. Juveniles mottled. Head small. Snout rounded, with nostrils near tip. Preopercle with about 12 spines at margin. Dorsal fin is continuous, originates anterior to pectoral-fin base, and has a few weak spines. HABITAT: Grand Banks to NC. Also in northeastern Atlantic and Mediterranean Sea. Oceanic. BIOLOGY: Juveniles associated with jellyfishes. Adults in deep water.

Nomeidae - Driftfishes

Bigeye cigarfish - *Cubiceps pauciradiatus* Günther, 1872

FEATURES: Blackish to blackish brown. Snout blunt, rounded. Nostrils near tip of snout. Mouth small. Eyes very large and surrounded by adipose tissue. Bony keel on breast. Pectoral fins long, wing-like. Pelvic fins insert posterior to pectoral-fin base. Scales on most of head. Lateral line on upper back; extends to below posterior base of second dorsal fin. HABITAT: Worldwide in tropical to warm temperate seas. Nova Scotia to FL, and Gulf of Mexico to Argentina. Occur at depths from about 330 to 2,700 ft. Reported to be pelagic.

Man-of-war fish - *Nomeus gronovii* (Gmelin, 1789)

FEATURES: Juveniles iridescent blue dorsally, silvery below. Sides with blackish blue bars, blotches, and spots. Pelvic fins entirely or mostly bluish black. Adults (not shown) are described as dark bluish or brown. Snout short, squared-off at tip. Nostrils near tip of snout. Mouth small. Eyes moderately sized; surrounded by adipose tissue. Pectoral fins large, fan-like. HABITAT: Worldwide in tropical to warm temperate seas. Rare in eastern Atlantic. Newfoundland to FL, Gulf of Mexico, Bahamas, Caribbean Sea to Brazil. Oceanic. BIOLOGY: Juveniles associated with Portuguese man-of-war.

Freckled driftfish - *Psenes cyanophrys* Valenciennes, 1833

FEATURES: Silvery with distinct dark spots forming narrow stripes on sides. Observed specimens blue dorsally, with dark fins. Others described as pale brown. Snout blunt, rounded. Eyes moderately large. Body oval in profile, deep, and compressed. Scales easily shed. HABITAT: Worldwide in tropical to subtropical seas. MA to FL, and northern Gulf of Mexico to northern South America. Also Bahamas and Bermuda. Associated with jellyfishes and *Sargassum* seaweed. Also reported to associate with flotsam. NOTE: This species may comprise several species.

Medusafishes

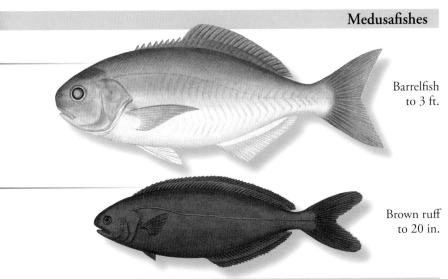

Barrelfish
to 3 ft.

Brown ruff
to 20 in.

Driftfishes

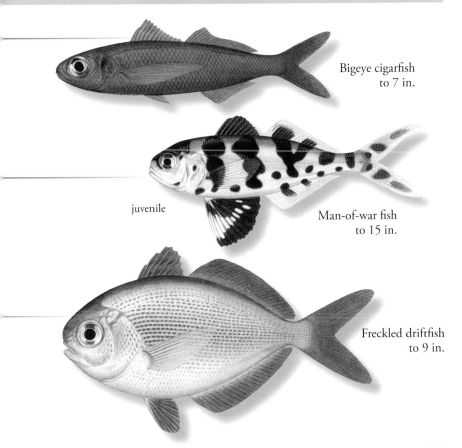

Bigeye cigarfish
to 7 in.

juvenile

Man-of-war fish
to 15 in.

Freckled driftfish
to 9 in.

Nomeidae - Driftfishes, *cont.*

Bluefin driftfish - *Psenes pellucidus* Lütken, 1880

FEATURES: Adults uniformly bluish black; may be paler on abdomen. May have pale bars. Body elongate and laterally compressed. Snout blunt, rounded. Nostrils at tip of snout. Mouth small. Eyes surrounded by adipose tissue. HABITAT: Worldwide in tropical to warm temperate seas. Georges Bank to FL, Gulf of Mexico, possibly to South America. Exact distribution uncertain.

ALSO IN THE AREA: Cape fathead, *Cubiceps capensis*; Driftfish, *Cubiceps gracilis*; Banded driftfish, *Psenes arafurensis*, p. 431.

Ariommatidae - Ariommatids

Silver-rag - *Ariomma bondi* Fowler, 1930

FEATURES: Bluish dorsally, silvery on sides and below. Others described as bluish brown dorsally. Lining of body cavity pale. Snout blunt. Mouth small. Eyes large and surrounded by adipose tissue. Caudal peduncle with two weak keels. Body elongate, laterally compressed. HABITAT: In tropical and warm temperate waters of Atlantic Ocean. Nova Scotia to FL, Gulf of Mexico, Caribbean Sea to Uruguay. Over continental shelves from about 80 to 2,000 ft. More common above 900 ft.

Brown driftfish - *Ariomma melanum* (Ginsburg, 1954)

FEATURES: Brownish dorsally and on sides, may be paler below. Others described as blackish. Lining of body cavity dark. Snout moderately blunt. Mouth small. Eyes moderately large, surrounded by adipose tissue. Caudal peduncle with two weak keels. Body elongate, laterally compressed. HABITAT: In tropical to warm temperate waters of Atlantic Ocean. NY to FL, northern Gulf of Mexico, and Caribbean Sea. Over continental slopes from about 460 to 2,400 ft. BIOLOGY: Juveniles near surface.

Spotted driftfish - *Ariomma regulus* (Poey, 1868)

FEATURES: Silvery to pale brown. Numerous dark spots on upper sides. Opercle, first dorsal fin, and pelvic fins blackish. Snout blunt, rounded. Mouth small. Eyes moderately large, surrounded by adipose tissue. Caudal peduncle with two weak keels. Body deep, oval in profile. HABITAT: NJ to FL, Gulf of Mexico, Caribbean Sea to Guyana. Also Great Bahama Bank. Occur at depths from about 600 to 1,600 ft.

Tetragonuridae - Squaretails

Bigeye squaretail - *Tetragonurus atlanticus* Lowe, 1839

FEATURES: Uniformly dark brown. Lower jaw with knife-like teeth; fits completely into upper jaw when mouth is closed. Head and eyes large. First dorsal fin low. Caudal peduncle with two low keels. Scales keeled. Body elongate. HABITAT: Worldwide in tropical to warm temperate seas. Nova Scotia to FL, eastern Gulf of Mexico, Greater Antilles to Panama. Oceanic in waters from about 68°–77°F.

Driftfishes

Bluefin driftfish
to 2.6 ft.

Ariommatids

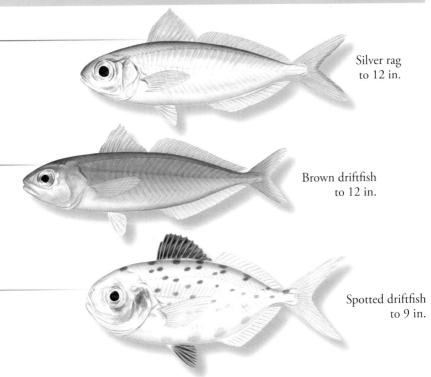

Silver rag
to 12 in.

Brown driftfish
to 12 in.

Spotted driftfish
to 9 in.

Squaretails

Bigeye squaretail
to 2 ft.

Stromateidae - Butterfishes

Gulf butterfish - *Peprilus burti* Fowler, 1944

FEATURES: Iridescent blue dorsally, silvery below. Fins dusky. Body rarely mottled. Mouth small. Snout short, blunt. Dorsal and anal fins long-based. Pelvic fins absent. A series of 17–25 pores present under anterior dorsal fin. Body deep and laterally compressed. HABITAT: In Gulf of Mexico from S FL to Yucatán. Pelagic. Occur over sandy and muddy bottoms of continental shelf from about 6 to 900 ft. More abundant below 500 ft. BIOLOGY: Gulf butterfish form large, loose schools. Juveniles associated with floating seaweeds and jellyfish.

Harvestfish - *Peprilus paru* (Linnaeus, 1758)

FEATURES: Iridescent blue to green dorsally, silvery below. Fins dusky. Tips of dorsal, caudal, and anal fins may be blackish. Mouth small. Snout short, blunt. Dorsal and anal fins long-based; anterior lobes elongate to very elongate. Pelvic fins absent. Pores under dorsal fin absent. Body very deep, laterally compressed. HABITAT: VA to FL. Gulf of Mexico to Argentina and Greater Antilles. Strays rarely to Gulf of Maine. Pelagic. Along coast in bays and inshore waters from about 165 to 230 ft. BIOLOGY: Form large schools. Juveniles associated with floating seaweeds and jellyfish.

Butterfish - *Peprilus triacanthus* (Peck, 1804)

FEATURES: Iridescent blue dorsally, silvery below. Sides with diffuse, dark blotches. Fins pale dusky. Mouth small. Snout short, blunt. Dorsal and anal fins long-based. A series of 17–25 pores present under anterior dorsal fin. Body moderately deep, laterally compressed. HABITAT: Gulf of St. Lawrence to FL. Most abundant from ME to Cape Hatteras. Pelagic. Occur over sandy and muddy bottoms in brackish estuaries to continental shelf to about 180 ft. BIOLOGY: Form large schools.

Caproidae - Boarfishes

Deepbody boarfish - *Antigonia capros* Lowe, 1843

FEATURES: Pinkish to reddish dorsally, silvery pink below. May display a broad, faint stripe on midside. Mouth small, almost vertical. Snout short, upturned. Dorsal fin with seven to nine strong spines. Anal fin with three strong spines. Body rhomboid in profile with depth exceeding length. HABITAT: Worldwide in tropical to subtropical seas. MA to FL, Gulf of Mexico, and northern South America. Adults occur near bottom from about 200 to 1,900 ft. Juveniles are pelagic.

Shortspine boarfish - *Antigonia combatia* Berry & Rathjen, 1959

FEATURES: Pinkish to reddish dorsally, silvery pink below. Mouth small, almost vertical. Snout short, somewhat upturned. Spiny dorsal fin with nine or ten strong spines. Anal fin with three strong spines. Body oval in profile, with body depth not exceeding length. HABITAT: NJ to FL, E Gulf of Mexico, Caribbean Sea to Brazil. Adults occur near bottom from about 220 to 1,900 ft. Juveniles are pelagic. BIOLOGY: Shortspine boarfish feed on small invertebrates.

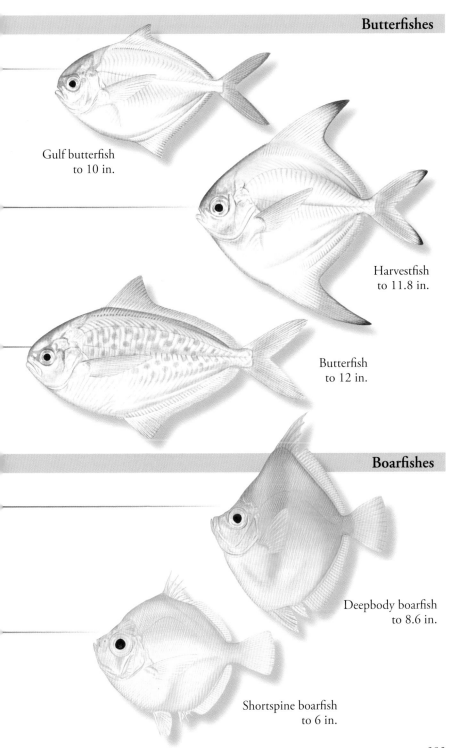

Gulf butterfish
to 10 in.

Harvestfish
to 11.8 in.

Butterfish
to 12 in.

Boarfishes

Deepbody boarfish
to 8.6 in.

Shortspine boarfish
to 6 in.

Scophthalmidae - Turbots

Windowpane - *Scophthalmus aquosus* (Mitchill, 1815)

FEATURES: Eyed side shades of tan, olivaceous, or reddish; somewhat translucent. Numerous dark spots on body and fins. May also display white spots and mottling. Blind side usually white. Mouth strongly oblique. First several dorsal-fin rays free, branched. Pelvic fin on eyed side originates near throat and is almost continuous with anal fin. Body nearly round in profile. Body and fins form a rhomboid shape. HABITAT: Gulf of St. Lawrence to FL. On sandy to muddy bottoms from shore to about 650 ft. BIOLOGY: Windowpane are migratory. Spawning takes place almost year-round. Feed on crustaceans, shrimps, and fishes.

Paralichthyidae - Sand flounders

Three-eye flounder - *Ancylopsetta dilecta* (Goode & Bean, 1883)

FEATURES: Eyed side shades of brown with dark flecks and some mottling. Three large, symmetrically arranged, ocellated spots on posterior body: one on upper body; a second on lateral line; the third on lower body. Blind side is white. Upper jaw extends to middle of lower eye. Anterior dorsal-fin rays elongate, with the third spine tallest. Pelvic fin on eyed side elongate. HABITAT: NC to FL, and Gulf of Mexico to Yucatán. Also Tobago. Demersal from about 165 to 1,200 ft.

Ocellated flounder - *Ancylopsetta ommata* (Jordan & Gilbert, 1883)

FEATURES: Eyed side shades of brown with numerous pale spots and rings. Four large, ocellated spots on posterior body: two on upper body; a third on lateral line; the fourth on lower body. Blind side is white. Upper jaw extends beyond pupil of lower eye. Anterior dorsal-fin spines somewhat elongate. HABITAT: NC to Jupiter, FL, and Gulf of Mexico to Campeche, Mexico. Occur from shallow bays to about 300 ft.

Gulf Stream flounder - *Citharichthys arctifrons* Goode, 1880

FEATURES: Eyed side tannish. Scales with dark margins. Observed specimens with dark body cavity that is visible from above. Body otherwise unmarked. Blind side is white. Horn-like projection on snout. Upper jaw extends to middle of lower eye. Eyes relatively large, with lower eye anterior to upper eye. Body somewhat elongate. Lateral line almost straight above pectoral fin. HABITAT: Georges Bank to S FL, in Gulf of Mexico from FL to Yucatán. Occur over soft bottoms, usually from about 150 to 1,200 ft.

Horned whiff - *Citharichthys cornutus* (Günther, 1880)

FEATURES: Eyed side tan to brown with a dark blotch at pectoral-fin base. Body and fins may have some faint mottling. Upper jaw extends to middle of lower eye. Eyes large, closely set in juveniles and females, widely separated in mature males. Males also with a prominent spine on snout. Body deep, tapering posteriorly. HABITAT: NC to FL, Gulf of Mexico to Brazil. Also Bahamas and Greater Antilles. Found over soft bottoms of continental shelves from about 65 to 1,200 ft.

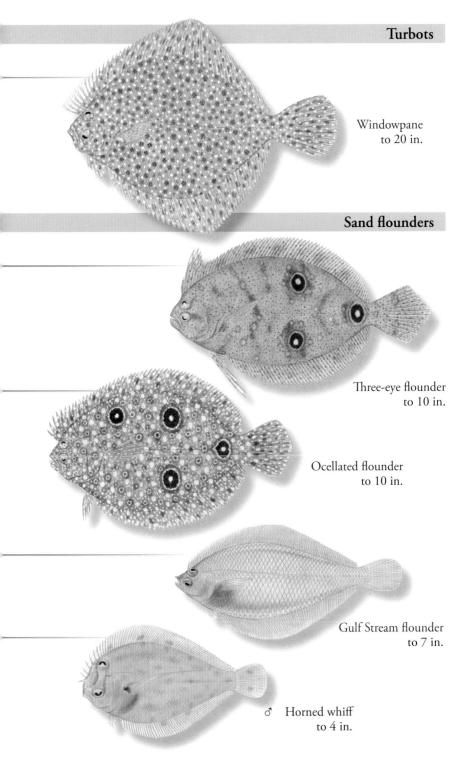

Windowpane
to 20 in.

Three-eye flounder
to 10 in.

Ocellated flounder
to 10 in.

Gulf Stream flounder
to 7 in.

♂ Horned whiff
to 4 in.

Paralichthyidae - Sand flounders, *cont.*

Anglefin whiff - *Citharichthys gymnorhinus* Gutherz & Blackman, 1970

FEATURES: Eyed side tannish, with several somewhat symmetrically arranged, dark-ringed blotches. Caudal fin with two dark blotches near base. Eyes large. Dorsal and anal fins expanded, angular. Males with several spiny projections on head and two large dark blotches on dorsal and anal fins. Females with reduced spines on head and dark flecks on dorsal and anal fins. HABITAT: FL Keys to AL. Bahamas, Antilles to Guyana, and Nicaragua. Range may extend north to NC. Bottom-dwelling from about 120 to 660 ft. More common in shallower portion of range.

Spotted whiff - *Citharichthys macrops* Dresel, 1885

FEATURES: Eyed side tan to brown with prominent dark spots and blotches on body and fins. Caudal fin with two dark spots near base. Blind side is white. Upper jaw extends to about middle of lower eye. Leaf-like cirri on margin of lower opercle on blind side. Pelvic-fin bases very short. Scales easily shed. HABITAT: NC to FL, Gulf of Mexico to Brazil. Found on hard sand and coarse shelly bottoms of continental shelves from shore to about 60 ft. Occasionally found to about 300 ft. Demersal.

Bay whiff - *Citharichthys spilopterus* Günther, 1862

FEATURES: Eyed side pale to dark brown, with or without numerous spots and blotches on body and fins. Dark spot on caudal peduncle. May display obscure, dark chevron-shaped mark on posterior body. Blind side is white. Upper jaw extends nearly to posterior portion of lower eye. Eyes separated by a low, narrow, concave ridge. Edge of opercle on blind side lacks cirri. Pelvic-fin bases short. HABITAT: NJ to FL, Gulf of Mexico to Brazil. Also Greater Antilles. Occur on muddy bottoms in estuaries and lagoons to continental shelves from shore to about 250 ft. Demersal.

Mexican flounder - *Cyclopsetta chittendeni* Bean, 1895

FEATURES: Eyed side brown with a blackish blotch under pectoral fin. Dorsal and anal fins with two large, dark blotches that are partially enclosed by a pale C-shaped mark; also with several small, dark, partially ocellated spots. Caudal fin with three dark, partially-ocellated blotches at margin. Blind side white. Upper jaw extends past rear margin of lower eye. Eyes separated by a moderately narrow, concave ridge. Pelvic fin bases short. HABITAT: NE Gulf of Mexico to Brazil. Also Jamaica. Absent from Gulf Coast of FL. Found on inner continental shelves from about 60 to 490 ft.

Spotfin flounder - *Cyclopsetta fimbriata* (Goode & Bean, 1885)

FEATURES: Eyed side is brown with numerous small, pale and dark speckles that form loose clusters. Pectoral fin with a broad, dark margin. Dorsal and anal fins with two large, partially ocellated dark blotches. Center of caudal fin with a large, partially ocellated spot; sometimes with dark blotches at margin. Blind side white. Upper jaw extends to rear margin of lower eye. Eyes close-set and separated by a narrow, concave ridge. Pelvic-fin bases short. HABITAT: NC to FL, Gulf of Mexico to Brazil. Also Greater Antilles. On soft bottoms from about 65 to 750 ft.

♂ Anglefin whiff
to 3 in.

Spotted whiff
to 8 in.

Bay whiff
to 8 in.

Mexican flounder
to 12.6 in.

Spotfin flounder
to 13 in.

Fringed flounder - *Etropus crossotus* Jordan & Gilbert, 1882

FEATURES: Eyed side pale to dark brown; others described as olive brown. Two or three dusky blotches sometimes present on lateral line. Observed specimen with a dense covering of minute flecks on body. Caudal fin sometimes with dark margin. Mouth very small, extending to anterior portion of lower eye. Eyes set close together, separated by a slightly elevated ridge. Upper margin of head slightly concave. Scales singular, lack smaller accessory scales. Body very deep. HABITAT: VA to FL, Gulf of Mexico to northern South America. Also Antilles. Demersal from shore to about 100 ft., sometimes to about 300 ft. BIOLOGY: Enter bays during summer months.

Smallmouth flounder - *Etropus microstomus* (Gill, 1864)

FEATURES: Eyed side pale brown. Area around mouth and eyes dark. May have dusky blotches along lateral line. A series of dusky spots is often present along base of dorsal and anal fins. Mouth very small. Eyes are set close together and separated by an elevated ridge. Upper margin of head slightly concave. Primary scales have smaller, accessory scales. Body deep. HABITAT: NY to NC. Strays to FL. Demersal from near shore to about 120 ft. BIOLOGY: Reported to spawn from summer to fall north of Cape Hatteras and year-round south of Cape Hatteras.

Gray flounder - *Etropus rimosus* Goode & Bean, 1885

FEATURES: Eyed side grayish to dusky gray brown. Three dark, obscure blotches on lateral line. Mouth small, with upper jaw extending to anterior portion of lower eye. Eyes are set close together and separated by an elevated, scaled ridge. Upper margin of head slightly concave. Primary scales have many smaller, accessory scales. Body very deep. HABITAT: From just north of Cape Hatteras to NW FL in Gulf of Mexico. Demersal from about 65 to 195 ft. in warm waters of 62°F or above. BIOLOGY: Gray flounder spawn in winter.

Shrimp flounder - *Gastropsetta frontalis* Bean, 1895

FEATURES: Eyed side tan to brown with three ocellated spots on body: two on upper side, one on lower side. Several dark lines on and between eyes. Body variably flecked and marbled. Anterior dorsal-fin spines tall, with third spine tallest. Pelvic fin on eyed side longer than pelvic fin on blind side. HABITAT: NC to FL, Gulf of Mexico, Bahamas, and Nicaragua to Panama. Demersal from about 115 to 605 ft. BIOLOGY: Shrimp flounder feed on crustaceans.

Gulf flounder - *Paralichthys albigutta* Jordan & Gilbert, 1882

FEATURES: Eyed side pale to dark brown with three ocellated spots on body: one on upper side; a second on posterior lateral line; the third on lower side. Spots may be faint in large specimens. Remainder of body with numerous small spots and flecks. Jaws large, with upper jaw extending beyond rear margin of lower eye. Upper eye located directly above lower eye. HABITAT: NC to FL, Gulf of Mexico to Panama. A few recorded from Bahamas. Demersal over hard or sandy bottoms from about 60 to 425 ft. BIOLOGY: Adults migrate offshore in fall and winter to spawn.

Fringed flounder
to 7.8 in.

Smallmouth flounder
to 6 in.

scale detail

Gray flounder
to 5 in.

Shrimp flounder
to 10 in.

Gulf flounder
to 2.3 ft.

Paralichthyidae - Sand flounders, *cont.*

Summer flounder - *Paralichthys dentatus* (Linnaeus, 1766)

FEATURES: Eyed side shades of brown, gray to greenish brown. May also be whitish, bluish, reddish to almost black. May or may not have ocellated spots. When present, five ocellated spots are symmetrically arranged on posterior body: two on upper side below dorsal profile; a third on lateral line; the fourth and fifth on lower side above ventral profile. Blind side white. Upper jaw extends to rear margin of lower eye. Lateral line arched over pectoral fin. HABITAT: Nova Scotia to FL. Strays to Canada. Most common from MA to NC. Occur in channels, inlets, and lower estuaries and over sandbars and seagrass beds. From near shore to about 600 ft.; usually to about 130 ft.

Southern flounder - *Paralichthys lethostigma* Jordan & Gilbert, 1884

FEATURES: Eyed side shades of pale to dark brown, with or without diffuse pale and dark spots. Pale spots may appear to merge into lines. Large specimens may be almost uniformly colored. Ocellated spots absent. Blind side white to dusky. Upper jaw extends past rear margin of lower eye. Lateral line arched over pectoral fin. Body depth about 36% of total length. Dorsal fin with 80–95 rays. HABITAT: NC to FL, and Gulf of Mexico to TX. Absent from southwestern coast of FL. In estuaries and coastal waters over soft bottoms from shore to about 140 ft. Adults also enter rivers.

Fourspot flounder - *Paralichthys oblongus* (Mitchill, 1815)

FEATURES: Eyed side brown to gray; may be mottled or display broad, pale areas. Four distinct, symmetrically-arranged, ocellated spots present on posterior body. Blind side may be as dark as eyed side. Upper jaw extends to posterior portion of lower eye. Body moderately deep. HABITAT: Georges Bank to Dry Tortugas, FL. Demersal in bays and sounds of northern part of range; in water of about 900 ft. or more off FL.

Broad flounder - *Paralichthys squamilentus* Jordan & Gilbert, 1882

FEATURES: Eyed side shades of brown, with or without diffuse, non-ocellated spots and blotches. Blind side is white in small specimens, dusky in large specimens. Upper jaw extends slightly beyond posterior margin of lower eye. Lateral line arched over pectoral fin. Body deep, about 44% of total length. Dorsal fin with 76–85 rays. HABITAT: NC to FL and in Gulf of Mexico. Demersal from near shore to about 755 ft. BIOLOGY: Juveniles occur in shallow water, move to deeper water with increasing size.

Shoal flounder - *Syacium gunteri* Ginsburg, 1933

FEATURES: Eyed side tannish, with or without round or ocellated spots or blotches. Usually with a large, dark, diffuse blotch on caudal peduncle. Bases of dorsal and anal fins with a series of dark spots. Blind side white. Upper jaw extends to middle of lower eye. Lower eye distinctly anterior to upper eye. Eyes widely separated in both sexes, more so in males. Males also with elongate upper pectoral-fin rays. Body deep. HABITAT: FL, Gulf of Mexico, northern South America, and Antilles. Occur over sandy and muddy bottoms from about 55 to 415 ft. Also over shelly bottoms.

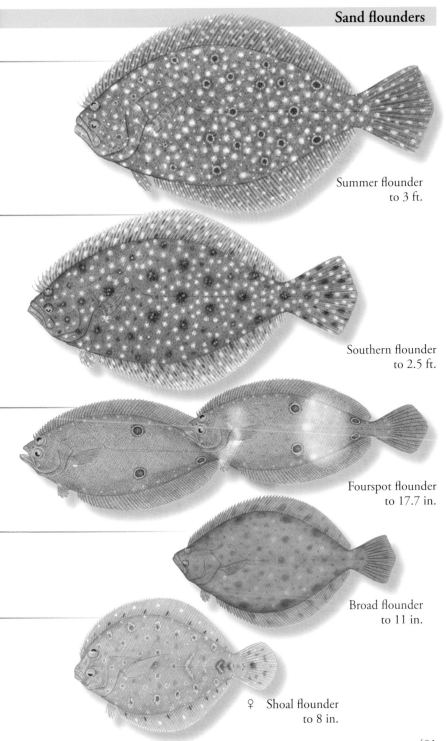

Summer flounder
to 3 ft.

Southern flounder
to 2.5 ft.

Fourspot flounder
to 17.7 in.

Broad flounder
to 11 in.

♀ Shoal flounder
to 8 in.

Paralichthyidae - Sand flounders, *cont.*

Channel flounder - *Syacium micrurum* Ranzani, 1842

FEATURES: Eyed side tan to brown with small dark and pale spots that form loose, ocellated clusters. Dark clusters under pectoral fin, along lateral line, and at center of caudal fin. Markings may be faint. Bases of dorsal and anal fins with a series of small, dark spots. Blind side white. Upper jaw extends to middle portion of lower eye. Males with moderately separated eyes and elongate pectoral-fin rays on eyed side. Body comparatively shallow. HABITAT: E FL, Bahamas, Caribbean Sea to Brazil. Possibly Gulf of Mexico. On soft bottoms from about 300 to 1,300 ft.

Dusky flounder - *Syacium papillosum* (Linnaeus, 1758)

FEATURES: Eyed side brown with small, dark and pale spots that may form loose, ocellated clusters. Distinct, dark clusters absent. Bases of dorsal and anal fins with a series of dark spots. Two dark lines from snout to upper eye. Blind side white to slightly dusky. Upper jaw extends to middle portion of lower eye. Males with widely separated eyes, iridescent lines on head, elongate pectoral-fin rays. HABITAT: NC to FL, Gulf of Mexico, Caribbean Sea to Brazil. On soft bottoms from about 33 to 460 ft.

ALSO IN THE AREA: Sand whiff, *Citharichthys arenaceus*; Shelf flounder, *Etropus cyclosquamus*, p. 431.

Pleuronectidae - Righteye flounders

Witch flounder - *Glyptocephalus cynoglossus* (Linnaeus, 1758)

FEATURES: Eyed side brown to grayish brown with or without darker bars. Pectoral fin with pale margin. Blind side is white with small, dark specks. Mouth very small. Lateral line not arched above pectoral fin. Dorsal fin with 97–117 rays. HABITAT: Gulf of St. Lawrence and Grand Banks to VA. To NC in deeper water. Also in eastern Atlantic. On soft, muddy bottoms from about 150 to 900 ft.

American plaice - *Hippoglossoides platessoides* (Fabricius, 1780)

FEATURES: Eyed side reddish to grayish brown. Smaller specimens with vague, dark blotches. Blotches fade with age. Tips of dorsal and anal fins white. Blind side white to bluish white. Upper jaw extends to below middle of lower eye. Lateral line slightly arched over pectoral fin. HABITAT: Greenland to Gulf of Maine. May stray to RI. Also in eastern North Atlantic. Over soft bottoms from shore to about 4,500 ft.

Atlantic halibut - *Hippoglossus hippoglossus* (Linnaeus, 1758)

FEATURES: Eyed side dark brown to olivaceous or slaty brown. Small specimens paler and more or less mottled. Large specimens darker, more uniformly colored. Blind side pure white in small specimens, blotched in large specimens. Mouth moderately large. Upper jaw extends to anterior half of lower eye. Lateral line arched over pectoral fin. HABITAT: SW Greenland to NY. Strays to VA. Also in northeastern Atlantic. On sandy, gravel, and clay bottoms from about 80 to 3,200 ft. Sought as foodfish. Currently listed as Endangered.

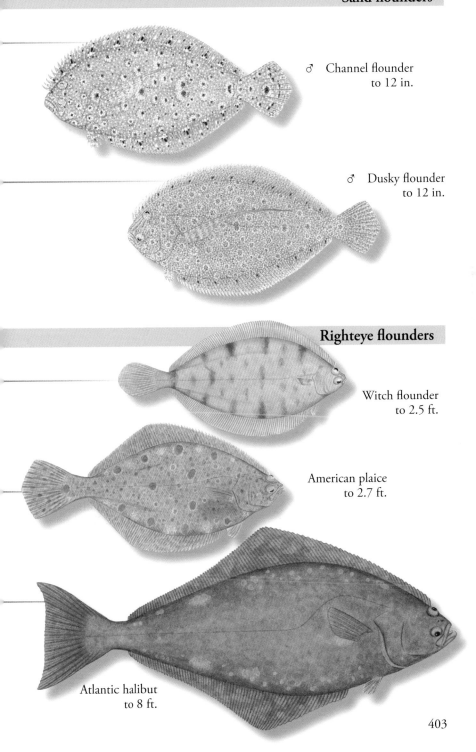

♂ Channel flounder
to 12 in.

♂ Dusky flounder
to 12 in.

Witch flounder
to 2.5 ft.

American plaice
to 2.7 ft.

Atlantic halibut
to 8 ft.

Pleuronectidae - Righteye flounders, *cont.*

Yellowtail flounder - *Limanda ferruginea* (Storer, 1839)

FEATURES: Eyed side brown to olive gray with rusty red spots. May also display pale spots. Tips of dorsal and anal fins and margin of caudal fin yellow. Blind side white with yellow along posterior body margins and on caudal peduncle and caudal fin. Mouth small, fleshy. Snout short, upturned. Eyes relatively large, almost touch each other. Dorsal fin with 73–91 rays. Lateral line arched over pectoral fin. HABITAT: Newfoundland to VA. Over sandy and sandy mud bottoms from about 33 to 330 ft.

Smooth flounder - *Pleuronectes putnami* (Gill, 1864)

FEATURES: Eyed side gray, muddy to slaty brown, or almost black. May be uniformly colored or variably mottled. Blind side is white. Mouth is small, snout short. Eyes separated by a scaleless ridge. Dorsal fin with 48–59 rays. Lateral line straight over pectoral fin. HABITAT: Labrador to RI. May occur southward to CT and northward to Siberia. Found coastally in bays, harbors, estuaries, and river mouths over soft mud bottoms to about 88 ft.

Winter flounder - *Pseudopleuronectes americanus* (Walbaum, 1792)

FEATURES: Color variable. Tannish, reddish brown, olive green, to dark slaty or black. May be uniformly colored, mottled, blotched, or partially white. Smaller specimens are paler and more marked than large specimens. Blind side is usually white; some with yellow on lower peduncle; some with dark blotches. Mouth small, snout short. Area between eyes scaled. Dorsal fin with 59–76 rays. Lateral line straight over pectoral fin. HABITAT: Labrador to GA. Over a variety of soft and hard bottoms from low-salinity rivers to moderately deep offshore waters.

Greenland halibut - *Reinhardtius hippoglossoides* (Walbaum, 1792)

FEATURES: Eyed side blackish, grayish, or yellowish brown. Paler on blind side. Mouth large. Upper jaw extends to posterior margin of lower eye. Eyes widely separated. Lateral line straight over pectoral fin. HABITAT: In Arctic waters of the Atlantic and Pacific oceans and the Bering Sea. Greenland to Gulf of Mexico. Demersal. Occur from about 295 to 5,250 ft. in cold waters of about 32°F.

Bothidae - Lefteye flounders

Peacock flounder - *Bothus lunatus* (Linnaeus, 1758)

FEATURES: Color changes with mood and habitat. Tan to gray to white. May be mottled or almost uniformly pale or dark. May display numerous pale to dark blue circles, semicircles, and spots on body. Usually with two or three variable, dark clusters on lateral line. Eyes widely separated, more so in mature males. Males with elongate pectoral-fin rays. Both with snout notched above mouth and fleshy tentacles on eyes. HABITAT: E FL, Bermuda, Bahamas, Caribbean Sea to Brazil. Also recorded from Flower Garden Banks. Found from shore to about 215 ft. over sandy bottoms near reefs. Also over seagrass beds and around mangroves.

Yellowtail flounder
to 2 ft.

Smooth flounder
to 12 in.

Winter flounder
to 2 ft.

Greenland halibut
to 4 ft.

Lefteye flounders

♂ Peacock flounder
to 17.8 in.

Eyed flounder - *Bothus ocellatus* (Agassiz, 1831)

FEATURES: Eyed side pale tan to gray with whitish rings and rosettes. Rings and rosettes may appear ocellated. Three dark clusters on lateral line. Caudal fin with a dark cluster on upper and lower lobes. Mouth small. Eyes widely separated, more so in males. Males also with a spine at tip of snout and elongate upper pectoral-fin rays. Both with a fleshy ridge on eyes. Dorsal fin with 76–91 rays. HABITAT: NY to W FL, Bahamas, Caribbean Sea to Brazil. Also Bermuda and S Gulf of Mexico. Recorded from Flower Garden Banks. Over soft bottoms from about 60 to 300 ft.

Twospot flounder - *Bothus robinsi* Top & Hoff, 1972

FEATURES: Eyed side tannish with numerous brown and pale spots. Others described as dark brown, without spots. Caudal fin with two dark spots on middle rays. Mouth small. Eyes widely separated, more so in males. Males also with a sharp spine on snout and elongate upper pectoral-fin rays. Eyes lack fleshy ridge. Dorsal fin with 78–90 rays. HABITAT: NC to FL, Gulf of Mexico, Bahamas, Caribbean Sea to Brazil. Also Bermuda. Demersal over soft bottoms from about 33 to 330 ft. BIOLOGY: Feed on a wide variety of invertebrates.

Pelican flounder - *Chascanopsetta lugubris* Alcock, 1894

FEATURES: Eyed side brownish, yellowish brown, to grayish. May be spotted or blotched. Fin membranes may be dusky. Lining of body cavity black; visible from above. Blind side is white. Mouth very large and oblique. Lower jaw extends past upper jaw. Snout short. Eyes are relatively large and close-set in both sexes. Body elongate, tapering posteriorly. HABITAT: Tropical Atlantic, Pacific, and Indian oceans. E FL, Gulf of Mexico to Brazil. Demersal. Occur from about 750 to 1,800 ft.

Spiny flounder - *Engyophrys senta* Ginsburg, 1933

FEATURES: Eyed side pale tan to grayish. Other specimens described as brownish. Three dark clusters along lateral line; middle cluster darkest. Small, dark clusters along dorsal and ventral profiles, often with narrow, dark rings. Blind side of females white. Blind side of males banded anteriorly. Spiny ridge present between eyes. Females and juveniles with a single, moderately long tentacle on eyes. Tentacles reduced to absent in large males. Dorsal fin with 74–83 rays. HABITAT: NC to FL, Gulf of Mexico, Nicaragua to Brazil. Also Bahamas. Over soft bottoms from about 100 to 600 ft.

Deepwater flounder - *Monolene sessilicauda* Goode, 1880

FEATURES: Eyed side tan to brownish with dark blotches loosely arranged into bands. Pectoral fin banded on eyed side. Caudal fin may have a dark, central blotch. Mouth small, eyes closely set. Arched portion of lateral line squarish. Pectoral fin absent on blind side. Body elongate. HABITAT: NY to FL, Gulf of Mexico, Colombia to Brazil. Over soft bottoms from about 255 to 1,800 ft.

ALSO IN THE AREA: Angry pelican flounder, *Chascanopsetta danae*; Spotfin sash flounder, *Trichopsetta melasma*; Sash flounder, *Trichopsetta ventralis*, p. 431.

♂ Eyed flounder
to 6.3 in.

♂ Twospot flounder
to 9.8 in.

Pelican flounder
to 12 in.

♀ Spiny flounder
to 4 in.

Deepwater flounder
to 7 in.

Achiridae - American soles

Lined sole - *Achirus lineatus* (Linnaeus, 1758)

FEATURES: Eyed side shades of brown to olive brown with small and large dark spots on body and fins. About eight narrow, faint to dark bars may be present. Blind side white anteriorly, becoming brownish posteriorly. Fimbriae present around mouth on eyed side and are broadly distributed on blind side of head. Pectoral fins present, very small. Tufts of dark cirri are randomly distributed on both sides of body. HABITAT: SC to FL, Gulf of Mexico, Caribbean Sea to Argentina. Absent from Bahamas. Demersal. Found coastally in high-salinity lagoons and over sandy mud bottoms in brackish water. BIOLOGY: Lined sole feed on invertebrates.

Naked sole - *Gymnachirus melas* Nichols, 1916

FEATURES: Eyed side brown to pale tan with 15–32 (usually 20–30) brown to black bars. Pale interspaces as wide as, or slightly wider than, dark bars. Caudal fin with three to five bars. Blind side pale creamy white. Juveniles dark, with no bars. Dermal folds and fimbriae on blind side of head and margin of body. Pectoral fin on eyed side rudimentary, occasionally hidden. Body scaleless. HABITAT: Cape Cod to FL. Also W FL and Bahamas. Demersal. Found in coastal waters from about 3 to 600 ft.

Fringed sole - *Gymnachirus texae* (Gunter, 1936)

FEATURES: Eyed side brown to pale tan with 25–49 (usually more than 30) brown to blackish bars. Pale interspaces almost twice as wide as dark bars. Caudal fin with three to six whole or broken bars. Blind side pale creamy white. Juveniles may be dark, with no bars. Dermal folds and fimbriae on blind side of head. Dermal cirri present on upper eyed-side interspaces. Pectoral fin on eyed side rudimentary, occasionally hidden. Body scaleless. HABITAT: Gulf of Mexico from NW FL to Campeche Bank. Bottom-dwelling over muddy bottoms from about 56 to 610 ft.

Scrawled sole - *Trinectes inscriptus* (Gosse, 1851)

FEATURES: Eyed side tannish to grayish with a dark, reticulating pattern from head to caudal fin. Several small, dark, scattered blotches may be present. Fimbriae present around mouth on eyed side and are broadly distributed on blind side of head. Pectoral fins are rudimentary. HABITAT: S FL, Bahamas, Greater and Lesser Antilles to Venezuela. Occur over soft bottoms in clear, coastal waters including estuaries, bays, and mangrove-lined lagoons.

Hogchoker - *Trinectes maculatus* (Bloch & Schneider, 1801)

FEATURES: Color and pattern variable. Shades of brown with darker, wavy spots, blotches, and lines. Usually with seven to eight widely separated dark bars. Bars and lines extend into fins. Blind side white to pale tan, occasionally blotched or spotted. Fimbriae present around mouth on eyed side and are broadly distributed on blind side of head. Pectoral fins absent. Cirri are scattered on body, but do not form tufts. HABITAT: Gulf of Maine to FL, Gulf of Mexico to Panama. In turbid inshore waters. Found in rivers, estuaries, and high-salinity coastal waters to about 195 ft.

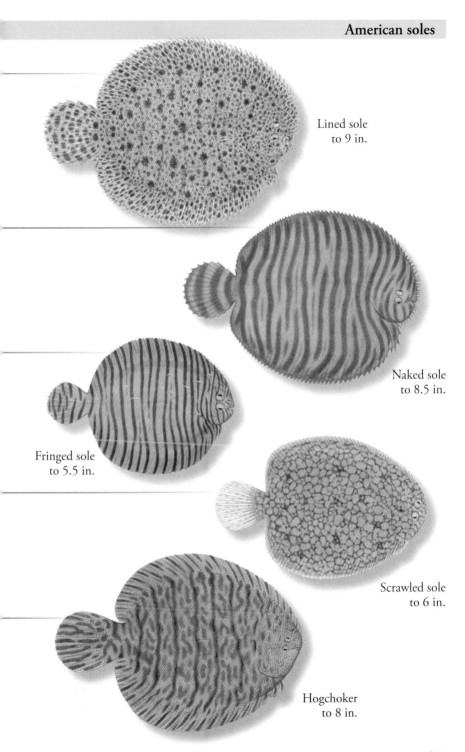

Lined sole
to 9 in.

Naked sole
to 8.5 in.

Fringed sole
to 5.5 in.

Scrawled sole
to 6 in.

Hogchoker
to 8 in.

Cynoglossidae - Tonguefishes

Caribbean tonguefish - *Symphurus arawak* Robins & Randall, 1965

FEATURES: Eyed side grayish to yellowish with two to seven (usually four to five) oblique brown bars. Bars may be complete or incomplete and may extend into fins. Sometimes with variable dark blotches on midline; when present, blotches more developed posteriorly. Narrow, dark bar above upper eye. Blind side flecked. Dorsal fin with 70–76 rays. Caudal fin with 11–14 (usually 12) rays. HABITAT: S FL (rare) and Bahamas. Scattered in Bahamas and Caribbean Sea. Occur over sandy bottoms around coral reefs from about 20 to 125 ft.

Offshore tonguefish - *Symphurus civitatium* Ginsburg, 1951

FEATURES: Eyed side pale to dark brown. May have 6–14 faint to distinct bars. Bars do not extend into fins. Opercle often with a dark patch. Posterior third of dorsal and anal fins dark. Caudal fin dark. Blind side whitish. Dorsal fin with 86–93 rays. HABITAT: Cape Hatteras to FL. In Gulf of Mexico from Dry Tortugas, and Florida panhandle to Yucatán. Over sandy and silty bottoms from about 3 to 240 ft.

Spottedfin tonguefish - *Symphurus diomedeanus* (Goode & Bean, 1885)

FEATURES: Eyed side usually uniformly dark brown. Specimens from light-colored bottoms usually pale brown or yellowish on eyed side. May display pale blotches and incomplete bars. Posterior dorsal and anal fins usually with one to five distinct, round spots. Diffuse spots may be present along anterior fin base. Blind side creamy white to yellowish. Dorsal fin with 86–96 rays. HABITAT: Cape Hatteras to FL, Gulf of Mexico to Brazil. Also Jamaica. Rarely Antilles. Demersal over shelly mud and sand bottoms from about 20 to 600 ft.

Blackcheek tonguefish - *Symphurus plagiusa* (Linnaeus, 1766)

FEATURES: Eyed side dark brown to dull tan, or whitish with a variable number of dark bars on head and body; bars may be absent. Dark blotch at upper opercle; may be absent in smaller specimens. Fins blotched; others described lacking spots or blotches on fins. Opercular cavity and throat region dark, visible from above. Body cavity unpigmented. Blind side uniformly white in adults. Snout short, rounded. Lower eye smaller than upper eye. Dorsal fin with 81–91 rays. Caudal fin usually with 10 rays. HABITAT: NY to FL Keys, Gulf of Mexico to Bay of Campeche. Also Cuba and Bahamas (rare). Most common from Chesapeake Bay to FL. Occur from shallow estuaries and tidal creeks to over the continental shelf from shore to about 98 ft.

Spottail tonguefish - *Symphurus urospilus* Ginsburg, 1951

FEATURES: Eyed side usually dark brown with 4–11 (usually 6–11) dark bars. Bars do not extend into fins. Caudal fin with a distinct, black, ocellated spot. Blind side creamy white. Dorsal fin with 82–90 rays. HABITAT: Cape Hatteras to FL. In the Gulf of Mexico from FL Keys, Dry Tortugas, W FL, AL, TX, and western Yucatán. One report from Cuba. Over vegetated bottoms from about 16 to 130 ft.

NOTE: There are eight other Tonguefishes in the area, p. 431.

Caribbean tonguefish
to 2.2 in.

Offshore tonguefish
to 6.5 in.

Spottedfin tonguefish
to 8.2 in.

Blackcheek tonguefish
to 9 in.

Spottail tonguefish
to 7 in.

Triacanthodidae - Spikefishes

Jambeau - *Parahollardia lineata* (Longley, 1935)

FEATURES: Dusky pale pink dorsally. Others described as pale yellow. Paler below. About nine reddish brown stripes on sides. Snout moderately long and blunt. Mouth small. Eyes large. Spiny dorsal fin with six stout spines; first spine very tall. Pelvic fins with one long, stout spine and two short rays. Body oblong in profile. HABITAT: VA to FL, Gulf of Mexico to Yucatán. Occur from about 390 to 1,300 ft.

ALSO IN THE AREA: Spotted spikefish, *Hollardia meadi*, p. 431.

Balistidae - Triggerfishes

Gray triggerfish - *Balistes capriscus* Gmelin, 1789

FEATURES: Somewhat variable. Grayish to greenish gray with about three obscure to very dark, irregular bars on upper sides. Whitish below. Bright blue spots or lines along dorsal profile and on dorsal and upper caudal fins. Second dorsal and anal fins spotted; spots form lines and rows. Caudal-fin margin convex in juveniles; upper and lower rays long and trailing in adults. HABITAT: Nova Scotia to FL, Gulf of Mexico, Bahamas, Caribbean Sea to Argentina. Also Bermuda and eastern Atlantic Ocean. Found over coral reefs and rocky, sandy, and grassy bottoms. Also reported from over shoals far inshore. BIOLOGY: Feed on invertebrates. Taken as foodfish.

Queen triggerfish - *Balistes vetula* Linnaeus, 1758

FEATURES: Greenish to bluish gray, or yellowish brown dorsally. Lower portion of head and abdomen yellowish. Two curved, blue bands on cheeks. Dark lines radiate from eyes. Upper body with dark, diagonal lines. Dorsal, caudal, and anal fins with blue inner margins. May pale, darken, or change color. Anterior dorsal-fin ray, upper and lower caudal-fin rays long and trailing in large specimens. HABITAT: MA to FL, Gulf of Mexico, Bahamas, Caribbean Sea to Brazil. Also E Atlantic. Reef associated.

Rough triggerfish - *Canthidermis maculata* (Bloch, 1786)

FEATURES: Blackish with irregular, pearly blue spots on body. Others described as brown to gray with pale spots. Spots larger and fewer in juveniles. A deep groove is present in front of eyes. Caudal fin rounded in juveniles. Scales keeled. Body comparatively shallow. HABITAT: Circumglobal in tropical to warm temperate seas. NJ to FL, Gulf of Mexico, Bahamas, and Caribbean Sea to Argentina. Also Bermuda. Pelagic. Usually associated with flotsam. Sometimes found over deep reefs.

Ocean triggerfish - *Canthidermis sufflamen* (Mitchill, 1815)

FEATURES: Brownish to grayish dorsally and on sides. Paler below. May be very pale overall. Usually with a black to brownish black blotch at pectoral-fin base. Margins of dorsal, caudal, and anal fins may be dark. Dorsal- and anal-fin lobes very tall. Caudal fin margin straight in juveniles; upper and lower margins slightly elongate in adults. HABITAT: MA to FL, Gulf of Mexico, Bahamas, and Caribbean Sea to northern South America. Also Bermuda. Pelagic. Occur over offshore reefs in clear water.

Jambeau
to 8 in.

Gray triggerfish
to 12 in.

Queen triggerfish
to 20 in.

Ocean triggerfish
to 2 ft.

Rough triggerfish
to 20 in.

Balistidae - Triggerfishes, *cont.*

Black durgon - *Melichthys niger* (Bloch, 1786)

FEATURES: Dark blackish brown to almost uniformly blackish. Cheeks may have yellowish to orange tinge, and blue lines may radiate from eyes. Scale edges often edged in blue, forming a distinct cross-hatched and zigzag pattern. All with a bright, pale blue band at dorsal- and anal-fin bases. Scales on posterior body are keeled. HABITAT: S FL, Bahamas, Caribbean Sea to Brazil. Also Bermuda. Recorded from Flower Garden Banks. Found in clear water over coral reefs to about 98 ft.

Sargassum triggerfish - *Xanthichthys ringens* (Linnaeus, 1758)

FEATURES: Variable. Shades of blue with blackish spots on sides to brownish gray with dark brownish spots on sides. May pale or darken. All with three narrow, blue to blackish, grooved lines on cheeks. Upper, lower, and posterior caudal-fin margin reddish to brownish. HABITAT: SC south to Brazil. Also Gulf of Mexico and Bermuda. Usually over deep reefs. Juveniles with floating *Sargassum* seaweed.

Monacanthidae - Filefishes

Dotterel filefish - *Aluterus heudelotii* Hollard, 1855

FEATURES: Olivaceous to brownish. May have a pale lateral band and pale blotches anteriorly and ventrally. Bright blue spots and wavy lines on head and body. Second dorsal fin with 36–41 rays. Body comparatively deep. HABITAT: MA to FL, Gulf of Mexico, Antilles to Brazil. Also Bermuda. Over seagrass beds and sandy and muddy bottoms to about 165 ft. BIOLOGY: Feed on a variety of plants and invertebrates.

Unicorn filefish - *Aluterus monoceros* (Linnaeus, 1758)

FEATURES: Grayish to brownish. May be uniformly colored or may display a reticulating pattern of small, pale spots and lines. Second dorsal and anal fins yellowish. Second dorsal fin with 46–50 rays. HABITAT: MA to FL, Gulf of Mexico to Brazil. Also Bermuda. Rare in Antilles. Over continental shelves to about 490 ft.

Orange filefish - *Aluterus schoepfi* (Walbaum, 1792)

FEATURES: Pale to metallic gray. Or, whitish with dark brownish bands. All with few to numerous orange to yellowish spots on head and body; spots may cluster. Second dorsal fin with 32–40 rays. HABITAT: Nova Scotia to FL, Gulf of Mexico, Bahamas, Caribbean Sea to Brazil. Also Bermuda. Over seagrass beds and sandy and muddy bottoms to about 165 ft. BIOLOGY: Feed on a variety of plants.

Scrawled filefish - *Aluterus scriptus* (Osbeck, 1765)

FEATURES: Highly variable. Greenish, yellowish, bluish to brownish. Ground color may be uniform or blotched and banded. Black spots and bright blue spots and lines on head and body. May pale, darken, or change color. Second dorsal fin with 43–49 rays. Body comparatively shallow. HABITAT: Worldwide in warm seas. Nova Scotia south to Brazil. Also Gulf of Mexico and Bermuda. Coral reef associated.

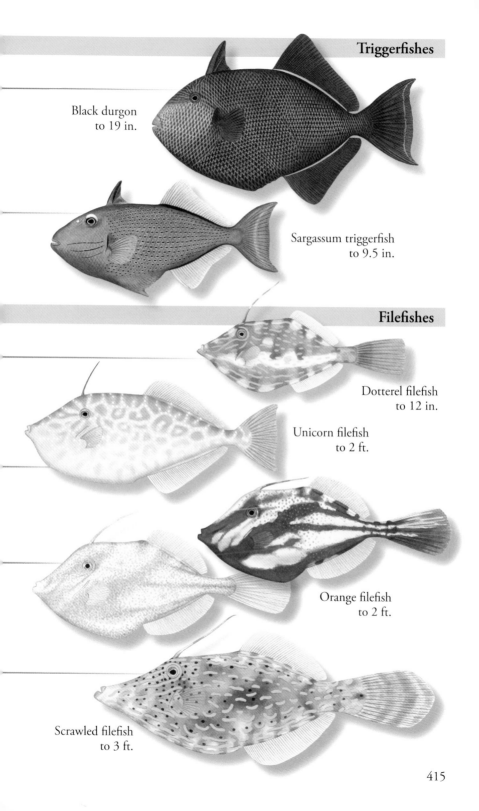

Triggerfishes

Black durgon
to 19 in.

Sargassum triggerfish
to 9.5 in.

Filefishes

Dotterel filefish
to 12 in.

Unicorn filefish
to 2 ft.

Orange filefish
to 2 ft.

Scrawled filefish
to 3 ft.

Monacanthidae - Filefishes, *cont.*

Whitespotted filefish - *Cantherhines macrocerus* (Hollard, 1853)

FEATURES: Variable. May be grayish anteriorly and orangish posteriorly or almost entirely grayish or brownish. Most with a faint to distinct, yellowish to white saddle on mid-back. Large, diffuse to distinct pale spots may cover body. Caudal fin always dark; may have a pale inner bar. Capable of quick color change. Adults with two or three pairs of enlarged spines on caudal peduncle. HABITAT: FL, Bahamas, Antilles to Brazil. Also Bermuda. Found in clear water over coral reefs.

Orangespotted filefish - *Cantherhines pullus* (Ranzani, 1842)

FEATURES: Highly variable. Shades of brown to gray. May have pale mottling. Most with two or more faint to distinct pale stripes and orange spots on sides. Spots have dark centers. Blue lines may be present on snout. One or two distinct white spots on caudal peduncle; some reported with faint or absent spots. Second dorsal fin with 33–36 rays. HABITAT: MA to FL, Gulf of Mexico, Bahamas, Caribbean Sea to Brazil. Also Bermuda and E Atlantic. Adults reef associated. Juveniles are pelagic.

Fringed filefish - *Monacanthus ciliatus* (Mitchill, 1818)

FEATURES: Highly variable. Pale to dark green, or tannish to brown. Body variably marked with a combination of pale and dark spots, lines, and blotches. May be uniformly dark or pale. Snout somewhat upturned. First dorsal-fin spine barbed. Second dorsal fin with 29–37 rays. Body comparatively deep. HABITAT: Newfoundland to FL, Gulf of Mexico, Bahamas, Caribbean Sea to Argentina. Also Bermuda and eastern Atlantic.

Slender filefish - *Monacanthus tuckeri* Bean, 1906

FEATURES: Highly variable. Tannish, grayish, brownish to greenish. Body variably marked with a combination of spots, blotches, bands, and pale reticulating pattern. Snout long, slightly upturned. Second dorsal fin with 32–37 rays. Body comparatively elongate. HABITAT: NC to FL, Bahamas, Antilles to northern South America. Also Bermuda. Occur in grassy coves and around reefs.

Planehead filefish - *Stephanolepis hispidus* (Linnaeus, 1766)

FEATURES: Variable. Pale tan to brown or olivaceous. Usually brownish with dark and pale oblique bars and blotches. Second dorsal fin with 29–35 (usually 31–34) rays; second ray filamentous in males. HABITAT: Nova Scotia to FL, Gulf of Mexico, Caribbean Sea to Brazil. Also Bermuda and eastern Atlantic. Possibly absent from Bahamas. Adults near soft and weedy bottoms; juveniles with floating seaweeds.

Pygmy filefish - *Stephanolepis setifer* (Bennett, 1831)

FEATURES: Variable. Pale tan to brown or olivaceous with irregular oblique bars and blotches. May have dark, oblong blotches and pale spots on sides. Caudal fin with two dark bands. Second dorsal fin with 27–29 (rarely 30) rays; second ray filamentous in males. HABITAT: NC to FL, Gulf of Mexico, Caribbean Sea to Brazil. Also Bermuda. Adults over soft and weedy bottoms; juveniles with floating seaweeds.

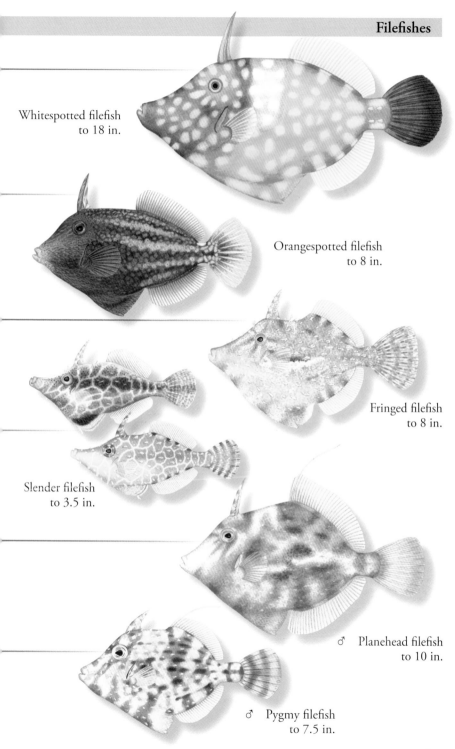

Whitespotted filefish
to 18 in.

Orangespotted filefish
to 8 in.

Fringed filefish
to 8 in.

Slender filefish
to 3.5 in.

♂ Planehead filefish
to 10 in.

♂ Pygmy filefish
to 7.5 in.

Ostraciidae - Boxfishes

Honeycomb cowfish - *Acanthostracion polygonia* Poey, 1876

FEATURES: Color and pattern variable. Yellowish, greenish, to bluish with brownish to blackish hexagonal marks on body. Centers of hexagons generally pale. Cheeks with a reticulating pattern. May change color with mood and habitat. Scales in front of eyes and anterior to anal fin expanded into spine-like projections. Pectoral fins usually with 12 rays (rarely 11). HABITAT: NJ to FL, Gulf of Mexico, Bahamas, and Caribbean Sea to Brazil. Also Bermuda. Occur primarily over coral reefs.

Scrawled cowfish - *Acanthostracion quadricornis* (Linnaeus, 1758)

FEATURES: Color and pattern variable. Yellowish, greenish, brown, or gray with bright blue to black spots and wavy lines on body. Lines may be broad to narrow and few to many. Usually with an unbroken line or a series of spots along dorsal and ventral profiles. Spots and lines extend onto caudal fin. Scales in front of eyes and anterior to anal fin expand into spine-like projections. Pectoral fin usually with 11 rays (rarely 10 or 12). HABITAT: MA to FL, Gulf of Mexico, Bahamas, and Caribbean Sea to Brazil. Also Bermuda. Rare strays to South Africa. Found primarily over shallow seagrass beds. BIOLOGY: Feed on soft corals, sponges, tunicates, and shrimps.

Trunkfish - *Lactophrys trigonus* (Linnaeus, 1758)

FEATURES: Color and pattern variable. Tannish, olivaceous, to grayish. Smaller specimens with small, pale spots and a reticulating pattern. Mature specimens with few spots and a dark, chain-like pattern behind pectoral fins. May also display a second dark area on sides. Scales anterior to anal fin prolonged into spine-like projections. Pectoral fins usually with 12 rays (rarely 11 or 13). HABITAT: MA to FL, Bahamas, Caribbean Sea to Brazil. Also Bermuda. Associated with seagrass beds. BIOLOGY: Feed on worms, bivalves, crabs, and tunicates. Other name: Buffalo trunkfish.

Spotted trunkfish - *Rhinesomus bicaudalis* (Linnaeus, 1758)

FEATURES: Color is variable. Tannish to pale gray with evenly spaced, relatively large, brown to blackish spots on head and body. Spots extend onto caudal fin. Large specimens with three or four white spots on upper body. May display diffuse, dark areas. Scales anterior to anal fin prolonged into spine-like projections. Pectoral fins with 12 rays. HABITAT: FL, Gulf of Mexico, Bahamas, and Caribbean Sea to Brazil. Reported from over shallow coral reefs. BIOLOGY: Spotted trunkfish feed on tunicates, sea cucumbers, brittle stars, and sea urchins.

Smooth trunkfish - *Rhinesomus triqueter* (Linnaeus, 1758)

FEATURES: Variable. Blackish to blackish brown with numerous small to relatively large pale spots on head and body. Spots may be white to yellowish. Pectoral, dorsal, and anal fins yellowish. Bases of pectoral and dorsal fins blackish. Specimens from Flower Garden Banks recorded as golden with darkly ringed pale spots. Body lacks prolonged spines. Pectoral fins with 12 rays. HABITAT: MA to FL, Bermuda, Gulf of Mexico, Bahamas, Caribbean Sea to Brazil. Coral reef associated.

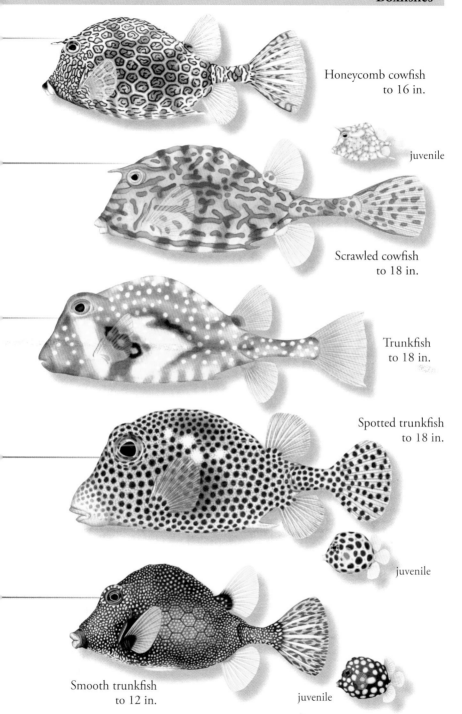

Honeycomb cowfish
to 16 in.

juvenile

Scrawled cowfish
to 18 in.

Trunkfish
to 18 in.

Spotted trunkfish
to 18 in.

juvenile

Smooth trunkfish
to 12 in.

juvenile

Tetraodontidae - Puffers

Goldface toby - *Canthigaster jamestyleri* Moura & Castro, 2002

FEATURES: Tan to brown dorsally. Sides pale yellow, fading to white below. Two dark brown to black stripes on sides. Dark blotch at dorsal-fin base. Dark scrawls on back. Blue lines on snout and blue spots on sides and abdomen. Caudal fin yellowish with blue spots. Pectoral fins with 15–16 rays. HABITAT: NC to northern Gulf of Mexico. Occur over hard bottoms and reefs from about 130 to 320 ft.

Sharpnose puffer - *Canthigaster rostrata* (Bloch, 1786)

FEATURES: Dark tan to brown dorsally. Sides yellowish to white. Bright blue lines radiate from eyes. Blue lines and spots on snout, sides, and abdomen. Caudal fin yellowish with blue lines and dark upper and lower margins. Dark areas on caudal fin extend slightly anteriorly onto body. Snout comparatively long. Pectoral fins with 16–18 rays. HABITAT: NC to FL, Gulf of Mexico, Bahamas, Caribbean Sea to northern South America. Coral reef associated from shallow water to about 320 ft.

Smooth puffer - *Lagocephalus laevigatus* (Linnaeus, 1766)

FEATURES: Greenish gray, gray, or brownish gray dorsally. Silvery on sides; may have a golden luster. White below. Small specimens with three or four dark saddles on upper sides. Snout blunt, rounded. Pectoral fins usually with 17–18 rays. Caudal-fin lobes about equal in length. HABITAT: MA to FL, Gulf of Mexico, Bahamas, and Caribbean Sea to Argentina. Also in eastern Atlantic. Adults pelagic over continental shelves. Juveniles occur near shore and over offshore banks.

Oceanic puffer - *Lagocephalus lagocephalus* (Linnaeus, 1758)

FEATURES: Dark bluish black dorsally. Others described as greenish dorsally. Silvery on sides, white below. Dark and pale areas distinctly lineated. Upper portion of pectoral fins dark. Several rows of dark spots under pectoral-fin base. Snout blunt, rounded. Lower caudal-fin lobe longer than upper lobe. HABITAT: Circumglobal in tropical to temperate seas. Oceanic. BIOLOGY: Feed on crustaceans and squids.

Marbled puffer - *Sphoeroides dorsalis* Longley, 1934

FEATURES: Variable. Mottled to uniformly brown to gray with small, pale speckles dorsally. One to five dark blotches on lower sides. Caudal fin dark at base and along margin. Some with a network of pale blue lines, others with yellow scrawls on lower head and sides. All with pair of dark lappets on back. HABITAT: NC to FL, Gulf of Mexico, Bahamas, Caribbean Sea to Surinam. Demersal from about 30 to 300 ft.

Northern puffer - *Sphoeroides maculatus* (Bloch & Schneider, 1801)

FEATURES: Olive green to gray or brown dorsally with six to eight irregular grayish black bars or blotches. Numerous small black spots pepper upper body. Sides yellowish with a row of dark blotches. Blotch under pectoral fin darkest. Blackish bar between eyes. Abdomen white. Prickles on abdomen extend past anus. HABITAT: Newfoundland to FL. In shallow coastal and inshore waters in a variety of habitats.

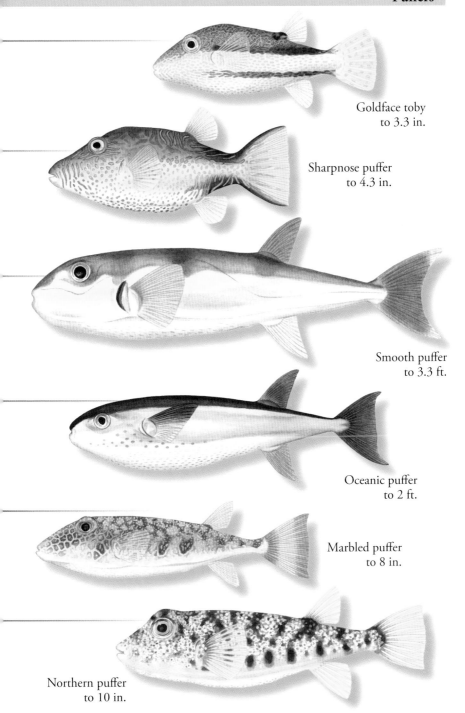

Goldface toby
to 3.3 in.

Sharpnose puffer
to 4.3 in.

Smooth puffer
to 3.3 ft.

Oceanic puffer
to 2 ft.

Marbled puffer
to 8 in.

Northern puffer
to 10 in.

Southern puffer - *Sphoeroides nephelus* (Goode & Bean, 1882)

FEATURES: Brown dorsally, with darker spots and blotches. Pale bluish white spots loosely arranged into rings and lines cover dorsal surface. Dark bar between eyes. Ventral area white. Some males with small, bright red spots. Caudal fin uniformly colored to vaguely barred. Prickles on abdomen end anterior to anus. HABITAT: E FL to AL, western Yucatán to Central America, Bahamas, and Antilles. Occur in shallow bays and estuaries and around bridges and piers. BIOLOGY: Southern puffer form lose aggregations around bridges and piers. Flesh reported to be toxic.

Blunthead puffer - *Sphoeroides pachygaster* (Müller & Troschel, 1848)

FEATURES: Uniformly brown or gray dorsally and on sides. Usually with obscure, dark blotches. White ventrally. Caudal fin dark with pale tips in large specimens. Snout very blunt, rounded in profile. Dorsal and anal fins short-based. Prickles entirely absent from body. HABITAT: SC to FL, Gulf of Mexico to Argentina. Possibly also in eastern Atlantic and Indian oceans. Demersal from about 180 to 600 ft. Reported from over sandy, muddy, and rocky bottoms. Juveniles pelagic.

Least puffer - *Sphoeroides parvus* Shipp & Yerger, 1969

FEATURES: Pale brown to gray with darker spots and blotches dorsally and on sides. Tiny, pale blue green spots pepper upper body. Blotch behind pectoral fins, when present, not darker than other blotches. White ventrally. Snout blunt. Prickles cover upper body from snout to dorsal fin, and ventrally to near anus. Dorsal and anal fins short-based. Lappets absent. HABITAT: In Gulf of Mexico from Pensacola, FL, to Bay of Campeche. Occur along coast and in bays and estuaries. Common over open, sandy mud bottoms.

Bandtail puffer - *Sphoeroides spengleri* Bloch, 1785

FEATURES: Highly variable. Usually dark brown dorsally, becoming paler on sides. May be tannish with some pale and dark mottling or dark blackish brown with pale mottling. All with a distinct row of large, dark spots from chin to caudal peduncle; spots do not usually merge with dark color above. White ventrally; may become dark. Caudal fin distinctly banded. Snout moderately pointed. Upper body with many small pale lappets. Prickles are present or absent dorsally and on cheeks. Abdomen with prickles almost to anus. HABITAT: MA to FL, Gulf of Mexico, Bahamas, Caribbean Sea to Brazil. Also Bermuda. Over reefs and seagrass beds to about 130 ft.

Checkered puffer - *Sphoeroides testudineus* (Linnaeus, 1758)

FEATURES: Dorsal area with a network of dark blotches and bands. Large, darker spots overlay blotches and bands. Interspaces tannish to grayish. Bands loosely arranged into a bull's-eye pattern on mid-back. Lower sides spotted. Abdomen pale. Caudal fin may be barred. Snout moderately blunt. Prickles present from snout to dorsal fin and ventrally to about anus; usually imbedded. HABITAT: RI to FL Keys, Campeche Bank, Bahamas, Caribbean Sea to Brazil. Rare to absent in northern Gulf of Mexico. In shallow water over seagrass beds, around mangrove areas, and in estuaries.

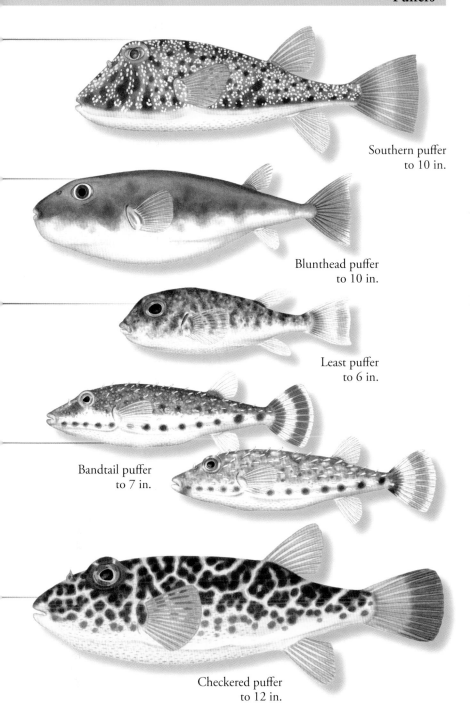

Southern puffer
to 10 in.

Blunthead puffer
to 10 in.

Least puffer
to 6 in.

Bandtail puffer
to 7 in.

Checkered puffer
to 12 in.

423

Diodontidae - Porcupinefishes

Bridled burrfish - *Chilomycterus antennatus* (Cuvier, 1816)

FEATURES: Tannish with small, close-set, black to brown spots on head and body. A large, dark brownish to greenish blotch above pectoral-fin base and on back at dorsal-fin base. A dark blotch may be present on upper head. Caudal fin may be spotted. A short to tall, fleshy tentacle present above eyes. Caudal peduncle lacks spines. HABITAT: S FL, Bahamas, Antilles to northern South America. Also reported from Honduras and Flower Garden Banks. Found over seagrass beds and coral reefs. BIOLOGY: Bridled burrfish feed on hard-shelled invertebrates.

Web burrfish - *Chilomycterus antillarum* Jordan & Rutter, 1879

FEATURES: Variable. Whitish to pale tan with a brownish to reddish reticulating pattern on head and body. Pale interspaces may be large to very small or appear as pale spots. Dark bar below eyes. All with a large, brown to black blotch above and behind pectoral fins and at dorsal-fin base. May also have a dark blotch above anal-fin base. Short to tall, fleshy tentacle present above eyes. Spines on abdomen with fleshy filaments. HABITAT: FL, Bermuda, Bahamas, Antilles to northern South America. Also reported from Flower Garden Banks. Near reefs and over seagrass beds.

Spotted burrfish - *Chilomycterus reticulatus* (Linnaeus, 1758)

FEATURES: Tannish dorsally and on sides. May have darker areas on upper body. Relatively small, close-set, blackish spots cover head, upper body, and fins. Anal fin may lack spots. Ventral area whitish. Juveniles are blue dorsally, white below, with blackish spots dorsally and on sides. Eyes lack a tentacle. Spines on head and body short, erect. Peduncle with one spine dorsally. HABITAT: Circumtropical. NJ to FL, western Gulf of Mexico, to Brazil. Demersal in coastal waters.

Striped burrfish - *Chilomycterus schoepfii* (Walbaum, 1792)

FEATURES: Pale yellowish brown dorsally and on sides with numerous brown, oblique, and wavy lines. Lines are narrow to broad but are generally evenly spaced. Five to seven blackish blotches present on back and sides. Ventral area white to yellow, sometimes blackish. Four to six fleshy tentacles present on chin; one tentacle above eyes. Caudal peduncle lacks spines. HABITAT: Nova Scotia (rare) to FL, Gulf of Mexico, and Bahamas (rare) to Brazil. Occur over seagrass beds in protected bays, estuaries, and coastal lagoons. Adults are demersal, juveniles are pelagic. BIOLOGY: Solitary. Feed on hard-shelled invertebrates.

Pelagic porcupinefish - *Diodon eydouxii* Brisout de Barneville, 1846

FEATURES: Blue dorsally and on sides. White below. Moderately large, blackish blue spots on head and body. Spots may be present on fin bases. Spines on head and body relatively long. One spine present on upper caudal peduncle. Dorsal and anal fins with bluntly pointed tips. Body relatively slender. HABITAT: Worldwide in tropical to warm temperate seas. Recorded from NC, northern and western Gulf of Mexico. Also northern South America. Records are scattered; probably more widespread. Pelagic.

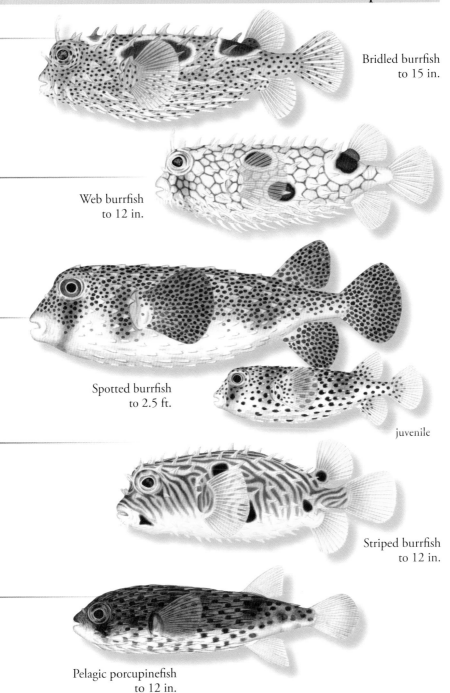

Bridled burrfish
to 15 in.

Web burrfish
to 12 in.

Spotted burrfish
to 2.5 ft.

juvenile

Striped burrfish
to 12 in.

Pelagic porcupinefish
to 12 in.

Diodontidae - Porcupinefishes, *cont.*

Balloonfish - *Diodon holocanthus* Linnaeus, 1758

FEATURES: Tannish dorsally, pale below with moderately-sized, dark spots on upper body. A dark bar extends between and below eyes. Several dark blotches on body. Fins usually unspotted; if spots are present, only on bases. Spines long; those on forehead are longer than those on body. Caudal peduncle lacks spines. HABITAT: Worldwide in tropical to warm temperate seas. MA to FL, Gulf of Mexico, Bahamas, Antilles to Brazil. Occur around mangrove areas and over shallow reefs and soft bottoms.

Porcupinefish - *Diodon hystrix* Linnaeus, 1758

FEATURES: Pale brownish yellow to olivaceous dorsally with numerous small, dark spots on head, upper body, and fins. Abdomen whitish. May have a darkly pigmented ring around abdomen. Spines long; those on forehead shorter than those on body. Spines present on peduncle. HABITAT: Worldwide in tropical to warm temperate seas. MA to FL, Gulf of Mexico, Bahamas, Caribbean Sea to Brazil. Reef associated.

Molidae - Molas

Sharptail mola - *Masturus lanceolatus* (Liénard, 1840)

FEATURES: Pale to dark brown dorsally. Others described as gray. White below. Observed specimens with mottling and scattered white spots. Mouth small. Gill opening round. Pectoral fins rounded. Dorsal and anal fins tall, nearly vertical. Posterior margin of body with a blunt to pointed 'tail.' Body oblong in profile. HABITAT: Worldwide in tropical to temperate seas. MA to FL and northern Gulf of Mexico. Also recorded from TX. Oceanic. At surface or in deep water. BIOLOGY: Sharptail mola are poor swimmers. Swim by waving dorsal and anal fins side to side.

Ocean sunfish - *Mola mola* (Linnaeus, 1758)

FEATURES: Variable. Pale to dark grayish to brownish dorsally, whitish to dusky below. Some are fairly uniformly colored. Males spotted. Mouth and nostrils small. Gill opening round. Pectoral fins rounded. Dorsal and anal fins tall, nearly vertical. Posterior margin of body rounded and somewhat to distinctly scalloped. Body round to oval in profile. HABITAT: Worldwide in tropical to warm temperate seas. Newfoundland to FL, Gulf of Mexico, Caribbean Sea to northern South America. From far offshore to inshore. Occur at surface or in deep water. BIOLOGY: Ocean sunfish are poor swimmers. Swim by waving dorsal and anal fins from side to side.

Slender mola - *Ranzania laevis* (Pennant, 1776)

FEATURES: Dark blackish blue dorsally, iridescent silvery on sides and below. Head and body with wavy, silver bars and spots. Bars on head outlined with dark lines and spots. Mouth small, vertically oriented. Gill openings round. Dorsal and anal fins tall, nearly vertical. Posterior margin of body almost straight. Body at least twice as long as it is deep. Small specimens very elongate. HABITAT: Worldwide in tropical to warm temperate seas. In western Atlantic from FL, eastern Gulf of Mexico to Brazil. Pelagic, oceanic. Reported from surface to about 460 ft.

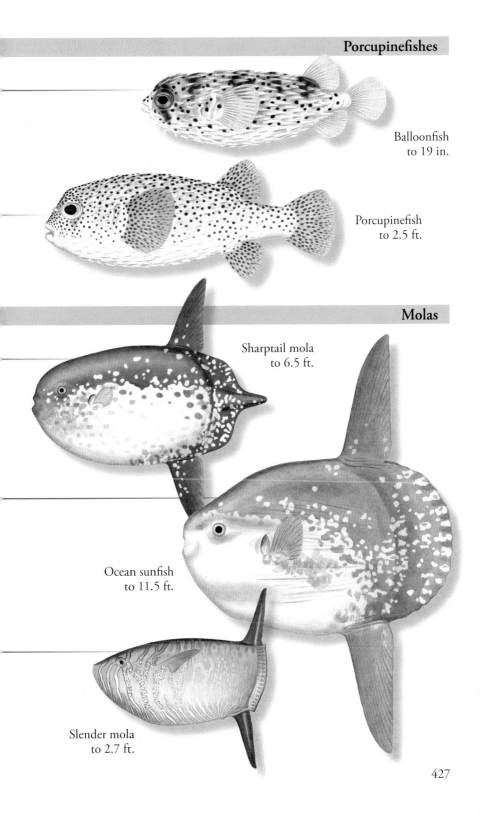

Balloonfish
to 19 in.

Porcupinefish
to 2.5 ft.

Molas

Sharptail mola
to 6.5 ft.

Ocean sunfish
to 11.5 ft.

Slender mola
to 2.7 ft.

Glossary

Below are common technical terms used in this book to describe fishes and their environment. Many of the terms referring to color patterns, anatomical features, and topography are illustrated in the front of this book.

Adipose eyelid: A fatty, transparent tissue that covers the eyes in some fishes.

Adipose fin: A fleshy, rayless fin on the dorsal midline between the dorsal and caudal fin in some fishes. In some species it is fused to the tail and separated from it by a slight notch.

Anadromous: A fish is anadromous when it lives primarily in the ocean and migrates into fresh water to spawn.

Anal fin: The fin on the ventral surface of fishes located between the anus and the caudal fin. See pp. 17, 18.

Anterior: At or toward the front, or head, of a fish.

Band: An oblique or curved marking. See p. 16.

Barbel: A fleshy, tentacle-like appendage, usually on the chin or snout or near the mouth of some fishes, that is usually sensitive to chemicals or to touch.

Base: The point or area where a fin emerges from the body. See p. 19.

Bioluminescence: Light emitted from a living organism.

Brackish: Water that is a mixture of fresh and salt water and is less salty than ocean water and more salty than fresh water.

Buckler: A bony structure, often with a projecting spine, located on the sides or dorsal surface or along the abdomen of some fishes.

Canine: A prominent, conical and pointed tooth, usually larger than other teeth.

Catadromous: A fish is catadromous when it lives primarily in fresh water and migrates to the ocean to spawn.

Caudal fin: The fin at the rear end of the body, also called the tail fin. See pp. 17, 18.

Caudal peduncle: The portion of the fish body between the end of the anal fin and the base of the caudal fin. See p. 18.

Cephalic disk: A modified dorsal fin on the head of remoras that forms an oval-shaped adhesive disc.

Cephalic fin: A portion of modified pectoral fin on the head of manta rays that is elongated and flexible and used to direct water into the mouth.

Chest: The ventral portion of the body in front of the pelvic fins; sometimes referred to as the breast. See p. 19.

Chevron: A diffuse, V-shaped mark that may be dark or pale.

Circumglobal: Occurring around the world, in all oceans.

Cirrus (pl. cirri): A fleshy filament or tab on the head or body, sometimes occurring as a fringe-like series.

Clasper: An elongate modification of the pelvic fin of chimaeras, sharks, skates, and rays that is used for internal fertilization.

Commensal: Describes a relationship between two species in which one species benefits and the other species is neither helped nor harmed.

Compressed: Describes a fish body that is flattened from side to side.

Concave: Arched or rounded inward.

Continental shelf: The submerged portion of a continent that is relatively flat and gently sloping and that extends from shore to a depth of about 660 ft. (200 m).

Glossary

Continental slope: The submerged portion of a continent that is sloping and extends from a depth of about 660 ft. (200 m) to 13,200 ft. (4,000 m).

Convex: Arched or rounded outward.

Corselet: Region behind the head and surrounding the pectoral fins that is covered in specially modified scales.

Demersal: Living unattached at or near the bottom.

Denticle: A small tooth-like or thorn-like projection.

Depressed: Describes a fish body that is flattened from top to bottom.

Detritus: Disintegrating organic material; usually refers to particulate plant or animal debris.

Dorsal fin: A fin located on the back of a fish, often notched or separated into two or more fins. See pp. 17, 18.

Dorsally: Refers to the back or upper portion of a fish.

Estuary: An area partially enclosed by land where fresh water combines with salt water.

Filamentous: Refers to a long, slender, thread-like structure.

Filter-feed: Describes a method of feeding by removing plankton from the water by using gill rakers.

Finlet: A short, isolated, separate fin segment between the dorsal and caudal fins and between the anal and caudal fins.

Flotsam: Floating wreckage or debris, usually man-made.

Hermaphroditic: A fish is hermaphroditic if it possesses both male and female reproductive organs.

Initial (phase): The color phase before sex change in a species that is hermaphroditic, with female and male reproductive organs at different periods in its life. See also **Terminal**.

Inshore: Refers to relatively shallow waters along the coast and around islands to depths of about 660 ft. (200 m).

Keel: A raised ridge on a scale or on either side of the caudal peduncle.

Laminae: Thin, flat structures of the cephalic disc on remoras.

Lateral line: A sensory organ along the side of the body that perceives vibrations and is formed by a series of tubes, canals, or pores. The lateral line is complete when it extends from the head to the caudal fin; it is incomplete when it does not reach the caudal fin.

Laterally: Refers to the side of a fish.

Lure: A fleshy appendage at the tip of a modified dorsal-fin spine that is used to attract prey.

Margin: The outermost edge of a fin.

Melanophore: A pigmented cell that contains dark brown or black pigment.

Nape: The part of the back immediately behind the head, or the area on the dorsal surface between the head and the dorsal-fin origin. See p. 18.

Oceanic: Living in open ocean waters beyond continental shelves.

Ocellated: Having a pattern of a single or multiple ocelli.

Ocellus (pl. ocelli): An eye-like spot in which the central color is surrounded by one or more differently colored rings. See p. 16.

Offshore: Refers to waters beyond the edge of the continental shelf.

Omnivore: An animal that feeds on both animals and plants.

Opercle: The portion of the side of the head, typically composed of large, flattened bones, that covers the gills. See p. 18.

Origin: Refers to the front of a fin, where the first spine or ray emerges from the body. See p. 19.

Pectoral fins: The paired fins behind the gill opening on each side of the body.

Pelagic: Living in open waters or open ocean, away from the bottom.

Pelvic fins: The paired fins on the lower part of the body and in front of the anus. The position of the pelvic fins varies. See pp. 17, 18.

Planktivore: An animal that feeds exclusively on planktonic (drifting) matter.

Posterior: At or toward the rear, or tail, of a fish.

Preopercle: The portion of the side of the head composed of a bone lying anterior to the opercle. The preopercle is often defined by a groove in the skin.

Profile: In side view, describes the outline of a fish body.

Protrusible: Refers to a mouth that is able to extend and lengthen forward to catch prey.

Ray: Supporting element of the fins. Soft rays are segmented, flexible, and branched, while spiny rays are unsegmented, hard, and unbranched.

Reticulating: A pattern or network of irregular, repeating, and interconnecting dark or pale lines.

Rhomboid: A shape similar to a square, but with oblique angles.

Rostrum: The area of the snout that is sometimes flattened or elongate.

Saddle: A patch of color extending across the back and onto the upper sides; may form a V- or U-shape. See p. 16.

Saline: Containing salt.

Scute: A bony plate or modified scale that is usually keeled.

Serrated: Having saw-like notches along the edge.

Snout: The region of the head in front of the eyes and above the mouth.

Spine: A sharp, unbranched, unsegmented, rigid fin ray; also a sharp bony projection on the head or body.

Spiracle: The circular opening between the eye and first gill opening in some sharks, rays, and sturgeons.

Subtropical: Describes the regions between the tropical region and temperate regions.

Temperate: Describes the regions between subtropical regions and polar regions.

Terminal (phase): The color phase after sex change in a species that is hermaphroditic, with female and male reproductive organs at different periods in its life. See also **Initial**.

Thorn: A sharp, bony projection.

Tropical: The region around the world that is warmest and associated with the equator.

Tubercle: A hardened, conical projection on the surface of the body.

Turbid: Refers to water that is murky due to suspended silt or sediment.

Ventrally: Refers to the abdominal or lower portion of a fish.

Vermiculation: Cluster of wavy lines or blotches. See p. 16.

A full bibliography and descriptions of species not illustrated in this book are available online at www.press.jhu.edu.

References listed below will provide the reader with more detailed information about the species included in this book. These references should be available in bookstores and online or at public and institutional libraries.

Böhlke, J. E., and C. C. G. Chaplin. 1968. *Fishes of the Bahamas and Adjacent Tropical Waters.* Wynnewood: Livingston Publishing Company.

Boschung, H. T., Jr., and R. L. Mayden. 2004. *Fishes of Alabama.* Washington: Smithsonian Books.

Campagno, L., and S. Fowler. 2005. *Sharks of the World.* Princeton: Princeton University Press.

Carpenter, K. E., ed. 2002. *FAO Species Identification Guide for Fishery Purposes; The Living Marine Resources of the Western Central Atlantic.* Vol. 1–3. Rome: Food and Agriculture Organization of the United Nations.

Collette, B., and G. Klein-MacPhee, ed. 2002. *Bigelow and Schroeder's Fishes of the Gulf of Maine.* 3rd ed. Washington: Smithsonian Institution Press.

Hoese, H. D., and R. Moore. 1977. *Fishes of the Gulf of Mexico: Texas, Louisiana, and Adjacent Waters.* 2nd ed. College Station: Texas A&M University Press.

McEachran, J. D., and J. D. Fechhelm. 2005. *Fishes of the Gulf of Mexico.* Vol. 1 and 2. Austin: University of Texas Press.

Murdy, E. O., R. S. Birdsong, and J. A. Musick. 1997. *Fishes of Chesapeake Bay.* Washington: Smithsonian Institution Press.

Nelson, J. S. 2006. *Fishes of the World.* 4th ed. Hoboken: John Wiley & Sons.

Nelson, J. S., et al. 2004. *Common and Scientific Names of Fishes from the United States, Canada, and Mexico.* 6th ed. Bethesda: American Fisheries Society, Special Publication 29.

Randall, J. E. 1996. *Caribbean Reef Fishes.* 3rd ed. Neptune: T. F. H. Publications, Inc.

Scott, W. B., and M. G. Scott. 1988. *Atlantic Fishes of Canada.* Toronto: University of Toronto Press.

The websites listed below have searchable databases that are useful sources of taxonomic and ichthyological information:

California Academy of Sciences, Eschmeyer's Catalog of Fishes: http://research .calacademy.org/ichthyology/catalog

Fishbase: www.fishbase.org/

Florida Museum of Natural History, Ichthyology: www.flmnh.ufl.edu/fish/

Gobies: http://gobiidae.com/

Integrated Taxonomic Information System: www.itis.gov/

IUCN Red List of Threatened Species: www.iucnredlist.org/

Smithsonian National Museum of Natural History, Research and Collections, Vertebrate Zoology: http://vertebrates.si.edu/fishes/

Index

Index

Index

Index

Index

Index

437

Index

Index

Index

Index

Index

Index

Index

Index

About the Authors

Val Kells is a widely recognized freelance Marine Science Illustrator. She completed her formal training in scientific communication and illustration at the University of California, Santa Cruz. Since then, she has worked with designers, educators, and curators to produce a wide variety of illustrations for educational and interpretive displays in numerous public aquariums, museums, and nature centers. Her artwork has been displayed at the North Carolina Aquariums, Texas State Aquarium, Long Beach Aquarium, and Monterey Bay Aquarium, among others. Her work has also been published in many books and periodicals. She most recently contributed to *Sea Turtles: A Complete Guide to Their Biology, Behavior, and Conservation*, and is currently completing the illustrations for *Field Guide to the Fishes of Chesapeake Bay*. Val is an avid fisherman and naturalist. She lives in Virginia with her family and spends her off time exploring and fishing the aquatic habitats of the coasts.

Kent Carpenter is a well-known Ichthyologist and Professor in Biological Sciences at Old Dominion University in Norfolk, Virginia. He completed his graduate studies at the University of Hawaii, Manoa and studied fishes overseas for over 20 years in the Philippines, Thailand, Saudi Arabia, Kuwait, and Italy. He is the author of numerous scientific articles on fishes and wrote and edited identification guides for fishery purposes for the Food and Agriculture Organization of the United Nations, including *The Living Marine Resources of the Western Central Atlantic*. His current work concentrates on evolution of fishes and marine conservation, and currently he is the Director of the Global Marine Species Assessment for the International Union for Conservation of Nature and Conservation International. Kent is an enthusiastic SCUBA diver and lives with his family in Virginia Beach, Virginia.

About the Art. The illustrations were produced with watercolor and gouache on 300 lb. coldpress Arches® watercolor paper. Brushes included sizes 2 to 14, round, oval, and flat. Techniques included the use of liquid rubber mask, blotters, and gesso.

For more information please visit www.fieldguidetofishes.com